Seismic Behaviour and Design of Irregular and Complex Civil Structures

GEOTECHNICAL, GEOLOGICAL AND EARTHQUAKE ENGINEERING

Volume 24

For further volumes:
http://www.springer.com/series/6011

Oren Lavan • Mario De Stefano
Editors

Seismic Behaviour and Design of Irregular and Complex Civil Structures

Springer

Editors
Oren Lavan
Faculty of Civil and Environmental Engineering
Technion – Israel Institute of Technology
Haifa
Israel

Mario De Stefano
Department of Construction and Restoration
University of Florence
Florence
Italy

ISSN 1573-6059
ISBN 978-94-007-5376-1 ISBN 978-94-007-5377-8 (eBook)
DOI 10.1007/978-94-007-5377-8
Springer Dordrecht Heidelberg New York London

Library of Congress Control Number: 2013930359

Printed on acid-free paper

Springer is part of Springer Science+Business Media (www.springer.com)

Preface

Real structures are almost always irregular as perfect regularity is an idealization that very rarely occurs. Structural irregularities may vary dramatically in their nature, and, in principle, the concept of irregularity is a *fuzzy* one. Because of the complex behaviour of such structures under earthquake excitations, it is not surprising that, in spite of the large research efforts in plan irregular building structures dating back to the 1970s, even in recent years, many papers have been devoted to a better understanding of seismic response both of simplified one-storey and of multi-storey building models.

Of course, research interest has shifted from investigating elastic response to inelastic response and, subsequently, to developing passive control design strategies, which appear to be a suitable alternative to traditional design in order to mitigate irregularity effects. Additionally, a number of studies have dealt with adequacy of design specifications subscribed by major seismic codes and with issues arising from use of analysis methods, such as pushover analysis; only few studies still are devoted to investigating experimental response and effects of rotational components of input ground motions.

Given the above-summarized research scenario, this book collects state-of-the-art versions of paper presented at the 6th European Workshop on the Seismic Behaviour of Irregular and Complex Structures, held under the auspices of Task Group 8 (TG8) of the European Association for Earthquake Engineering in Haifa at the Technion – Israel Institute of Technology during September 2011.

The book provides for the most updated review of the issue, including contributions coming from 13 countries and giving a complete and international outlook. It is an essential tool for understanding the problem of structural irregularities and provides for the most updated review of this issue, starting from ground rotations, angle of incidence of seismic waves and their effects on structures, through the study of the seismic behaviour, the analysis and design of such structures, up to control and monitoring of irregular buildings. The topic of ground rotations is new in the context of irregular buildings and was added under the umbrella of TG8 in the workshop and in the book for the first time.

The editors wish to thank all contributing authors for the efforts they made to revise and upgrade their workshop papers to book chapters, Dr. Valentina Mariani, TG8 Secretary, for supervising book processing, Prof. A. Rutenberg, founder of TG8, for his continuous dedication to TG8, Prof. A. Ansal, EAEE Secretary General, who has encouraged us to edit this book and last, but not least, the publisher for its very professional support throughout all the editing phases.

The editors
Oren Lavan and Mario De Stefano

Contents

Part I Ground Rotations, Angle of Incidence of Seismic Waves, and Their Effect on Structures

1 An Investigation of an External Impact Conversion into the Strained Rotation Inside Ancient Boulder Structures (Solovky Islands, White Sea) 3
Felix Yudahin, Galina Antonovskaya, Nataly Kapustian, Egor Egorov, and Aleksey Klimov

2 Numerical Analyses of Seismic Ground Rotations from the Wave Passage Effects 15
Zbigniew Zembaty

3 Impact of Seismic Rotational Components on Symmetric Structures 29
Chao He and Qifeng Luo

Part II Seismic Behaviour of Irregular Structures

4 The Effect of Common Irregularities on the Seismic Performance of Existing RC Framed Buildings 47
Angelo D'Ambrisi, Mario De Stefano, Marco Tanganelli, and Stefania Viti

5 Influence of the Variability of Concrete Mechanical Properties on the Seismic Response of Existing RC Framed Structures 59
Angelo D'Ambrisi, Mario De Stefano, Marco Tanganelli, and Stefania Viti

6 Seismic Response of Irregular Industrial Steel Buildings 73
Julien Richard, Sanda Koboevic, and Robert Tremblay

7 **The Inter-storey Pounding Effect on the Seismic Behaviour of Infilled and Pilotis RC Structures** 87
Maria J. Favvata and Chris G. Karayannis

8 **Modal Irregularity in Continuous Reinforced Concrete Bridges. Detection, Effect on the Simplified Seismic Performance Evaluation and Ways of Solution** 103
Gustavo Ayala and Marco Antonio Escamilla

9 **Influence of Infill Panels on the Seismic Response of Existing RC Buildings: A Case Study** 119
Marco Tanganelli, Stefania Viti, Mario De Stefano, and Andrei M. Reinhorn

Part III Seismic Analysis of Irregular Structures

10 **Physically Based Prediction of the Maximum Corner Displacement of One-Storey Eccentric Systems** 137
Tomaso Trombetti, Stefano Silvestri, Giada Gasparini, and Michele Palermo

11 **Plan Irregular Structures: Simplified Approach** 155
Dietlinde Köber and Dan Zamfirescu

12 **Application of Nonlinear Static Method with Corrective Eccentricities to Multistorey Buildings: A Case Study** 173
Melina Bosco, Giovanna A. Ferrara, Aurelio Ghersi, Edoardo M. Marino, and Pier Paolo Rossi

13 **A Multimode Pushover Procedure for Asymmetric Buildings Under Biaxial Seismic Excitation** 189
Grigorios E. Manoukas, Asimina M. Athanatopoulou, and Ioannis E. Avramidis

14 **Three-Dimensional Modal Pushover Analysis of Unsymmetric-Plan Buildings Subjected to Two Components of Ground Motion** 203
Juan C. Reyes and Anil K. Chopra

15 **Estimating Torsional Demands in Plan Irregular Buildings Using Pushover Procedures Coupled with Linear Dynamic Response Spectrum Analysis** 219
Carlos Bhatt and Rita Bento

16 **Analysis of Collapse of Irregular Tall Structures Using Mixed Lagrangian Formulation** 235
Georgios Apostolakis, Andrei M. Reinhorn, Gary Dargush, Oren Lavan, and Mettupalayam Sivaselvan

Part IV Seismic Design of Irregular Structures

17 Improved Earthquake-Resistant Design of Irregular Steel Buildings ... 253
Miltiadis T. Kyrkos and Stavros A. Anagnostopoulos

18 Design of a Plan Irregular RC Frame Building by Direct Displacement-Based Design Method ... 269
Vittorio Capozzi, Gennaro Magliulo, and Roberto Ramasco

Part V Seismic Control and Monitoring of Irregular Structures

19 A Simple Methodology for the Seismic Passive Control of Irregular 3D Frames Using Friction Dampers ... 285
Yael Daniel, Oren Lavan, and Robert Levy

20 An Energy-Based Design Method for Buildings with Supplemental Damping and Nonlinear Behavior ... 297
Raul Barrón and A. Gustavo Ayala

21 Mitigation of the Seismic Response of Structures with Vertical Stiffness and Strength Irregularity Using Supplemental Dampers ... 309
Jack Wen Wei Guo and Constantin Christopoulos

22 Allocation and Sizing of Multiple Tuned Mass Dampers for Seismic Control of Irregular Structures ... 323
Yael Daniel and Oren Lavan

23 The Possibility of a Continuous Monitoring of the Horizontal Buildings' Rotation by the Fibre-Optic Rotational Seismograph AFORS Type ... 339
Leszek R. Jaroszewicz, Zbigniew Krajewski, and Krzysztof P. Teisseyre

24 Seismic Monitoring of Linear and Rotational Oscillations of the Multistory Buildings in Moscow ... 353
Nataly Kapustian, Galina Antonovskaya, Vadim Agafonov, Kostantin Neumoin, and Maksim Safonov

Index ... 365

Contributors

Vadim Agafonov Moscow Institute of Physics and Technology (State University), Dolgoprudnuy, Moscow, Russia

Stavros A. Anagnostopoulos Department of Civil Engineering, University of Patras, Patras, Greece

Galina Antonovskaya Institute of ecology problems of the North UrB, Russian Academy of Science (RAS), Arkhangelsk, Russia

Georgios Apostolakis Department of Civil, Structural and Environmental Engineering, University at Buffalo, The State University of New York, Buffalo, NY, USA

Asimina M. Athanatopoulou Aristotle University of Thessaloniki, Thessaloniki, Greece

Ioannis E. Avramidis Aristotle University of Thessaloniki, Thessaloniki, Greece

A. Gustavo Ayala Instituto de Ingeniería, Universidad Nacional Autónoma de México, Mexico City, Mexico

Raul Barrón Facultad de Ingeniería, Universidad Autónoma de Zacatecas, Zacatecas, Mexico

Rita Bento Department of Civil Engineering and Architecture, Instituto Superior Técnico, Technical University of Lisbon, Lisbon, Portugal

Carlos Bhatt Department of Civil Engineering and Architecture, Instituto Superior Técnico, Technical University of Lisbon, Lisbon, Portugal

Melina Bosco Department of Civil and Environmental Engineering, University of Catania, Catania, Italy

Vittorio Capozzi Dipartimento di Ingegneria Strutturale, Università di Napoli Federico II, Naples, Italy

Anil K. Chopra Civil and Environmental Engineering, University of California, Berkeley, Berkeley, CA, USA

Constantin Christopoulos University of Toronto, Toronto, ON, Canada

Angelo D'Ambrisi Dipartimento di Costruzioni e Restauro, Facoltà di Architettura, Università degli Studi di Firenze, Florence, Italy

Yael Daniel Faculty of Civil and Environmental Engineering, Technion – Israel Institute of Technology, Haifa, Israel

Gary Dargush Department of Mechanical and Aerospace Engineering, University at Buffalo, The State University of New York, Buffalo, NY, USA

Mario De Stefano Department of Construction and Restoration, University of Florence, Florence, Italy

Egor Egorov Moscow Institute of Physics and Technology, State University, Moscow, Russia

Marco Antonio Escamilla Instituto de Ingeniería, Universidad Nacional Autónoma de México, Ciudad Universitaria, México D.F, Mexico

Maria J. Favvata Division of Structural Engineering, Department of Civil Engineering, Democritus University of Thrace, Xanthi, Greece

Giovanna A. Ferrara Department of Civil and Environmental Engineering, University of Catania, Catania, Italy

Giada Gasparini Dipartimento di Ingegneria Civile, Ambientale e dei, Materiali (DICAM), Università di Bologna, Bologna, Italy

Aurelio Ghersi Department of Civil and Environmental Engineering, University of Catania, Catania, Italy

Jack Wen Wei Guo University of Toronto, Toronto, ON, Canada

Chao He Research Institute of Structural Engineering and Disaster Reduction, Tongji University, Shanghai, China

East China Electric Power Design Institute of China Power Engineering Consulting Group, Shanghai, China

Leszek R. Jaroszewicz Institute of Applied Physics, Military University of Technology, Warsaw, Poland

Nataly Kapustian Schmidt Institute of Physics of the Earth, Russian Academy of Science (RAS), Moscow, Russia

Institute of Ecology Problems of the North UrB, Russian Academy of Science (RAS), Arkhangelsk, Russia

Chris G. Karayannis Division of Structural Engineering, Department of Civil Engineering, Democritus University of Thrace, Xanthi, Greece

Aleksey Klimov Moscow State University of Civil Engineering, Moscow, Russia

Dietlinde Köber Reinforced Concrete Department, Technical University of Civil Engineering, Bucharest, Romania

Sanda Koboevic École Polytechnique de Montréal, Chemin de Polytechnique, Montreal, QC, Canada

Zbigniew Krajewski Institute of Applied Physics, Military University of Technology, Warsaw, Poland

Miltiadis T. Kyrkos Department of Civil engineering, University of Patras, Patras, Greece

Oren Lavan Faculty of Civil and Environmental Engineering, Technion – Israel Institute of Technology, Haifa, Israel

Robert Levy Department of Structural Engineering, Faculty of Engineering Sciences, Ben-Gurion University of the Negev, Beer-Sheva, Israel

Qifeng Luo Shanghai Institute of Disaster Prevention and Relief, Tongji University, Shanghai, China

Gennaro Magliulo Dipartimento di Ingegneria Strutturale, Università di Napoli Federico II, Naples, Italy

Grigorios E. Manoukas Aristotle University of Thessaloniki, Thessaloniki, Greece

Edoardo M. Marino Department of Civil and Environmental Engineering, University of Catania, Catania, Italy

Konstantin Neumoin Moscow Institute of Physics and Technology (State University), Moscow, Russia

Michele Palermo Dipartimento di Ingegneria Civile, Ambientale e dei, Materiali (DICAM), Università di Bologna, Bologna, Italy

Roberto Ramasco Dipartimento di Ingegneria Strutturale, Università di Napoli Federico II, Naples, Italy

Andrei M. Reinhorn Department of Civil, Structural and Environmental Engineering, University at Buffalo, The State University of New York, Buffalo, NY, USA

Juan C. Reyes Civil and Environmental Engineering, Universidad de los Andes, Colombia, Bogota, Colombia

Julien Richard Hatch and Associates, Montreal, QC, Canada

Pier Paolo Rossi Department of Civil and Environmental Engineering, University of Catania, Catania, Italy

Maksim Safonov LLC R-Sensors, Dolgoprudnuy, Moscow, Russia

Stefano Silvestri Dipartimento di Ingegneria Civile, Ambientale e dei, Materiali (DICAM), Università di Bologna, Bologna, Italy

Mettupalayam Sivaselvan Department of Civil, Structural and Environmental Engineering, University at Buffalo, The State University of New York, Buffalo, NY, USA

Marco Tanganelli Dipartimento di Costruzioni e Restauro, Facoltà di Architettura, Università degli Studi di Firenze, Florence, Italy

Krzysztof P. Teisseyre Institute of Geophysics, Polish Academy of Sciences, Warsaw, Poland

Robert Tremblay École Polytechnique de Montréal, Chemin de Polytechnique, Montreal, QC, Canada

Tomaso Trombetti Dipartimento di Ingegneria Civile, Ambientale e dei, Materiali (DICAM), Università di Bologna, Bologna, Italy

Stefania Viti Dipartimento di Costruzioni e Restauro, Facoltà di Architettura, Università degli Studi di Firenze, Florence, Italy

Felix Yudahin (deceased) Institute of ecology problems of the North UrB, Russian Academy of Science (RAS), Arkhangelsk, Russia

Dan Zamfirescu Reinforced Concrete Department, Technical University of Civil Engineering, Bucharest, Romania

Zbigniew Zembaty Faculty of Civil Engineering, Department of Structural Mechanics, Opole University of Technology, Opole, Poland

Part I
Ground Rotations, Angle of Incidence of Seismic Waves, and Their Effect on Structures

Chapter 1
An Investigation of an External Impact Conversion into the Strained Rotation Inside Ancient Boulder Structures (Solovky Islands, White Sea)

Felix Yudahin†**, Galina Antonovskaya, Nataly Kapustian, Egor Egorov, and Aleksey Klimov**

Abstract The experimental seismic study of ancient structures are discussed to find out constructive solutions forming and reasons of their high long-term stability against external impacts, including local non-catastrophic earthquakes. The theoretical background is based on the fact that any block media is capable to convert external deformational influence into the seismic radiation. One of the conversion processes is a strained rotation of blocks. The resulting seismic field contains eigen oscillations of the blocks and high-frequency signal generated at the blocks contacts. The experimental observations were performed with standard linear seismic sensors and direct rotational readout sensors. The experimentally observed response is compared with numerical models of the structure dynamics.

---

†(deceased)

G. Antonovskaya
Institute of Ecology Problems of the North UrB, Russian Academy of Science (RAS), 23 Naberezhnaya S. Dviny, 163000 Arkhangelsk, Russia

N. Kapustian (✉)
Institute of Ecology Problems of the North UrB, Russian Academy of Science (RAS), 23 Naberezhnaya S. Dviny, 163000 Arkhangelsk, Russia

Schmidt Institute of Physics of the Earth, Russian Academy of Science (RAS), IPE, 10 B. Gruzinskaya str., D-242, 123995 Moscow, Russia
e-mail: nkapustian@gmail.com

E. Egorov
Moscow Institute of Physics and Technology, State University, 9 Institutskii pereulok, Dolgoprudnuy, 141700 Moscow, Russia
e-mail: egorovev@mail.ru

A. Klimov
Moscow State University of Civil Engineering, 26 Yaroslavskoe highway, 129337 Moscow, Russia
e-mail: mgsu.klimov@gmail.com

O. Lavan and M. De Stefano (eds.), *Seismic Behaviour and Design of Irregular and Complex Civil Structures*, Geotechnical, Geological and Earthquake Engineering 24, DOI 10.1007/978-94-007-5377-8_1,

1.1 Introduction

The objective of the current investigation is to figure out the origin of the long-term high stability of some ancient buildings towards external impacts including local non-catastrophic earthquakes. This is done by studying the mechanic implemented in the structure of these buildings.

One of the methods found relies on construction formed by series of large monolith blocks with a characteristic dimension from 1/100 to 1/10 of the whole building. Glacial granite or basalt boulders are a construction material commonly used in Northern Europe. And the houses build with it proved to be the most durable towards both natural and anthropogenic influence including seismic. Despite the hard basic constructional material a large portion of the wall space belongs to gaps filled with mortar or brick masonry. This less rigid media allows the construction to adjust itself optimally to the load applied. One of the adjusting mechanisms is a strained rotation. Substantially, it is dominant at the external load processing by Earth crust blocks at platform areas (Kocharyan and Spivak 2003) so that it can possibly contribute to the geodynamical stability of the platforms.

We performed seismic experiments on natural objects to probe a boulder masonwork impact processing mechanism. Several masonry types were studied with varying boulder size, mortar type and external load properties. Basic impact types were studied including gradual slow, series of pulses and a prolonged vibration of a combined nature, being a wind pulses with microseisms. The objects examined all belong to Solovetsky monastery. They are: a sea and a lake boulder dams and the famous fortification – White Tower. The striking feature of objects location especially dams is an absence of constant anthropogenic vibrations, which is rarely found on the planet nowadays. Additionally, the dam locations are uninhabited thus greatly simplifying the detection of the media response to an external impact.

1.2 Experiment Description

1.2.1 *Methodology and Dams Description*

A theory behind observations is based on the fact that any blocked media is capable to convert external deformational influence into the seismic radiation. One of the conversion processes is a strained rotation of blocks. The resulting seismic field contains eigenmodes of blocks and a high-frequency signal radiated by joints between blocks. Theoretically this mechanism decreases the strength of construction but preserves its integrity against external impacts. The purpose of the field experiment was to establish observation schemes and data processing techniques capable of the characteristic seismic signal identification and confirming this type of structural behavior.

Fig. 1.1 A satellite view (*to the left*) and a photo (*to the right*) of the big sea dam at Solovky Islands

The observations were performed with standard linear seismic sensors (seismological accelerometers CMG-5T, velocimeters CMG3TD by Guralp and Russian SM-3) and direct readout rotational sensors. The direct registration of the rotational modes was carried out with the help of special sensor installed side by side to the linear sensors, which measures angular velocities (or angular accelerations) in XY, XZ and YZ planes. The model of the sensor is METR-03 by R-sensors Co (Lee et al. 2009).

The big sea dam in Fig. 1.1 is built in nineteenth century and dams the channel between the two islands in White sea. It is 1.1 km long 10 m width and its height is 6–7 m while 2 m of these are always underwater. The boulders forming the dam are averagely 1 m in diameter and are bind with sand-clay mortar. The dam has zigzag shape because of the several factors. Firstly, the basement of the dam is partially a morainal ridge since a plenty of such ridges is found at islands shores, which are clearly seen from satellite picture in Fig. 1.1. These ridges are responsible for the curvatures at the dam ends. Secondarily, curvatures at the central part are considered to be the most appropriate for the water flow of sea currents in the channel. This suggestion has to be proved by calculations though the shape of the central part is a shape of the second oscillation mode of a slab fixed at sides. Passages for small ships are found at the central part. Besides the natural loads including thermal and tidal cycles, gales and erosion processes dams were barbarian treated in 30-s of twentieth century.

The small (lake) dam is placed in an interconnected system of internal lakes of the Bolshoy Solovetsky Island. The dam serves as a backing wall (a shore part substantially) for a remote lake thus enclosing the system. The dam is 225 m length the width and height are both of 4 m. The same composition based on morainal boulders 1 m size is bonded with sand-clay mortar. We performed seismic prospecting profile along the dam. The results of data processing reveal two boundaries, first one is related to dam-bottom boundary and the second (13 m depth) is ascribed to a bedrock.

1.2.2 Observation of the Blocked Media Response to External Impacts: Pulses

To understand the possible mechanism responsible for the external load processing two "clean" experiments were carried out during two field seasons in 2008–2009 with the same shot points. Several stations with three-component accelerometers CMG-5T Guralp were placed in several points at the dam ridge (2008, 2009) and few additionally aside in 2009. The pulses were generated to probe the structure with a 20–80 Hz frequency impact with 100 and 10 pulses series in 2008 and 100 pulses in 2009. The record was maintained after the pulses till the batteries got empty.

Frequency-time plots are presented in Fig. 1.2 for components: Z, transversely (⊥) and parallel (II) to the dam at points 1, 2 and 3 (2008). The pulse series is clearly observed at 500s. The higher microseisms level of 30–40 Hz is present in the parallel component in respect to the transversal one. Notably this raised level is seen before and after pulses. This could be a result of the dam being fixed at edges in a shore. During 2008 and 2009 aftershot measurements the curves elevated at higher frequencies (>30 Hz). This change varies across horizontal components – the high-frequency radiation transversely is lower than the radiation along the dam. This effect reproduces in 2008 and in 2009. Additionally a maximum at low frequencies is detected in all components. Its position shifts with years is 2.5 Hz in 2008 and 3 Hz in 2009. In frequency-time plots in Fig. 1.2 this maximum is seen as a stable narrow line with stable frequency.

The observation of such a signal in an uninhabited region is very strange especially it being present at the dam with no sign of it outside. These all factors are the witness this peak to be an eigenfrequency of the dam. Given dam's geometry, ground bonding at bottom and along its sides, construction material (torn rubble stone) and the mortar type (the weakest one) the finite element model was constructed to estimate eigenfrequency of a slab similar to the dam. The low-frequency amplitude values are schematically shown in Fig. 1.3 for observation points along the dam. The first form oscillations fit the experimental data by the frequency and amplitude ration for points. The shift from 2.5 to 3 Hz in 2008–2009 is explained simply by a rainy summer in 2008 and thus a mortar being more elastic compared to the dry summer of 2009. The rigid mortar corresponds to a rigid construction with higher eigenfrequency.

The amplitude ratio of low (2–3 Hz) to high ($\sim$36 Hz) frequency signals varies considerably from parallel to transverse the dam (Fig. 1.2). Low-frequency signals dominate at the transverse component and high-frequency – along it. These two directions will be mentioned further as "basic" for the respective signal frequencies – Fig. 1.3. First oscillation mode is well seen on the low-frequency amplitudes envelope. The maximal high-frequency radiation is directed perpendicularly. Additionally the maxima for the both vibration types is detected at the center of the dam.

In order to understand these peculiarities let us consider the following model describing the processes taking place in the dam's centre (Fig. 1.3). Some of the

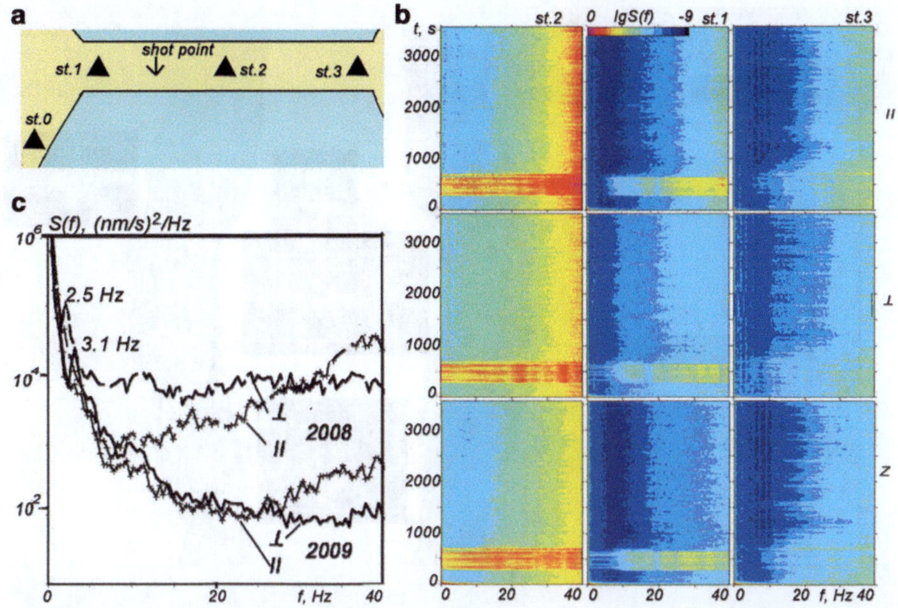

Fig. 1.2 The shot impact experiment at the lake dam: (**a**) – the measurement scheme, (**b**) – microseism frequency-time plots for observations: Z, perpendicular (⊥) and parallel (II) to the dam directions, (**c**) – microseism power spectra observed in 2008 and 2009

boulders move when the first mode oscillations take place. This deviation is larger for central ones than for the boulders at edges. The resulting "hooking" between the sides of boulders during the strained rotation leads to the microscopic defects formation in bonding material, e.g. micro cracks. The series of high-frequency pulse is witness of this mechanism created by the intense low-frequency oscillations. An example is shown in Fig. 1.3. It is known that shearing deformation result in the seismic radiation propagating transversely to the deformation direction (Myachkin et al. 1974), which is indeed in the case.

The following mechanism occupies some time required to process an external influence. Fig. 1.4 shows the amplitude-time plots of both high- and low-frequency signals at the components on both (⊥ and II) horizontal directions at point 2 in the center of the dam. The local minima ("calm") appears in all curves few minutes after the shot. Then it is followed by a rise of the amplitudes reaching the level higher than before shocks. The duration of such specific "excitations" reaches 1.5–2 h.

The series of intervals with magnitude surges are present with an important one at 40 min after shocks – Fig. 1.4. The time shift between magnitude curves of low- (A_{LF}) and high-frequency (A_{HF}) oscillations is the important fact. The rise of A_{HF} takes place at high A_{LF} values and the decrease of A_{LF} at high values of A_{HF} correspondingly. These peculiarities can be explained by a non-controversial model. It relies on construction eigenfrequency oscillations, which are caused

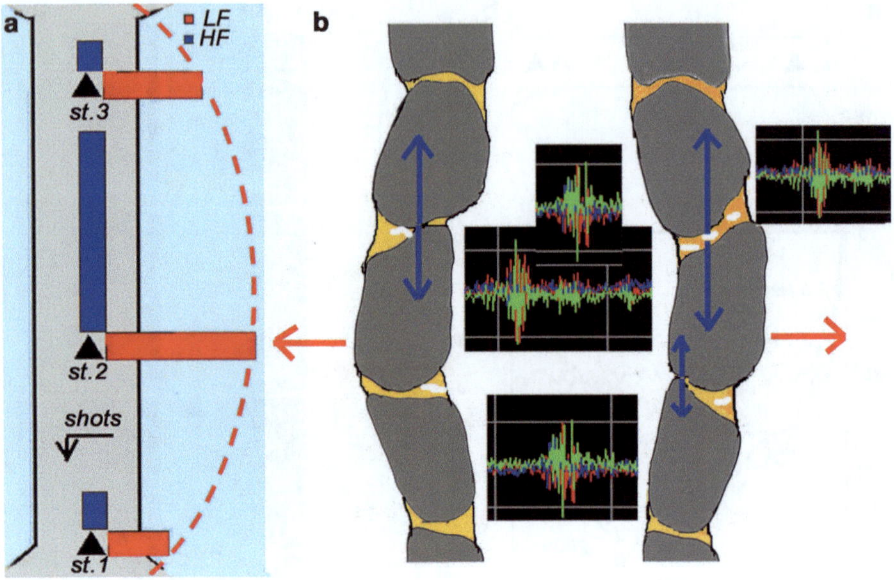

Fig. 1.3 The transformation of low-frequency eigenmode oscillations into high-frequency microseisms: (**a**) – an experimental scheme with sensors positions and amplitudes of dominant oscillations, (**b**) – the scheme of the transformation mechanism based on boulder steering at dam eigenmode oscillations (*red arrows*) and on micropulse formation at the boundaries of the boulders, *insets* – typical micro pulses waveforms along all three seismic components

by microseisms or wind pulses, being present at all times. The batch of shot impacts including some matching boulder's eigenfrequency (40 Hz according to Kocharyan and Spivak 2003) assisted severely the reorganization of the bonds between blocks. This resulted in a "calm" and in followed dam rocking resulting from the new bond structure. The strongest force affects boulders and mortar at high amplitude oscillations, especially in the centre of the dam. This is the case when the considered transformation mechanism of low-frequency oscillations into high-frequency radiation can take place. Thus presence of high-frequency component points at the crack formation. The bond between boulders weakens due to defects in mortar resulting in the dam oscillation damping.

A surprisingly time and spatial signal reproducibility is seen from the analysis of data obtained at all points during several years of study. The main trends are: high- and low-frequency resonance oscillations and their amplitude ratio, seismic radiation anisotropy (across and along the dam) and finally the maximum shot response at the center of the dam. Another important feature is the reproduction of time evolution peculiarities shown at Fig. 1.4: the radiation "calm" (low-frequency especially) followed with an anomalous surge long time (40 min in 2008 and 70 min in 2009) after the shot. The possible situation following these is a sort of an oscillation "reorganization" resulting in time evolution at different points being similar.

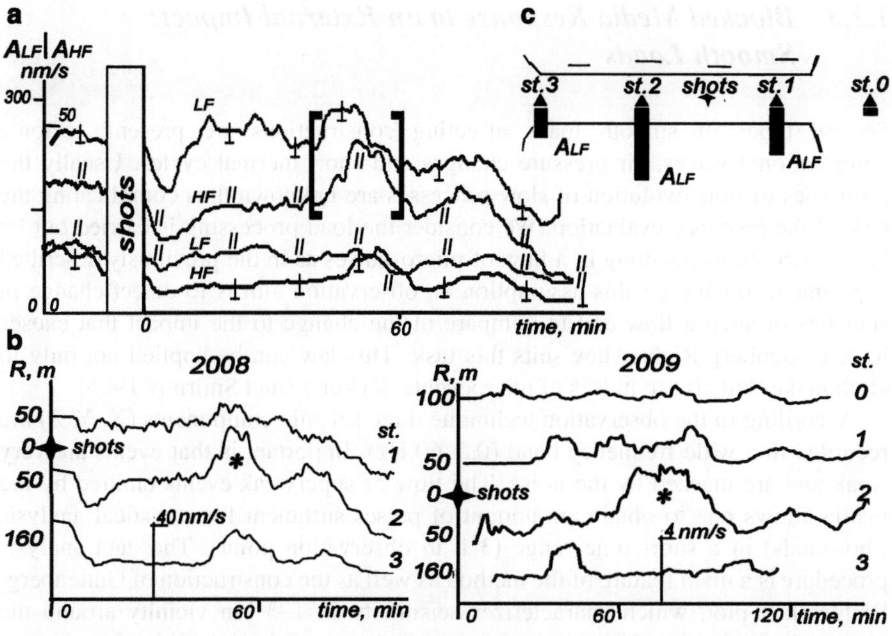

Fig. 1.4 The lake dam experiment: **(a)** – time behavior of high- and low-frequency amplitudes along axis parallel (II) and perpendicular (⊥) to the dam at point 2 in 2008, **(b)** – time behavior of high-frequency amplitudes after shots at various points in 2008 and 2009 (*curves* sorted by the distance from shot point and asterisks mark anomalous surges), **(c)** – an experimental scheme with amplitudes of the dam eigenmode oscillations

An analogous experiment with a shot impact carried out on a larger sea dam doesn't contradict qualitatively the former one. An abrupt surge of the amplitude-time evolution is detected at the center of both dams. This effect relates to an increase of the local strained deformed state at the observation points. The eigenfrequency oscillations are the cause at the small dam appearing as a 2–3 Hz peak in power spectrum. The cause for the sea dam is a strain concentrator, which is an ark in the dam body build to allow a passage for small ships.

Let us estimate the seismic conversion efficiency of an impact into response. Given $\sigma \sim v$, v – being an amplitude of displacement velocity and considering that response and impact have similar frequencies and an amount of phases, we get $v_{response}/v_{impact} \approx 4{*}10^{-8}$ m/s/$4{*}10^{-6}$ m/s $= 0.01$ with an efficiency of 1%. This very value of 1–2% was obtained in an additional experiment with a sea dam, which is described further. Since only up to 10% of an impact enters the solid the total efficiency is of an order of 0.1%. These values were obtained in our experiment with an "intermediate chain", which is a resonance dam excitation at low frequencies i.e. with a nonlinear impact transformation.

1.2.3 Blocked Media Response to an External Impact: Smooth Loads

Several types of smooth loads affecting constructions are present: tectonic deformational waves, air pressure changes and short thermal cycles. Usually the properties of time evolution of slow processes are unknown thus complicating the task of the response evaluation. We consider the load processing is carried out by blocks vibrations resulting in a flow of micro-pulses as in the previously described experiment. Basing on this assumption an observation aim is to detect change in statistics of such a flow and to compare of the change to the impact that causes it. A Guttenberg-Richter law suits this task. This law can be applied not only to earthquakes but also to pulses of microevents (Rykunov and Smirnov 1996).

According to the observation technique three seismic components (X, Y, Z) are recorded in a wide frequency band (0.5–60 Hz). Important is that events are very weak and are masked by the noise. The flow of superweak events emitted by the media allows one to obtain an amount of pulses sufficient for statistical analysis (thousands) in a short time range (3 h in observation point). The data analysis procedure is a main feature of the method as well as the construction of Guttenberg-Richter law plot, which characterizes seismicity in 1–3 km vicinity around the observation point. Analysis are based on calculation of the coherence function for different components of a seismic signal in an observation point $\kappa_{ij}(f)$:

$$\kappa_{ij}(f) = \frac{|S_{ij}(f)|}{\sqrt{S_{ii}(f) \cdot S_{jj}(f)}},$$

where i, j are pairs of (Z, X, Y) components, f – frequency, $S_{ii}(f)$, $S_{jj}(f)$ – power spectra for the i and j component respectively and $|S_{ij}(f)|$ is an averaged mutual spectrum for the i, j components couple. The coherence function is virtually a correlation coefficient for each frequency component of a signal.

It is easy to show that $\kappa_{ij}(f)$ is equal to 0 if the seismic noise is collected from the entire media volume. In the frequency band of a signal source coherence function is $0 < \kappa_{ij}(f) \leq 1$ if the preferential direction is present to the source within media or on its surface. Thus, a coherence function calculation allows picking out signals emitted from the compact volume. In addition, it permits not only suppressing of signals from large volume sources (e.g. surface microseisms) but also of signals generated by the moving transport. Comparison of $\kappa_{ij}(f)$ for component pairs in a horizontal plane with $\kappa_{ij}(f)$ estimation in vertical planes allows elimination of signal sources on the surface.

It's possible to calculate the set of $\kappa_{ij}(f)$ estimated for consecutive time intervals and to obtain the distribution (histograms) of $\kappa_{ij}(f)$. Let us consider that a microseism level is identical in different points of an observation area or that levels are easily compared. Then for very weak events (signal-to-noise ratio $S/N << 1$) it is possible to use this distribution of $\kappa_{ij}(f)$ as a Guttenberg-Richter plot in the vicinity of an observation point.

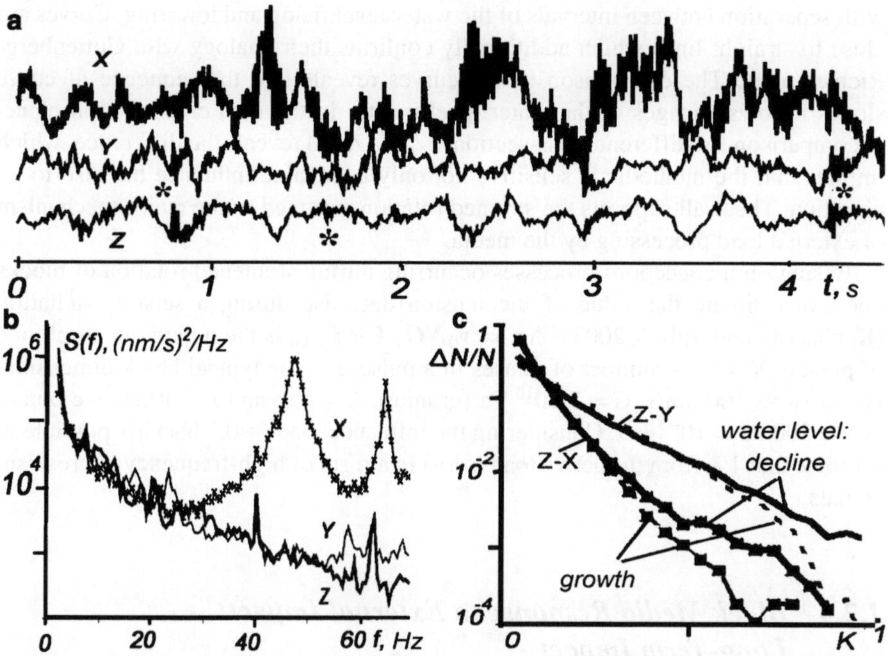

Fig. 1.5 The microseismic field at the sea dam: (**a**) – an example of a three-component signal recording with micropulses (*), (**b**) – the typical power spectra for Z, X and Y components, (**c**) – coherence diagrams (histograms) corresponding to the phases of sea tide

For the sea dam the main influence is the water level raising and lowering due to the sea tide, and a load estimation gives 0.1 bars (12 h tide period). Seismic measurements were put on hold during the phase of the tide when the water level had achieved its maximum and started to decrease. Hence, the load was alternating during the registration. Typical power spectra and a part of measured microseismic signal are shown in Fig. 1.5. Several micro-events, which are clearly distinguishable on the Z and Y components, are marked. The main frequency of these events is equal 40 Hz approximately. This fact is reflected in a local maxima present on the spectra for all components. Suffice to say, according to (Kocharyan and Spivak 2003) this frequency is an eigenfrequency for vibrations of blocks with the characteristic dimension of 1 m. The seismic signal in a frequency band of 45–55 Hz is the most intense for the X (⊥) component of microseisms. Thus, this radiation can be produced by processes caused by the water level change taking into account that the tide influence direction and X component coincides. There are maxima present at higher frequencies on Z and Y components and partly on X component (>60 Hz). Presence of such maxima can be attributed to a boulder "stirring" caused by the tide.

We calculated coherence functions for different pairs of components in the time interval sliding along the record. Using these values, we plot in Fig. 1.5 histograms with separation between intervals of the water level rising and lowering. Curves are close to straight lines which additionally confirms their analogy with Guttenberg-Richter plots. The comparison of the curves reveals that the sequence of curve slopes follows changes of the water level mostly in the dam cross-section plane. A comparison for different cross-sections (Z–X, Z–Y) reveals the difference, which implies that the methodic is sensitive not only to a value influence but also to its direction. These all supports the strained rotation assumed as the major mechanism of external load processing by the media.

Basing on a concept of processes occurring during straitened rotation of blocks one can estimate the value of the tension decrease during a seismic radiation (Kocharyan and Spivak 2003): $\Delta\sigma = v_0 NG/L\pi f$, v_0 is the maximum amplitude of pulses, $N = 5$ – number of phases in a pulse, L – the typical block dimension. Given $\nu_0 \approx 100$ nm/s, $G = 2{\cdot}10^{10}$ Pa (granite), $L = 1$ m and $f = 40$ Hz we obtain $\Delta\sigma \approx 100$ Pa $= 10^{-3}$ bar. Considering the influence load ($\approx$0.1 bar) it's possible to estimate that 1% from the active load is lost in a form of high-frequency microseism signals.

1.2.4 Block Media Response to External Impact: Long-Term Impact

All the constructions studied earlier were composed of rigid blocks with weak bonds between them – granite boulders with a clay-sand mortar. Let us consider the building with strong bonds. It is a tower of Solovetsky monastery (sixteenth century). It has a round shape and is built from boulders with a brick-lime filling (Fig. 1.6). The characteristic oscillations caused by microseisms and wind pulses were studied. Both linear and rotation sensors were used with placement positions at gun slots of the fourth floor and at the ground. In a horizontal plane the observation components were aligned along the radius and the tangent.

The comparison of power spectra at different points and for different types of seismometers is shown in Fig. 1.6. Peaks are present, which correspond to the eigenfrequency oscillations of the building. A model was made (Fig. 1.6) allowing a calculation of construction dynamics. The resulting data were the trajectories of various oscillation modes motion. The examples of these trajectories are shown in Fig. 1.6. Thus, basing on the comparison between the calculation and experiment the spectra peaks were identified – Fig. 1.6. The rotation of the whole tower appears in a rotation signal jointly with the rotation of the various parts of the tower. The 20 Hz peak on the spectra for a rotation around Z axis relates to the roof motion while the 40 Hz peak is attributed to the boulder rotation. Substantially, the peaks at these very frequencies are detected at the dams, which are composed of the similar boulders.

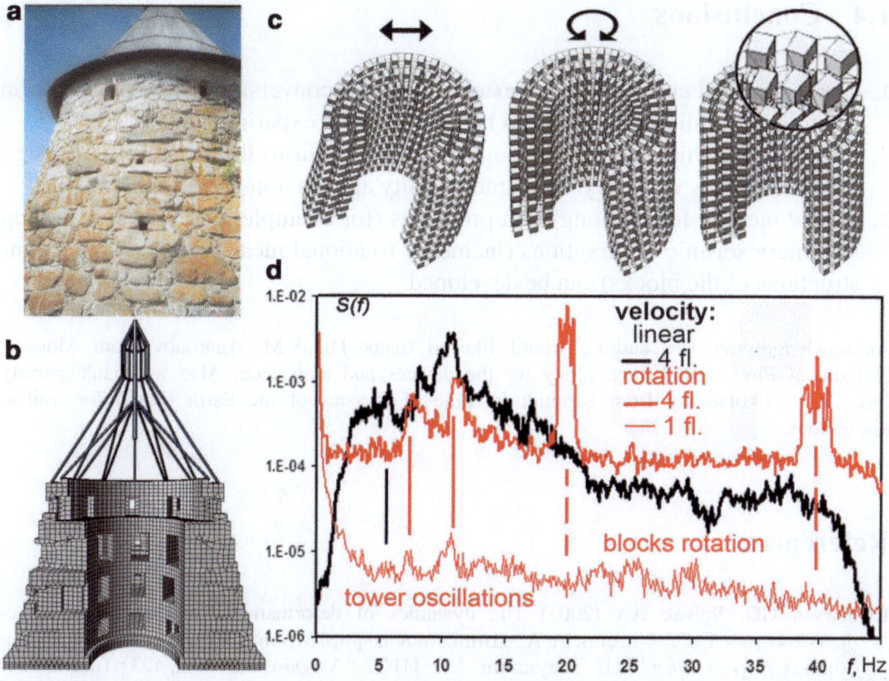

Fig. 1.6 The seismometric study of Solovky tower operation: (**a**) – photo, (**b**) – a constructional scheme used in modeling, (**c**) – blocks movement during eigenmode oscillations, (**d**) – power spectra of linear and rotational sensors records at various points

1.3 Results

1. For all the tested structures the response typical for strained rotation was observed. For the sea dam the variation of the micro pulses flow was correlated with a growth/decline of the water level on the dam walls. The more complex mechanism was found for the lake dam. The artificial strikes produced eigenmode oscillations of the dam structure, which are associated with strained rotations of the blocks. If the amplitude of the eigenmode oscillations exceeds some threshold the high-frequency seismic radiation is produced. For the tower the strained rotations are also associated with the eigenmode oscillations.
2. The intensity of the signal, frequency and time behavior of the seismic radiation depends on the type of excitation, design of the structure and properties of the constructive materials. Some features of the response can be predicted *a priori* using numerical models.
3. All external impacts during the experimental study are very weak (the pressure variance ~100 Pa), but can be easily detected by seismometers including the rotational ones.

1.4 Conclusions

1. The theoretical concepts of the external impacts conversion by the block media by strained rotations mechanisms has been proven experimentally.
2. The long-term history of the ancient structures confirms the possibility to design the similar ones with very long time stability against non-catastrophic impacts.
3. A new method for the long-term processes (for example, tectonic) study using stationary seismic observations (including rotational measurements on the constructions of the blocks) can be developed

Acknowledgments The authors would like to thank Dr. V.M. Agafonov from Moscow Institute of Physics and Technology for the devices and assistance. Also we thank warmly Prof. V.N. Nikolaevsky from Schmidt Institute of Physics of the Earth (RAS) for fruitful discussions.

References

Kocharyan GG, Spivak AA (2003) The dynamics of deformation for rock blocked media/(Кочарян Г.Г., Спивак А.А. Динамика деформирования блочных массивов горных пород. Ред. В.В. Адушкин. М., ИКЦ "Академкнига".), 423 p (in Russian)

Lee WHK, Igel H, Trifunac MD (2009) Recent advances in rotational seismology. Seismol Res Lett 80(3):479–490

Myachkin VI, Kostrov BV, Sobolev GA, Shamina OG (1974) Laboratory and theoretical investigations of earthquake process. Phys Earth (Fizika Zemli) 10:2526–2530

Rykunov LN, Smirnov VB (1996) Seismology in microscale (Рыкунов Л.Н., Смирнов В.Б. Сейсмология микромасштаба//Взаимодействие в системе литосфера-гидросфера-атмосфера. М.: Недра), pp 5–18 (in Russian)

Chapter 2
Numerical Analyses of Seismic Ground Rotations from the Wave Passage Effects

Zbigniew Zembaty

Abstract This chapter presents the problem of deriving rotational ground motions (torsion about vertical axis and rocking about horizontal axis) from the wave decompositions of translational ground motions. Respective formulas for the torsional and rocking power spectral densities are formulated in terms of translational acceleration seismic components. The numerical analysis shows that rotational components are shifted to higher frequencies compared to the respective translational ones and are inversely proportional to the shear wave velocity at a site, which means that they will be more pronounced for softer than hard sites.

2.1 Introduction

The presence of the rotational components in seismic surface ground motion or structural response was only occasionally discussed during the early period of seismic engineering (e.g., Immamura 1937; Richter 1958; Newmark and Rosenblueth 1971). Recent analyses give more and more arguments for treating seriously not only the rotations in structural seismic response, which come from structural irregularities, but also the ground motion rotations (see, e.g., recent Springer monograph edited by Teisseyre et al. (2007) or special issue of *Bulletin of the Seismological Society of America* from 2009, edited by Lee et al. (2009)).

The seismic rotations, as they appear on the ground surface, can be an effect of direct surface or gravity wave propagation and hypothetical rotational waves appearing in the near field of strong earthquakes or they may appear as an effect of upcoming body waves. The measurements of small, teleseismic rotations are

Z. Zembaty (✉)
Faculty of Civil Engineering, Department of Structural Mechanics,
Opole University of Technology, ul. Mikolajczyka 5, 45-271 Opole, Poland
e-mail: z.zembaty@po.opole.pl

O. Lavan and M. De Stefano (eds.), *Seismic Behaviour and Design of Irregular and Complex Civil Structures*, Geotechnical, Geological and Earthquake Engineering 24, DOI 10.1007/978-94-007-5377-8_2,

actually carried out with some success (Lee et al. 2009); however, the *strong* motion rotations were not measured so far. Thus, indirect methods of their assessment for engineering purposes are applied. Since the early 1980s, two groups of such methods have emerged in the literature:

- The first group of methods utilizes the synchronized measurements of translational ground motions at the short distances (e.g., Niazi 1986; Oliveira and Bolt 1989; Castellani and Zembaty 1996).
- The second group of methods is based on the analyses of the wave passage effects at a site, constructing respective wave field, and differentiating it with respect to a spatial coordinate. These methods emerged from the landmark paper by Trifunac from 1982 (Trifunac 1982) and were later developed by Lee and Trifunac (1985, 1987, 2009), Rutenberg and Heidebrecht (1985), Castellani and Boffi (1989), Zembaty et al. (1993), Zembaty (2009), and Li et al. (1997, 2002).

In this chapter, methods of the second group will be utilized. It is known (e.g., Trifunac 1982) that the surface torsion $\varphi(t)$ can be an effect of *SH* and *Love* wave propagation, while the rocking ground motion $\psi(t)$ results from *P* and SV waves as well as *Rayleigh* wave ground motion. The surface wave contribution in total strong ground motion can vary from earthquake to earthquake, and the actual amount of this contribution is still disputable. In this chapter, only effects of body waves on surface rotations will be studied.

2.2 Problem Statement

Any site on the ground surface can be subjected to six motions during earthquakes:

- Three translations $u(t)$, $v(t)$, and $w(t)$ along *x*-, *y*-, and *z*-axes
- Three rotations around these axes $\psi_X(t)$, $\psi_Y(t)$, and $\psi_Z(t)$

The rotation $\psi_Z(t)$ around vertical axis is usually called torsion and is denoted here as $\varphi(t)$, while the rotations $\psi_X(t)$ and $\psi_Y(t)$ around horizontal axes are called rockings. Directing axis *x* toward the epicenter defines the system of so-called principal axes (Fig. 2.1).

In 1975, Penzien and Watabe (1975) have shown that three translational components of seismic ground motion along the respective principal axes *x*, *y*, and *z* are uncorrelated. What is more, when spatial seismic effects at two distinct surface points *A* and *B* are analyzed, the respective coherence matrix transforms as a tensor when the system of coordinates is changed (see, e.g., Zembaty 1997). The decomposition of plane waves analyzed in the coordinate system of principal axes leads to two rotations only:

- Torsion $\varphi(t) = \psi_Z(t)$ around vertical axis
- Rocking $\psi(t) = \psi_Y$ around horizontal axis *y*, perpendicular to site-epicenter direction (*x*-axis)

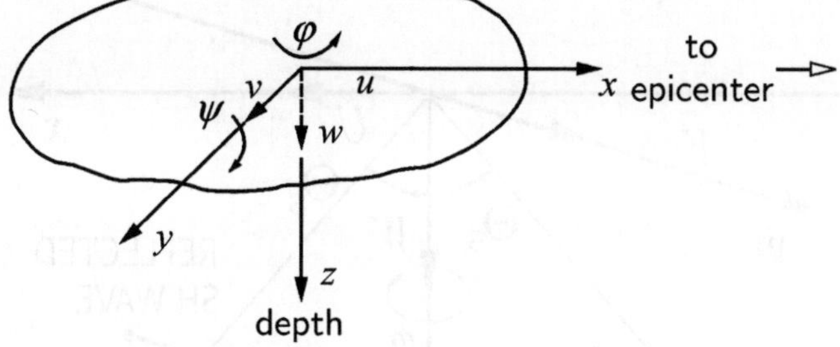

Fig. 2.1 System of principal axes at a site and two rotations φ and ψ

Consider now an elastic half-space. In this chapter, only the two surface rotations (φ and ψ), defined above, are considered (Fig. 2.1). These two rotations can be obtained using solid body mechanics as

$$\varphi(t) = \frac{1}{2}\left(\frac{\partial v(t,x,y)}{\partial x} - \frac{\partial u(t,x,y)}{\partial y}\right) \tag{2.1}$$

$$\psi(t) = \frac{\partial w(t,x,y)}{\partial x} \tag{2.2}$$

Now, the ground motion shall be decomposed onto respective wave components and differentiated with respect to spatial coordinates x and y. We shall start from the SH waves decomposition leading to respective torsion (Eq. 2.1).

2.3 Torsion from SH Waves

Consider an SH wave incident on the free surface at an angle Θ_S (Fig. 2.2).

This wave is reflected at the same angle Θ_S, and the same amplitude is kept for the reflected wave (Aki and Richards 1980). Thus, the amplitudes of the ground motions $u(t)$ and $w(t)$ along axes x and z, respectively, are equal to zero, while the amplitude of the ground motion $v(t)$ along axis y does not depend on the angle of incidence and equals

$$A_v = 2A_{SH} \tag{2.3}$$

From formula (2.1), it is evident that the torsional component (ground rotations around vertical axis) will be built by the derivatives of two horizontal motions $u(t)$ and $v(t)$. For plane waves and the principal coordinate system from Fig. 2.1, the SH component $v(t)$ along y-axis depends only on coordinate x, while P and SV contributions (along x-axis) do not depend on y; so Eq. 2.1 gives

$$\varphi = \frac{1}{2}\left(\frac{\partial v}{\partial x} - \frac{\partial u}{\partial y}\right) = \frac{1}{2}\frac{\partial v}{\partial x} \tag{2.4}$$

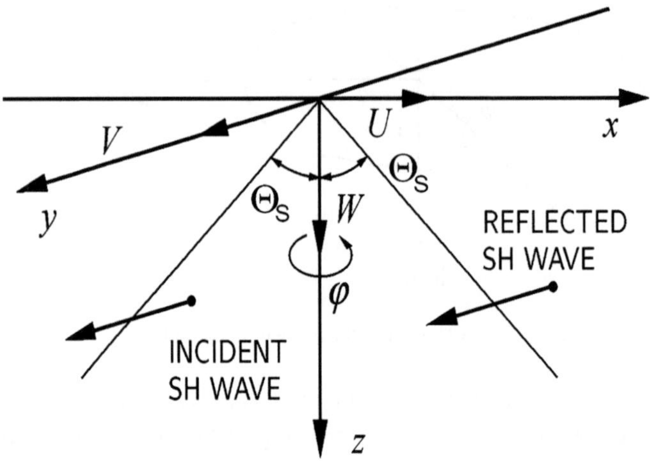

Fig. 2.2 *SH* wave incident on the ground surface

Now assume that the horizontal acceleration along *y*-axis can be written as a stochastic process in form of Stieltjes-Fourier integral

$$\ddot{v}(t) = \int_{-\infty}^{\infty} e^{i\omega t}\, d\hat{\ddot{v}}(\omega) \tag{2.5}$$

The process $\hat{\ddot{v}}(\omega)$ is a random function in frequency domain with orthogonal increments, i.e.,

$$\left\langle d\hat{\ddot{v}}(\omega_1) d\hat{\ddot{v}}(\omega_2)^* \right\rangle == \begin{cases} \left\langle \left| d\hat{\ddot{v}}(\omega) \right|^2 \right\rangle = S_{\ddot{v}}(\omega) d\omega \quad \text{for} \quad \omega_1 = \omega_2 = \omega \\ 0 \quad \text{for} \quad \omega_1 \neq \omega_2 \end{cases} \tag{2.6}$$

where ω is angular frequency [rad/s], symbol $< \ldots >$ denotes mathematical expectation, an asterisk denotes complex conjugate, and $S_{\ddot{v}}(\omega)$ represents power spectral density of the acceleration process $\ddot{v}(t)$. The acceleration $\ddot{v}(t)$ in a frequency band $(\omega, \omega + d\omega)$ can be written as a wave propagating along x direction:

$$d\ddot{v} = 2 \exp\left[i\omega \left(t - \frac{x \sin(\Theta_S)}{c_S} \right) \right] d\hat{\ddot{\Phi}}_{SH}(\omega) \tag{2.7}$$

coming from the horizontal component of the acceleration S wave $d\ddot{\Phi}(t) = e^{i\omega t} d\hat{\ddot{\Phi}}_{SH}(\omega)$ incident at the free surface with angle Θ_S and velocity c_S. For $x = 0$, Eq. 2.7 simplifies to

$$d\ddot{v} = 2e^{i\omega t} d\hat{\ddot{\Phi}}_{SH}(\omega) \tag{2.8}$$

From formulas 2.4, 2.7, and 2.8, one obtains the torsional component:

$$d\ddot{\varphi} = \frac{1}{2}\frac{\partial}{\partial x} d\ddot{v}(t,\omega,x)|_{x=0} = -i\omega\frac{\sin(\Theta_S)}{c_S}e^{i\omega t}d\hat{\ddot{\Phi}}_{SH}(\omega) \quad (2.9)$$

Taking into account Eq. 2.8, one obtains

$$d\ddot{\varphi} = -\frac{i\omega}{2}\frac{\sin(\Theta_S)}{c_S}d\hat{\ddot{v}}(\omega) \quad (2.10)$$

which can be integrated in the whole frequency domain

$$\ddot{\varphi}(t) = \int_{-\infty}^{\infty}\left(-\frac{i\omega}{2}\frac{\sin(\Theta_S)}{c_S}\right)d\hat{\ddot{v}}(\omega) \quad (2.11)$$

and using Eq. 2.6. can be applied to obtain any response statistics. For example, the mean square torsion equals

$$\sigma_{\ddot{\varphi}}^2 = \int_{-\infty}^{\infty}\frac{1}{(2c_S)^2}\sin^2(\Theta_S)\omega^2 S_{\ddot{v}}(\omega)\,d\omega \quad (2.12)$$

The integrand in this equation is power spectral density of the torsional component:

$$S_{\ddot{\varphi}}(\omega) = \frac{1}{(2c_S)^2}\sin^2(\Theta_S)\omega^2 S_{\ddot{v}}(\omega) \quad (2.13)$$

It can be seen from formula (2.13) that the resulting torsional spectrum will be frequency shifted (ω^2 factor), i.e., proportional to the third derivative of translational displacements. In Fig. 2.3, the coefficient $\sin^2(\Theta_S)/(2c_S)^2$ from Eq. 2.13 is plotted versus angle of incidence of shear waves Θ_S.

It can be seen from this figure and Eq. 2.13 that the torsional component increases with the increasing angle of incidence θ_S, and that it is inversely proportional to the shear wave velocity, which means that it will be more pronounced for softer soils.

2.4 Rocking from P and SV Wave Reflections

2.4.1 P Wave Reflection

Consider P waves incident on the ground surface (Fig. 2.4a). Each incident P wave generates a reflected P wave going down, under the same angle $\Theta_{PP} = \Theta_P$ and a

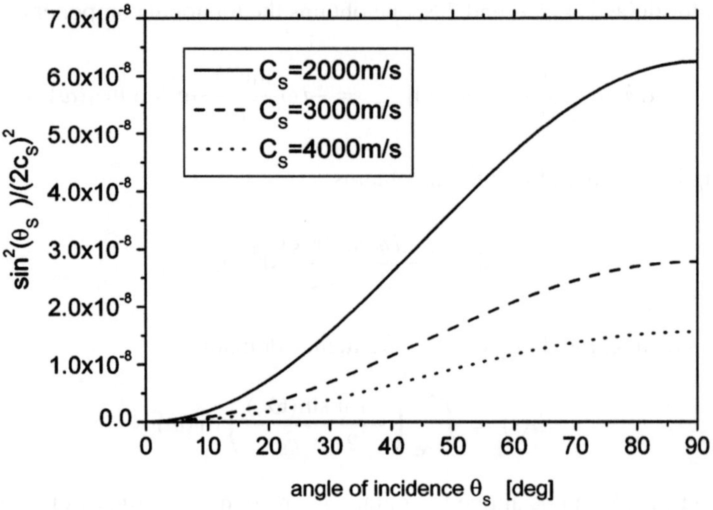

Fig. 2.3 Coefficient $[\sin(\Theta_S)/2c_S]^2$ of Eq. 2.13 versus angle of incidence Θ_S

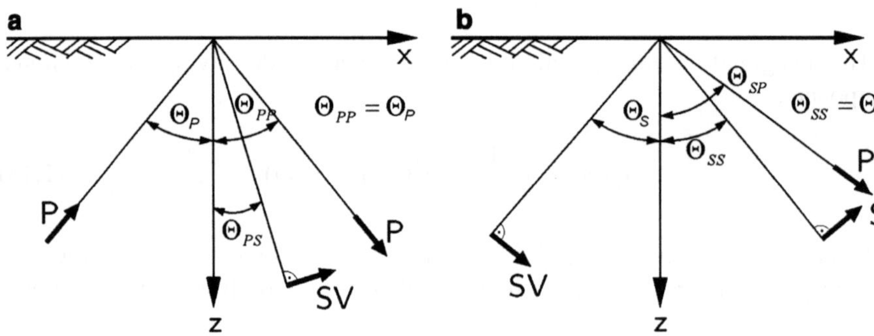

Fig. 2.4 *P* and *SV* waves incident on the ground surface at two different angles Θ_P and Θ_S

reflected SV wave at an angle Θ_{PS} which results from the familiar formula of the geometric optics (e.g., Aki and Richards 1980):

$$\frac{\sin(\Theta_P)}{\sin(\Theta_{PS})} = \frac{c_P}{c_S} = S \tag{2.14}$$

where c_P is the *P* wave velocity and *S* stands for the ratio of *P* to *S* wave velocity.

2.4.2 SV Wave Reflection

Consider next the *SV* wave reflection at the incidence angle Θ_S (Fig. 2.4b). Each wave SV is reflected as *SV* and as *P* wave under the angle which can be obtained from following equation:

$$\frac{\sin(\Theta_S)}{\sin(\Theta_{SP})} = \frac{c_S}{c_P} = \frac{1}{S} \tag{2.15}$$

2.4.3 Derivation of Rocking Spectral Density from Joint Reflections of P and SV Waves

Referring to Fig. 2.4a, b, one can derive amplitudes of horizontal (A_X) and vertical (A_Z) surface ground motions in terms of respective amplitudes A_P and A_{SV} of the *P* and SV waves with certain circular frequency ω [rad/s] impinging on the free surface

$$A_x = U_P A_P + U_S A_{SV} \tag{2.16}$$

$$A_z = W_P A_P + W_S A_{SV} \tag{2.17}$$

where coefficients U_P, U_S, W_P, W_S equal (see Fig. 2.4a, b)

$$U_P = (1 + P_P)\sin\Theta_P + P_P\cos\Theta_{PS} \tag{2.18}$$

$$W_P = (P_P - 1)\cos\Theta_P + P_S\sin\Theta_{PS} \tag{2.19}$$

$$U_S = (1 + S_S)\cos(\Theta_S) + S_P\sin(\Theta_{SP}) \tag{2.20}$$

$$W_S = (1 - S_S)\sin(\Theta_S) - S_P\cos(\Theta_{SP}) \tag{2.21}$$

The coefficients of the free surface reflections are given, in turn, by following formulas:

$$P_p = \frac{-(S^2 - 2\sin^2\Theta_P)^2 + 4\sin^2\Theta_P\cos\Theta_P\cos\Theta_{PS}}{(S^2 - 2\sin^2\Theta_P)^2 + 4\sin^2\Theta_P\cos\Theta_P\cos\Theta_{PS}} \tag{2.22}$$

$$P_S = \frac{4S\sin\Theta_P\cos\Theta_P(S^2 - 2\sin^2\Theta_P)}{(S^2 - 2\sin^2\Theta_P)^2 + 4S\sin^2\Theta_P\cos^2\Theta_P\cos\Theta_{PS})} \tag{2.23}$$

$$S_p = \frac{4 \sin \Theta_S \cos \Theta_S (1 - 2\sin^2 \Theta_S)}{S(1 - 2\sin^2 \Theta_S)^2 + 4\sin^2 \Theta_S \cos \Theta_{SP} \cos \Theta_S} \tag{2.24}$$

$$S_S = \frac{S(1 - 2\sin^2 \Theta_S)^2 - 4\sin^2 \Theta_S \cos \Theta_{SP} \cos \Theta_S}{S(1 - 2\sin^2 \Theta_S)^2 + 4\sin^2 \Theta_S \cos \Theta_{SP} \cos \Theta_S} \tag{2.25}$$

which were obtained from Chap. 5 of Aki, Richards monograph (Aki and Richards 1980). The amplitudes of both waves reflected without wave-type conversion ($P \to P$, $S \to S$) as well as with such the conversion ($P \to S$, $S \to P$) depend on the incident angles Θ_P and Θ_S in a complicated way. For some angles, there is almost no conversion; for another ones, the converted waves dominate. It should be noted that the S_P and S_S coefficients are becoming complex when the incidence angle becomes critical $\Theta_S = \Theta_{cr}$, and instead of being reflected, the S wave propagates as a surface wave (Aki and Richards 1980).

Assume now that the incident P and SV waves are random processes with the Stieltjes–Fourier representation as it was already shown for the torsional ground motion in Chap. 3. Then the horizontal acceleration signals $\ddot{u}(t)$ and vertical one $\ddot{w}(t)$ can be written as

$$\ddot{u}(t) = \int_{-\infty}^{\infty} e^{i\omega t} d\hat{\ddot{u}}(\omega); \quad \ddot{w}(t) = \int_{-\infty}^{\infty} e^{i\omega t} d\hat{\ddot{w}}(\omega) \tag{2.26}$$

where dashed symbols are random processes in the frequency domain with orthogonal increments. This time the orthogonality condition holds not only separately for each of the two processes $\ddot{u}(t)$ and $\ddot{w}(t)$ (Eq. 2.6) but also simultaneously

$$\left\langle d\hat{\ddot{u}}(\omega_1) d\hat{\ddot{w}}^*(\omega_2) \right\rangle = \begin{cases} \left\langle \left| d\hat{\ddot{u}}(\omega) d\hat{\ddot{w}}^*(\omega) \right| \right\rangle = S_{\ddot{u}\ddot{w}}(\omega) d\omega \quad \text{for} \quad \omega_1 = \omega_2 = \omega \\ 0 \quad \text{for} \quad \omega_1 \neq \omega_2 \end{cases} \tag{2.27}$$

where $S_{\ddot{u}\ddot{w}}(\omega)$ is the co-spectral density of processes $\ddot{u}(t)$ and $\ddot{w}(t)$. Now the horizontal and vertical accelerations along x- and z-axes (Fig. 2.4a, b) are written as infinitesimal contributions of two stochastic processes, horizontal and vertical, as follows

$$e^{i\omega\tau} d\hat{\ddot{u}}(\omega), \quad e^{i\omega\tau} d\hat{\ddot{w}}(\omega) \tag{2.28}$$

in frequency band (ω, $\omega + d\omega$). Following Eqs. 2.14 and 2.15, these two motions can be written as "P" and "SV" wave contributions:

$$\begin{cases} d\ddot{u} = d\ddot{u}_p + d\ddot{u}_s \\ d\ddot{w} = d\ddot{w}_p + d\ddot{w}_s \end{cases} \tag{2.29}$$

Procedure analogous to that described here in Chap. 3 leads to two waves propagating on the ground surface in vertical plane with apparent velocities $\sin\theta_P/c_P$ and $\sin\Theta_S/c_S$:

$$d\ddot{w}(t,\omega,x) = W_p \exp\left[i\omega\left(t - \frac{x\sin(\Theta_P)}{c_p}\right)\right] d\hat{\ddot{\Phi}}_p(\omega)$$

$$+ W_s \exp\left[i\omega\left(t - \frac{x\sin(\Theta_S)}{c_s}\right)\right] d\hat{\ddot{\Phi}}_s(\omega) \tag{2.30}$$

which are projections of the waves $d\ddot{\Phi}_P(t) = e^{i\omega t}\, d\hat{\ddot{\Phi}}_p(\omega)$ and $d\ddot{\Phi}_S(t) = e^{i\omega t}\, d\hat{\ddot{\Phi}}_S(\omega)$ with frequency ω, impinging the free surface at angles Θ_P and Θ_S. Differentiating Eq. 2.30 with respect to spatial coordinate *x* (see Eq. 2.2), spectral density of the rocking component can be obtained

$$S_{\ddot{\psi}}(\omega) = |W_X|^2\omega^2 S_{\ddot{u}}(\omega) + 2W_X W_Z^* S_{\ddot{u}\ddot{w}}(\omega) + |W_Z|^2\omega^2 S_{\ddot{w}}(\omega) \tag{2.31}$$

where the coefficients W_X and W_Z

$$W_x = \frac{1}{c_S}\frac{W_p W_s}{D}\left[\frac{\sin(\Theta_P)}{S} - \sin(\Theta_S)\right] \tag{2.32}$$

$$W_z = \frac{1}{c_S}\left[\frac{W_p U_s}{D}\frac{\sin(\Theta_P)}{S} - \frac{W_S U_P}{D}\sin(\theta_S)\right] \tag{2.33}$$

are written in terms of shear wave velocity c_S, the ratio of *P* to *S* wave velocity $S = c_P/c_S$, and the incidence angles of both waves. The typical engineering, local site (soil) parameter is the shear wave velocity c_S which equals to square root of the ratio of shear modulus of elasticity *G* to the soil density ρ ($c_S = \surd(G/\rho)$). This means that, as it was for the torsional component, the rocking ground motion is inversely proportional to stiffness of the soil at the site. If we follow the assumption of Penzien and Watabe (1975), of lack of correlation among principal axes (*x*, *y*, *z*), then Eq. 2.31 simplifies to

$$S_{\ddot{\psi}}(\omega) = |W_X|^2\omega^2 S_{\ddot{u}}(\omega) + |W_Z|^2\omega^2 S_{\ddot{w}}(\omega) \tag{2.34}$$

In Fig. 2.5, plots of modulus of coefficients W_X and W_Z, normalized with respect to shear wave velocity, are shown for three values of the *P* to *S* wave velocity coefficient $S = c_P/c_S$, representing deviations for typical soils, because the *S* coefficient directly depends on the Poisson coefficient ν of the soil with $S = \surd(2-2\nu)/\surd(1-2\nu)$. The plots are shown for equal angles of incidence $\Theta_P = \Theta_S$. It can be seen that the W_Z coefficient does not depend strongly on *S*. On the other hand, the coefficient W_X depends substantially on S only for overcritical angles where anyway it is much smaller than W_Z.

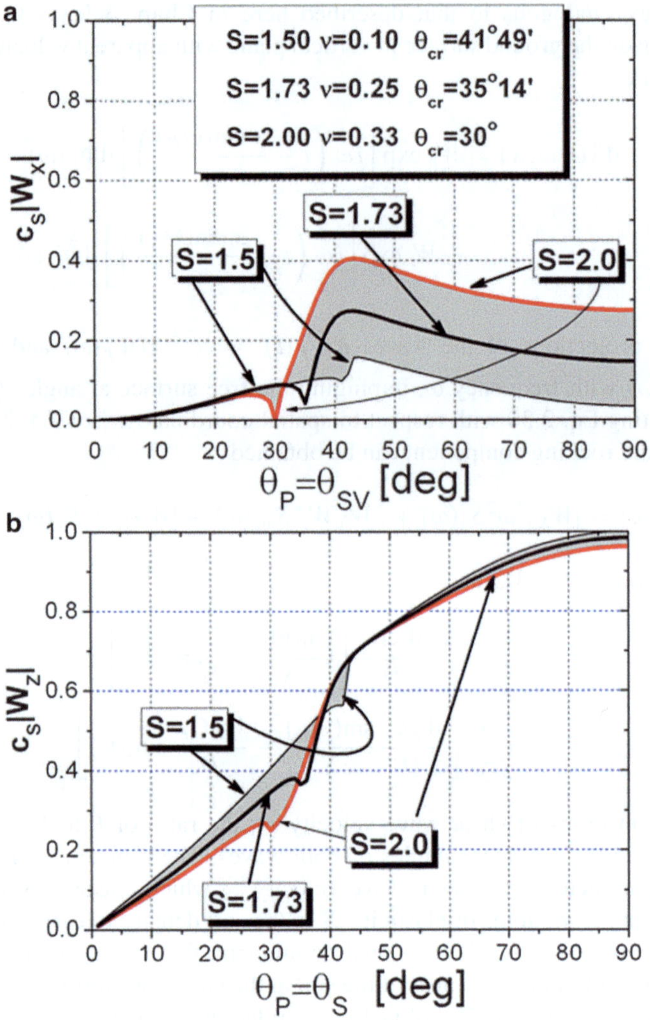

Fig. 2.5 Plots of modulus of coefficients W_X and W_Z (normalized with respect to shear wave velocity) for three values of $S = 1.5$, 1.73, and 2.0 and for equal incidence angles $\Theta_P = \Theta_S$

In Fig. 2.6, a 3D plot of the dependence of the modulus of W_X coefficient on two (assumed as separate) angles of incidence Θ_P and Θ_S is shown for $S = 1.73$, while in Fig. 2.7 respective 3D plot of the dependence of the modulus of W_Z coefficient on the two angles Θ_P and Θ_S is shown for $S = 1.73$.

The cross sections of these 3D plots, going diagonally for $\Theta_P = \Theta_S$, display the results presented in Fig. 2.5 for $S = 1.73$. The plot of $|W_X|$, shown in Fig. 2.6 as 3D plot, is rescaled due to its substantial increase in the area, where the *S* wave comes at a critical angle, while the *P* wave comes at a very small angle. Besides, the $|W_X|$ coefficient stays rather small compared to $|W_Z|$. Except for small Θ_S angles and

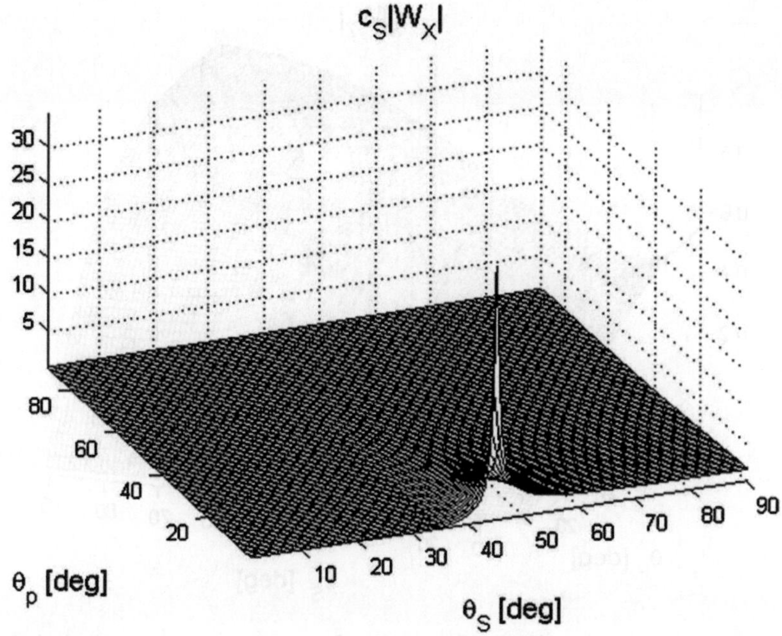

Fig. 2.6 Plot of modulus of coefficient W_X (normalized with respect to shear wave velocity) shown as a function of two, independent incidence angles Θ_P and Θ_S

large Θ_P angles, the $|W_Z|$ coefficient almost uniformly increases with increasing incidence angles. Some faster variations of $|W_Z|$ can be observed for small Θ_P angles and critical Θ_S angles. If one assumes that the surface of the ground overlays more stiff soils as is the typical situation, then the body waves will have a tendency to refract to the vertical direction. In this case, one has to deal mostly with low values of Θ_S and Θ_P. In this case, respective area of low Θ_S and low Θ_P values of the 3D plot from Figs. 2.6 and 2.7 will find practical application.

2.5 Conclusions

A concise numerical analysis of the effect body wave reflections from the free surface on the respective torsional (rotation about vertical axis) and rocking ground motion (rotation about horizontal axis) is presented. The torsional ground motion was obtained from *SH* wave reflection data and horizontal component of ground motion, while the rocking from the *P* and SV wave reflections as the horizontal and vertical translational components. Respective formulas were derived (Eqs. 2.13, 2.31, 2.32, 2.33, 2.34). The ω^2 coefficient results in a phase shift of the respective torsional and rocking power spectral densities compared to the translational ones. That is, the torsional and rocking accelerations are direct function

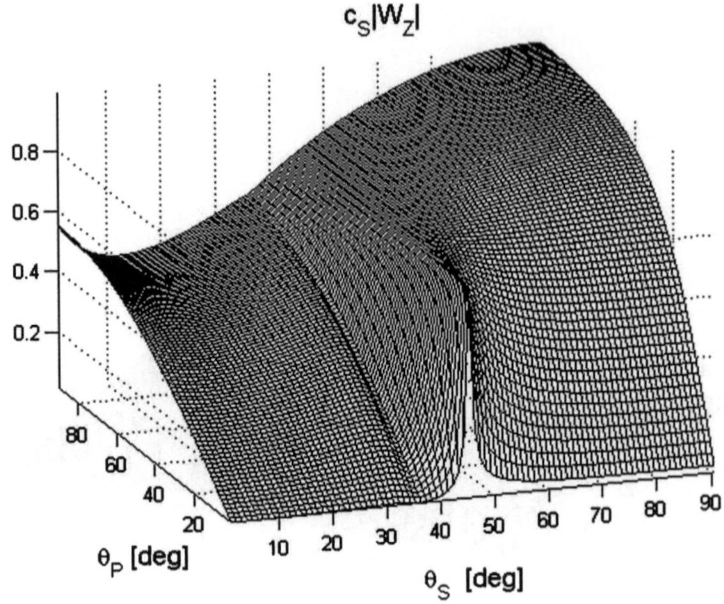

Fig. 2.7 Plot of modulus of coefficient W_Z (normalized with respect to shear wave velocity) shown as a function of two, independent incidence angles Θ_P and Θ_S

of third time derivative of displacements. Both torsional and rocking ground motions are inversely proportional to shear wave velocity at the site, which means that they will be more pronounced for compliant sites than for the hard ones.

When P and S waves propagate vertically, they neither produce surface torsion nor rocking. The torsional component increases with increasing angle of incidence of shear waves. When it comes to rocking component, it increases as the angles of incidence of both waves uniformly increase, but it is the vertical ground motion component which dominates in the resulting ground rocking.

It should be noted here that the actual torsion and rocking ground motions will depend not only on the body wave reflections but also on the surface, Love, and Rayleigh wave contributions. The problem of the presence of surface waves at a particular site during an earthquake is not a unique one. The final answer to the question of torsional and rocking strong ground motion will be solved, when sufficient number of *strong* rotational signals are measured and analyzed which requires further developments in seismic instrumentation.

References

Aki K, Richards PG (1980) Quantitative seismology. W.H Freeman, San Francisco

Castellani A, Boffi G (1989) Rotational components of seismic motion. Earthq Eng Struct Dyn 18:785–797

Castellani A, Zembaty Z (1996) Comparison between earthquake rotation spectra obtained through different experimental sources. Eng Struct 18:597–603

Immamura A (1937) Theoretical and applied seismology. Maruzen Co., Tokyo

Lee VW, Trifunac MD (1985) Torsional accelerograms. Soil Dyn Earthq Eng 4:132–139

Lee VW, Trifunac MD (1987) Rocking strong earthquake accelerations. Soil Dyn Earthq Eng 6:75–89

Lee VW, Trifunac MD (2009) Empirical scaling of rotational spectra of strong earthquake ground motion. Bull Seismol Soc Am 99(2B):1378–1390

Lee WHK, Celebi M, Todorovska M, Igel H (eds) (2009) Rotational seismology and engineering applications. Bull Seismol Soc Am 99(2B), Special issue

Li H-N, Suarez LE, Singh MP (1997) Rotational components of earthquake ground motions. Earthq Eng Eng Vib 17:37–50

Li H-N, Sun L-Y, Wang S-Y (2002) Frequency dispersion characteristics of phase velocities in surface wave for rotational components of seismic motion. J Sound Vib 258:815–827

Newmark NM, Rosenblueth E (1971) Fundamentals of earthquake engineering. Prentice-Hall, Englewood Cliffs

Niazi M (1986) Inferred displacements, velocities and rotations of a long rigid foundation located at El Centro differential array site during the 1979 imperial valley, California earthquake. Earthq Eng Struct Dyn 14:531–542

Oliveira CS, Bolt BA (1989) Rotational components of surface strong ground motion. Earthq Eng Struct Dyn 18:517–526

Penzien J, Watabe M (1975) Characteristics of 3-dimensional earthquake ground motion. Earthq Eng Struct Dyn 3:365–373

Richter CF (1958) Elementary seismology. W.H Freeman, San Francisco

Rutenberg A, Heidebrecht AC (1985) Response spectra for torsion, rocking and rigid foundations. Earthq Eng Struct Dyn 13:543–557

Teisseyre R, Takeo M, Majewski E (eds) (2007) Earthquake source asymmetry, structural media and rotation effects. Springer, Berlin

Trifunac MD (1982) A note on rotational components of earthquake motions for incident body waves. Soil Dyn Earthq Eng 1:11–19

Zembaty Z (1997) Vibrations of bridge structure under kinematic wave excitations. J Struct Eng ASCE 123:479–488

Zembaty Z (2009) Tutorial on surface rotations from the wave passage effects – stochastic approach. Bull Seismol Soc Am 99:1040–1049

Zembaty Z, Castellani A, Boffi G (1993) Spectral analysis of the rotational component of earthquake motion. Probab Eng Mech 8:5–14

Chapter 3
Impact of Seismic Rotational Components on Symmetric Structures

Chao He and Qifeng Luo

Abstract This chapter focuses on studying the impact of the rotational components of seismic excitation on different styles of buildings. The rotational components are synthesized from the translational components of the 2008 Wenchuan earthquake recorded at Mianzhu station. The analysis results show that the rotational components cause much larger torque in a symmetric structure than the translational components. The total seismic responses are comprised of the responses caused by the rotational components and the responses caused by the translational components. For a 30-m-tall symmetric building, the rotational components contribute about 10% of the total seismic responses. For a high-rise building, the rotational components can cause about 50% of the total responses, but for 160-m-high structure, when the rotational components are added, the responses increase by a much smaller level than that caused by the rotational components alone; for 443-m-high building, the responses caused by the translational components and by the rotational components can be directly added together. It implies that the rotational components effect on the structure may be closely related to the structure height. This chapter uses Eurocode8 and ANSYS method to analyze the model, and the results show that Eurocode8 might overrate the seismic rotational loads in some cases.

C. He
Research Institute of Structural Engineering and Disaster Reduction, Tongji University, Shanghai 200092, China

East China Electric Power Design Institute of China Power Engineering Consulting Group, Shanghai 200063, China
e-mail: hechao04@163.com

Q. Luo (✉)
Shanghai Institute of Disaster Prevention and Relief, Tongji University, Shanghai 20092, China
e-mail: luo@tongji.edu.cn

O. Lavan and M. De Stefano (eds.), *Seismic Behaviour and Design of Irregular and Complex Civil Structures*, Geotechnical, Geological and Earthquake Engineering 24, DOI 10.1007/978-94-007-5377-8_3,

3.1 Introduction

It has been a long time since the structural seismic torsional damage was observed. Some methodologies have been provided to analyze the seismic rotational phenomenon. However, the knowledge on the seismic rotational components and their effects on structures are still very limited for lack of measuring methods. Europe codes provided an empirical formula to calculate the rotational components of seismic excitation based on the velocity of shear wave. Considering that the shear wave is not the only factor for the rotational components, the existing formula is not adequate in estimating the rotational components and the level of uncertainty may be higher than desirable.

In recent years, we were shocked at the earthquake disasters. It is critical for us to understand earthquake comprehensively. It has been lately realized from seismic damage and numerical simulations that seismic rotational components coursed some structures damage. Unfortunately, the existing rotational components records are very limited. The study of seismic rotational phenomena has become a hot topic. References (Teisseyre et al. 2006, 2008; Lee et al. 2009; Yang et al. 2010; He et al. 2011) illustrate the early studies in this topic, methodologies development, and the achievements in detail. One estimation method of rotational components based on the theory of elastic plane wave propagation is developed by Newmark (1969) and Trifunac (1982). The rotational seismic motions are estimated from main-shock translational acceleration records of the 2008 Wenchuan earthquake with this method by Che and Luo (2010). On the other hand, in most of structure design, the rotational components are almost not considered. This chapter focuses on studying the impact of the rotational components of seismic excitation on different symmetric tall structures.

3.2 Estimation of the Rotational Components

3.2.1 Estimation Method for Rotational Components

Based on the elastic wave method (Trifunac 1982; Li et al. 1997), certain relationship exists between seismic translational components and rotational components. Figure 3.1 shows the schematic plastic deformation of soil under the seismic loading.

The angle that the soil element rotates around the axis Z:

$$\varphi_Z = \frac{1}{2}(\theta_1 - \theta_2) = \frac{1}{2}\left(\frac{\partial u_y}{\partial x} - \frac{\partial u_Z}{\partial y}\right). \tag{3.1}$$

Over the last 30 years, scientists have been improving the relationship formula with elastic wave method. A relatively reliable formula can be expressed as

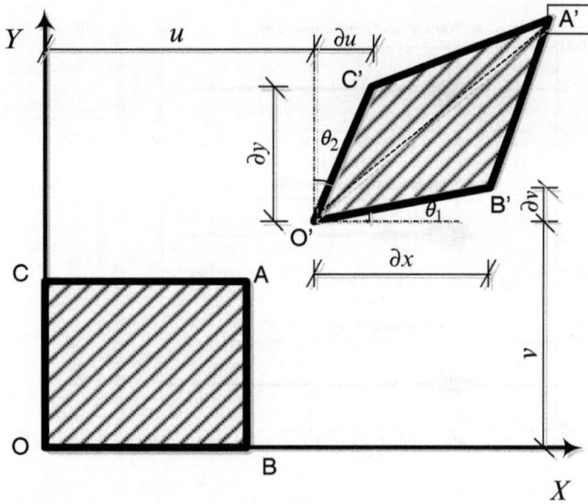

Fig. 3.1 Plastic deformation of soil (Li et al. 1997)

$$\begin{cases} \ddot{\varphi}_1(\omega) = \frac{i\omega}{C}\ddot{u}_3(\omega)\sin\theta \\ \ddot{\varphi}_2(\omega) = \frac{i\omega}{C}\ddot{u}_3(\omega)\cos\theta \\ \ddot{\varphi}_3(\omega) = \frac{1}{2}\cdot\frac{i\omega}{C}[\ddot{u}_2(\omega)\cos\theta - \ddot{u}_1(\omega)\sin\theta] \end{cases} \tag{3.2}$$

where ω is circular frequency; C is wave velocity; $\ddot{\varphi}_1$, $\ddot{\varphi}_2$, $\ddot{\varphi}_3$ are rotational components time histories on the orthogonal axes; θ is principal axis angle; and C is expressed as (Sun and Chen 1998; Hong 2010)

$$C = \frac{\nu_T}{\sin\alpha}. \tag{3.3}$$

Taking the frequency dispersion effect of incidence angle into account, the incidence angle of body wave in Sichuan area $\alpha = 28.4225°$ (Hong 2010). Based on 291 Wenchuan earthquake records from 97 stations in Sichuan province, China, ν_T can be written as (Hong 2010)

$$\begin{cases} \bar{\nu}_T(f_i) = 3.2711 + 0.6952\log_{10}f - 0.0908(\log_{10}f)^2 \\ \nu_T(f_i) = \bar{\nu}_T(f_i)(1 + 0.2\varsigma) \end{cases}. \tag{3.4}$$

From Eqs. 3.3 and 3.4, C can be written as (Hong 2010)

$$\begin{aligned} C &= \frac{\left(3.2711 + 0.6952\log_{10}f - 0.0908(\log_{10}f)^2\right)(1 + 0.2\varsigma)}{\sin(28.2445°)} \\ &= (6.912 + 1.469\log_{10}f - 0.192(\log_{10}f)^2)(1 + 0.2\varsigma). \end{aligned} \tag{3.5}$$

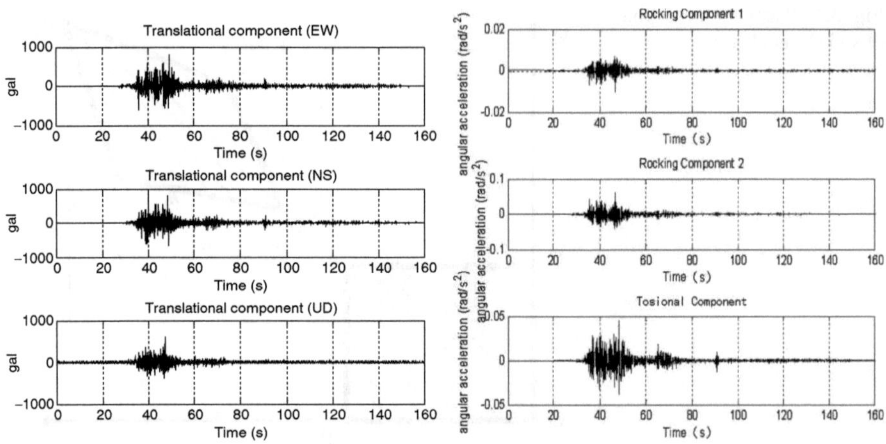

Fig. 3.2 Six components at Mianzhu station

From Eqs. 3.2 and 3.5, it is easy to calculate the rotational components from the translational components.

3.2.2 *Synthesized Rotational Components*

With the method illustrated above, we calculated three rotational components for the 2008 Wenchuan earthquake recorded at Mianzhu and Baoxing seismic stations. The recorded three translational components and synthesized three rotational components are shown in Figs. 3.2 and 3.3.

In the near field of Wenchuan earthquake, what we actually recorded was just the acceleration. Because the rotational components are based on the generation of transient or permanent displacement, the rotational acceleration based on the elastic wave theory is not accurate (Graizer 2010). However, Wang and Hu (1991) indicated that the precision of the method is acceptable from engineering opinion. We can use the estimated data with this method to study structural response, the sensitivity of structure styles to rotational components, and correlate the calculation results with the applicable seismic design codes.

3.3 Contribution of Rotational Motions to Structural Response

The seismic translational components acting on the bottom of the structure are in the form of translational acceleration, while the rotational components are in the form of angle acceleration. The torsional destruction of structure might attribute

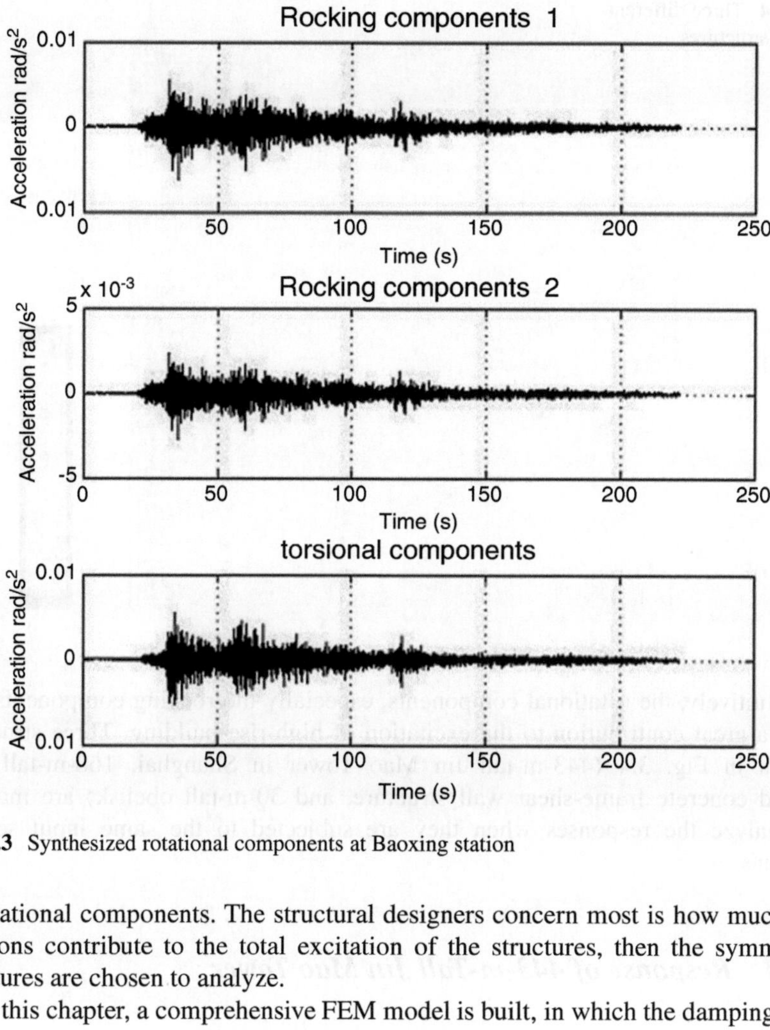

Fig. 3.3 Synthesized rotational components at Baoxing station

to rotational components. The structural designers concern most is how much the rotations contribute to the total excitation of the structures, then the symmetric structures are chosen to analyze.

In this chapter, a comprehensive FEM model is built, in which the damping is in the form of the Rayleigh damping shown as below:

$$[C] = \alpha[M] + \beta[K],$$

$$\alpha = \frac{2\left((\varsigma_i/\omega_i) - (\varsigma_j/\omega_j)\right)}{\left((1/\omega_i^2) - (1/\omega_j^2)\right)},$$

$$\beta = \frac{2\left(\varsigma_j\omega_j - \varsigma_i\omega_i\right)}{(\omega_j^2 - \omega_i^2)}, \tag{3.6}$$

where $[M]$ and $[K]$ are the mass matrix and stiffness matrix and ς_i and ω_i are the damping ratio and natural frequency of the ith mode.

Fig. 3.4 Three different height structures

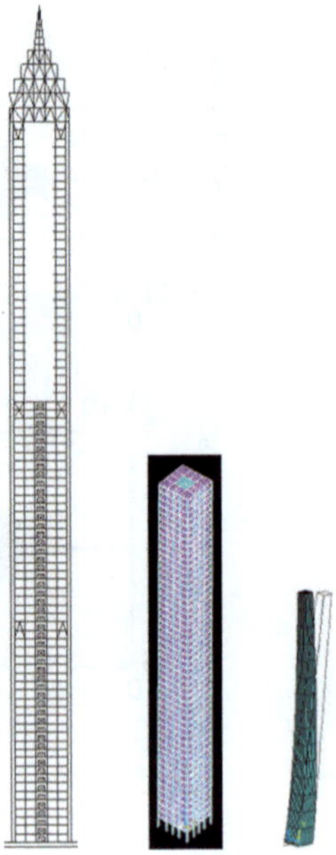

Intuitively, the rotational components, especially the rocking components, will have a great contribution to the excitation of high-rise building. Three structures shown in Fig. 3.4 (443-m-tall Jin Mao Tower in Shanghai, 160-m-tall reinforced concrete frame-shear wall structure, and 30-m-tall obelisk) are modeled to analyze the responses when they are subjected to the same input seismic motions.

3.3.1 Response of 443-m-Tall Jin Mao Tower

Figure 3.5 shows the horizontal displacement, the ratios of displacement caused by rotational components to the total displacement, and the ratios of displacement caused by translational components to the total displacement. The total displacement is defined as the displacement caused by both rotational and translational components.

From Fig. 3.5a, the displacement caused by the three rotational components is in better agreement with that caused by the three translational components. It is obvious that both of them are larger than half of the displacement caused by the six components.

Figure 3.6 shows the different rocking angles along the height direction and the ratios of rocking angles caused by two kinds of seismic loads to total rocking angle.

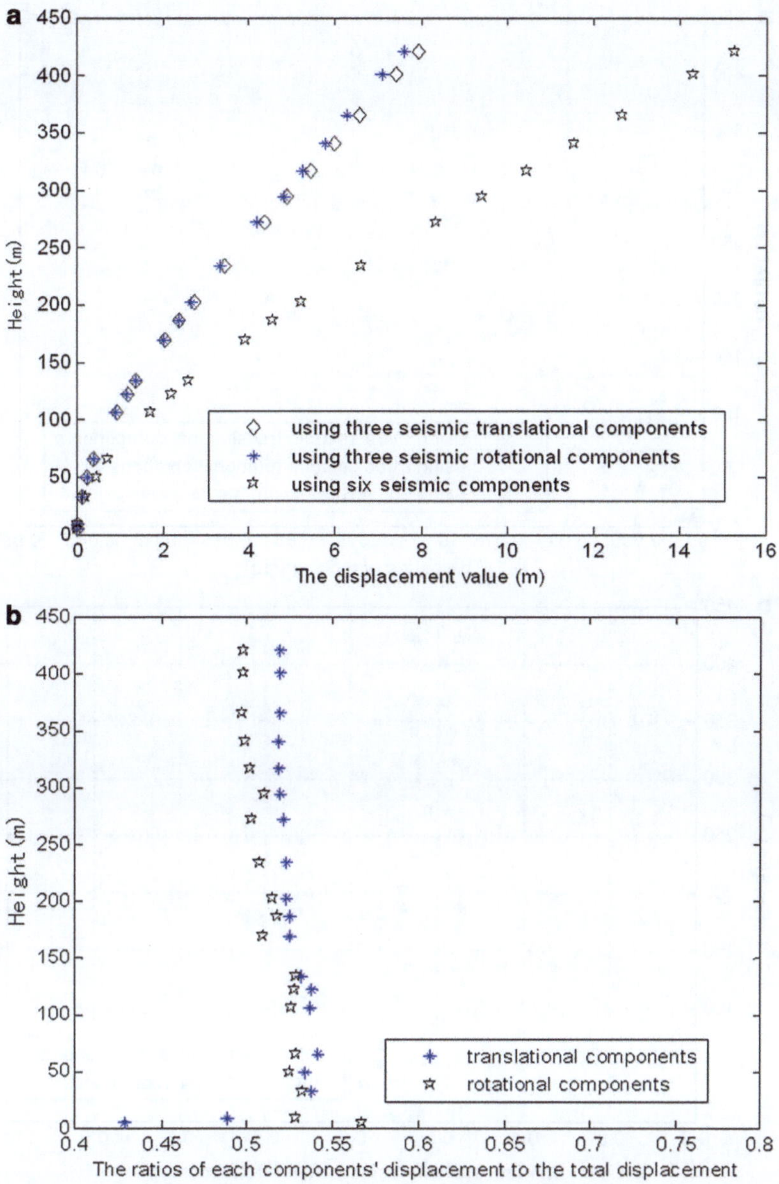

Fig. 3.5 (**a**) Translational displacements caused by translational and rotational components and (**b**) their contribution to the total structural translational displacement

The rocking angle caused by rotational components is larger than that caused by translational components. The ratio of rotational components is larger than 0.5, while the ratio of translational components is smaller than 0.5.

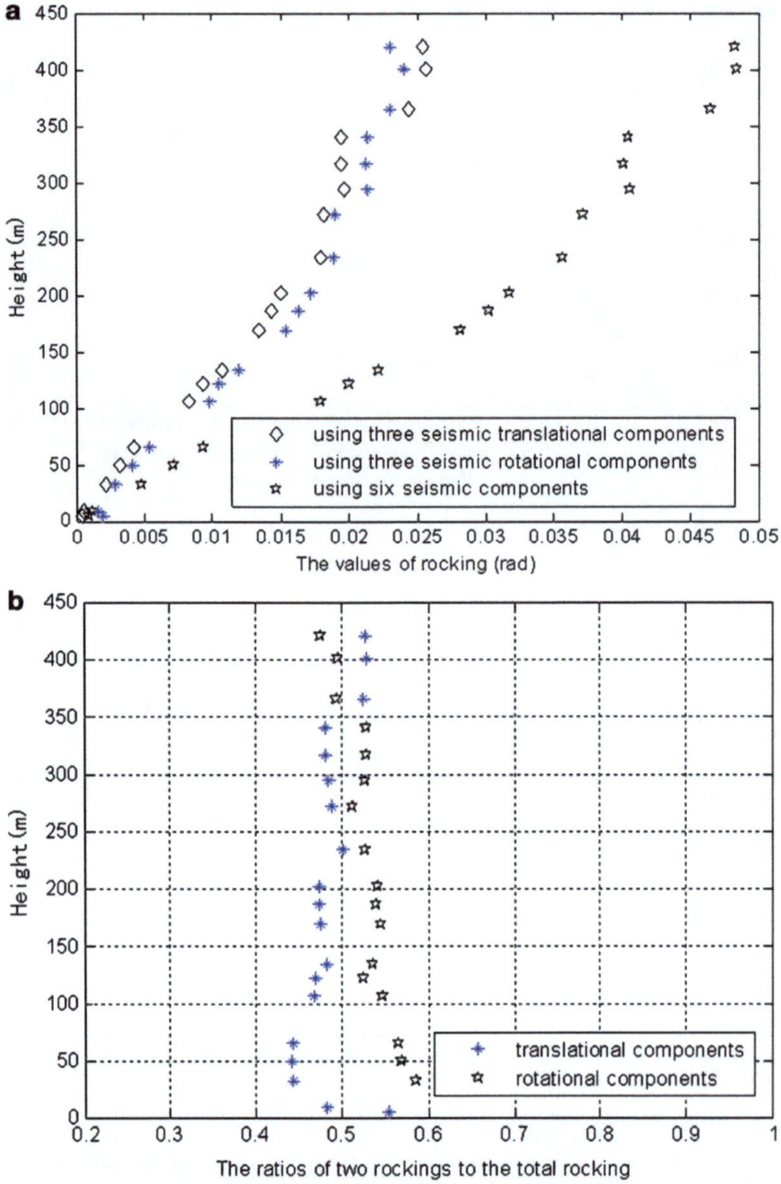

Fig. 3.6 (**a**) Rocking caused by rotational and translational components and (**b**) their contribution to the total structural rocking

In terms of the torsion of the structure, the translational components contribute about 10% and the rotational components contribute 90% to the total response (Fig. 3.7a, b). In other words, the torsion damage of the structure is mainly caused by the rotational components. However, for structures like Jin Mao Tower, the seismic rotational components are negligible.

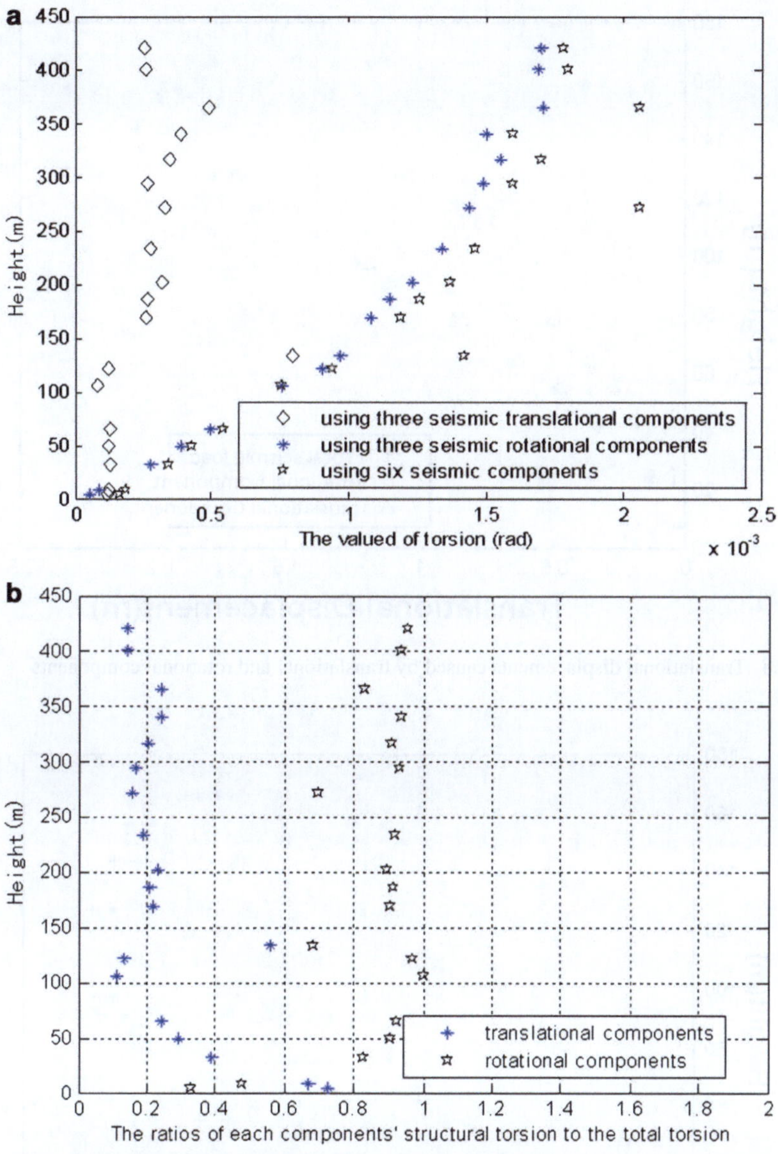

Fig. 3.7 (**a**) Structure torsion caused by translational and rotational components and (**b**) their contribution to the total torsion

3.3.2 Response of 160-m-Height Reinforced Concrete Frame-Shear Wall Structure

We also figure out the translational displacement, rock, and torsion of the 160-m-height reinforced concrete frame-shear wall structure, shown in Figs. 3.8, 3.9, and 3.10.

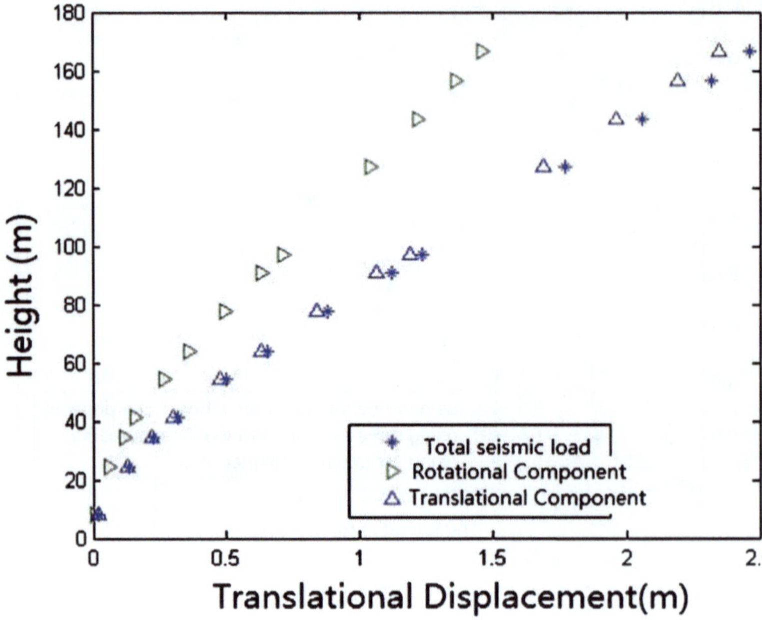

Fig. 3.8 Translational displacements caused by translational and rotational components

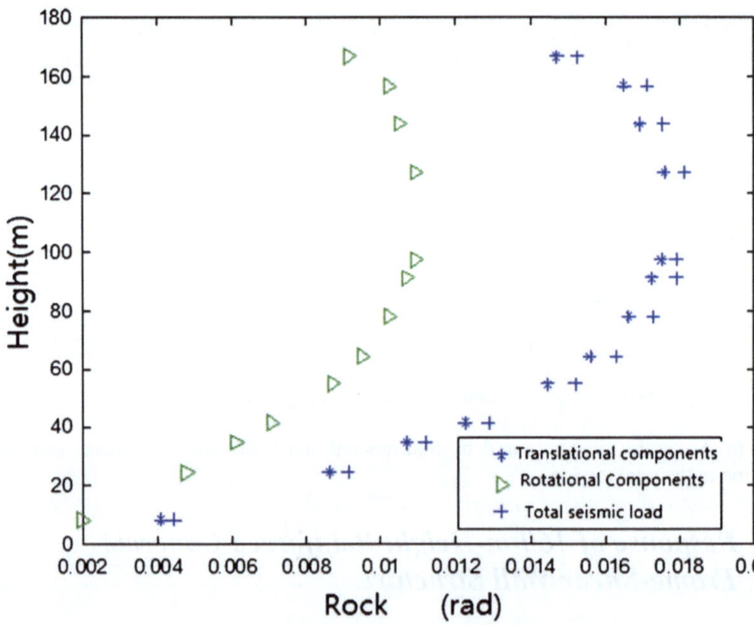

Fig. 3.9 Rocking caused by rotational and translational components

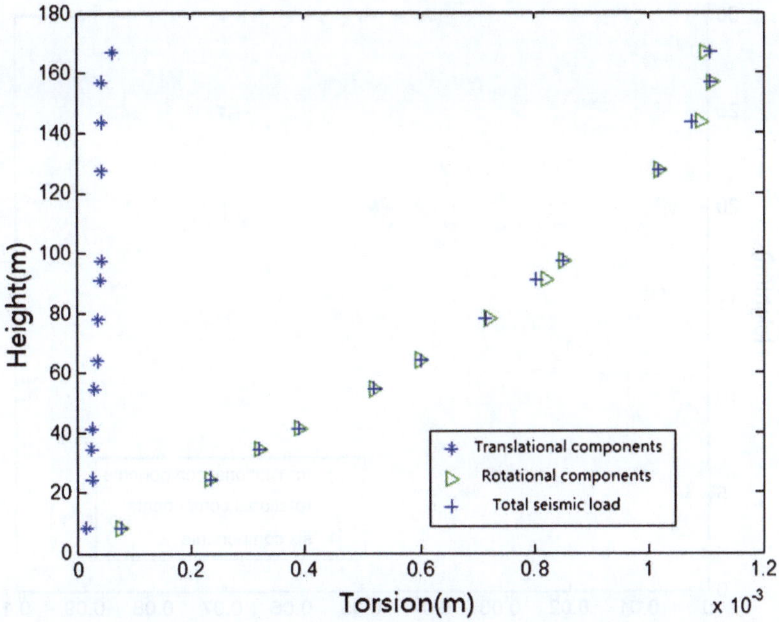

Fig. 3.10 Structure torsion caused by translational and rotational components

From Fig. 3.8, we can see that the displacement caused by rotational components only is more than half of the total displacement (translational and rotational). It needs to be noted that the total displacement is just a little larger than the displacement caused by the translational components only.

The change of structural rock is similar with its translational displacement; the rock caused by the rotational components is more than half of the total rock, while the total rock is just a little larger than that caused by the translational components.

Similar to the torsion of the Jin Mao Tower, most of the torsion phenomena are caused by the rotational components.

3.3.3 Response of 30-m-Height Monument

Because structural rocking is similar with the translational displacement, and the structural torsion is mostly caused by the rotational components, for the 30-m-height monument, we only calculate the translational displacement which is shown in Fig. 3.11. We can see from the figure that the input rotational components do not increase structural translational displacement response; the contrary, it makes the translational displacement smaller.

For the 30-m-tall symmetric structure, the response caused by the rotational components only is about 10% of the total responses (translational plus rotational).

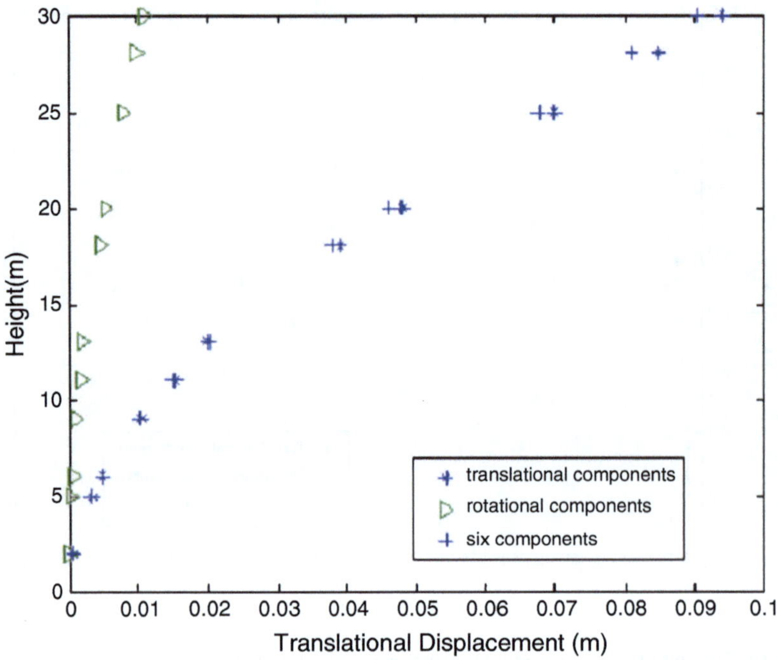

Fig. 3.11 Translational displacement caused by translational and rotational components

If we load the rotational components to the structure which has already been loaded with the translational components, the response of the structure will be reduced. For the high-rise building, the response caused by the rotational components is about 50% of the total responses (translational plus rotational). But when we load the rotational components to 160-m-high structure, the magnified effect is much smaller than the response caused by the rotational components only; for 443-m-high structure, the responses caused by the translational components and the rotational components can be directly added.

3.4 Comparison of Two Kinds of Analyzed Results with Different Method

3.4.1 Analyzed Result with Eurocode8 Method

Taking one reinforced concrete chimney as an example, Zembaty and Boffi calculated how much the rotational components account for in the whole seismic load according to the Eurocode8 (Zembaty and Boffi 1994). The simplified chimney

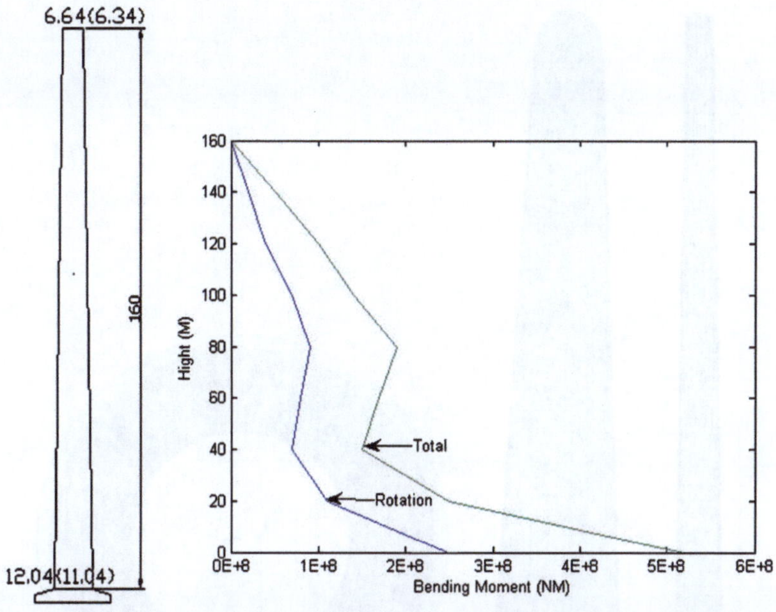

Fig. 3.12 Reinforced concrete chimney model and the bending moment (Zembaty and Boffi 1994)

model and the calculated results are shown in Fig. 3.12, in which we can see the inner and outer diameters of the chimney. Two different bending moments are caused by rotational components and by total seismic load.

From Fig. 3.12, we can see that the rotational components cause about half of the bending moment which caused by all of the ground motions. According to Eurocode8, the total structural seismic response equals the sums of responses caused by rotational components and translational components. We can deduce that the other half of structural response is caused by translational components. Moreover, the two kinds of seismic components almost play the same role in damaging structure, which makes the rotational component a serious concern.

3.4.2 *Analyzed Result with Theoretical Method (ANSYS Software)*

We use the same chimney (dimensions, material, etc.) as an example. The model is shown in Fig. 3.13. Loading seismic motions from Mianzhu seismic station in Wenchuan earthquake, we calculate the maximum structural translational displacement with ANSYS software, and the results are shown in Fig. 3.14. From the figure we can see the amount of rotational components account for in the total

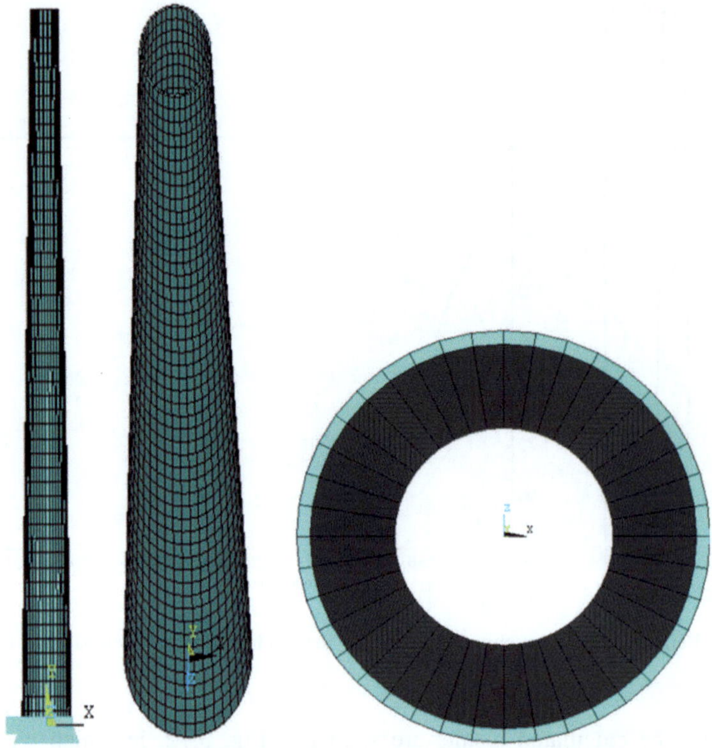

Fig. 3.13 Reinforced concrete chimney model

seismic load; also we can see the magnified displacement when we add the rotational components to the chimney which has been loaded with translational components. The difference between two seismic loadings can be analyzed.

3.4.3 Comparison of Two Kinds of Analyzed Results

Taking the structural translational results as baseline, Eurocode8 suggests that the rotational components are as important as the translational components, while the theoretical result indicates that the contribution of rotational components is much smaller than that of translational components. Eurocode8 could be too conservative in estimating the rotational components load.

Additionally, there is no evidence that the total earthquake effect can be equivalent to the sum of translational effect and rotational effect even for smaller rotational components response and larger translational components response. Except for the Jin Mao Tower, for other structures in this chapter, the increased response caused by adding rotational components to the structure, which has loaded translational components, is much smaller than that caused just by rotational components.

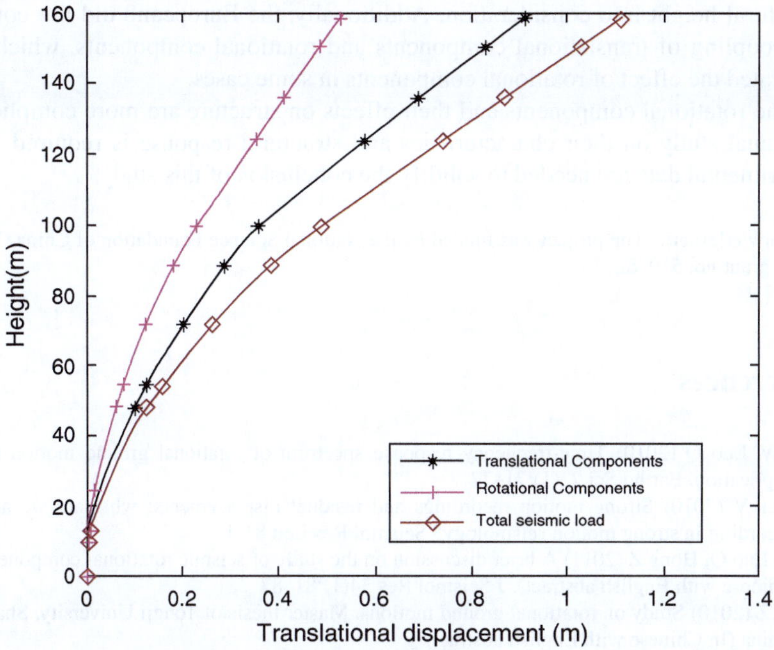

Fig. 3.14 Theoretical displacement response of chimney

3.5 Conclusions

The research in this chapter shows that the symmetric structure torsion caused by the rotational components is much larger than that caused by the translational components, and it is the rotational components that caused the symmetric structure damage. For the symmetric structures, the torsion caused by the translational components can be ignored.

For 30-m-tall symmetric structures, the responses caused by the rotational components are about 10% of the total responses (translational plus rotational components); the total response will be reduced when we add rotational components to the structures which have been loaded with translational components. For the high-rise building, the responses caused by the rotational components are about 50% of the total responses, but for 160-m-tall structure, when added the rotational components, the increased responses are much smaller than that just caused by the rotational components; for 443-m-tall structure, the total response is the sum of the responses caused by the translational components and the rotational components separately. The research implies that the rotational components effect on the structure is structure height dependent.

The calculation example indicates that European8 could be conservative in estimating the seismic rotational loads. Optimization could be made by taking the structural height into consideration. Additionally, the European8 did not consider the coupling of translational components and rotational components, which may overrated the effect of rotational components in some cases.

The rotational components and their effects on structure are more complicated; continual study on their characteristics and structural response is required. More experimental data are needed to solidify the conclusion of this study.

Acknowledgment The project was funded by the National Science Foundation of China (NSFC) under grant no. 51078273.

References

Che W, Luo Q (2010) Time-frequency response spectrum of rotational ground motion and its application. Earthq Sci 23(1):71–77

Graizer V (2010) Strong motion recordings and residual displacements: what are we actually recording in strong motion seismology? Seismol Res Lett 81:4

He C, Luo Q, Hong Z (2011) A brief discussion on the study of seismic rotational components (In Chinese with English abstract). J Seismol Res 34(1):81–87

Hong Z (2010) Study on rotational ground motions. Master thesis of Tongji University, Shanghai, China (In Chinese with English abstract)

Lee WHK, Celebi M, Todorovska MI, Igel H (2009) Introduction to the special issue on rotational seismology and engineering applications. Bull Seismol Soc Am 99(2B):945–957

Li H, Sunarez LE, Singh MP (1997) Rotational components of earthquake ground motions (In Chinese with English abstract). Earthq Eng Eng Vib 17(2):37–51

Newmark NM (1969) Torsion in symmetrical buildings. In: Proceedings of the 4th WCEE, vol 2, A–3, pp. 19–32. Santiago, Chile

Sun S, Chen G (1998) Synthesis method for estimation of rotation components of ground motion (in Chinese with English abstract). J Seismol 1:19–24

Teisseyre R, Takeo M, Majewski E (eds) (2006) Earthquake source asymmetry, structural media and rotation effects. Springer, Berlin

Teisseyre R, Nagahama H, Majewski E (eds) (2008) Physics of asymmetric continua: extreme and fracture processes: earthquake rotation and Soliton waves. Springer, Berlin/Heidelberg

Trifunac MD (1982) A note on rotational components of earthquake motions on ground surface for incident body waves. Soil Dyn Earthq Eng 1(1):11–19

Wang J, Hu Y (1991) Research on the seismic rotational components (In Chinese with English abstract). Earthq Eng Eng Vib 11(2):1–10

Yang F, Luo Q, Che W (2010) Torsional phenomena in the 2008 great Wenchuan earthquake. Earthq Sci 23:75–83

Zembaty Z, Boffi G (1994) Effect of rotational seismic ground motion on dynamic response of slender towers. Eur Earthq Eng 8:3–11

Part II
Seismic Behaviour of Irregular Structures

Chapter 4
The Effect of Common Irregularities on the Seismic Performance of Existing RC Framed Buildings

Angelo D'Ambrisi, Mario De Stefano, Marco Tanganelli, and Stefania Viti

Abstract This chapter deals with the seismic performance of irregular 3D RC existing framed structures subjected to seismic actions. More specifically, the effect of the noncoincidence between the mass and the stiffness centers on the seismic response of these structures is investigated. The analysis is performed on a 4-story 3D framed sample structure designed for vertical loads only. A very detailed nonlinear model of the structure is implemented under the computer code SeismoStruct. The seismic response of the structure is analyzed performing a nonlinear incremental dynamic analysis. The obtained response domain is compared with the limit values provided by FEMA 356 for the different limit states. The effect of the introduced irregularities on the seismic performance of the structure is not negligible despite the low value of the eccentricity. The performed analysis evidences that a particular attention has to be paid to the seismic behavior of RC buildings realized in the 1960s and 1970s, before the adoption of seismic codes, since even light irregularities can consistently affect their seismic performance.

4.1 Introduction

A significant part of the Italian building heritage is constituted by reinforced concrete buildings designed according to inadequate rules and realized with concrete having poor mechanical properties during the intense constructive activity experienced by the country in the in the 1960s and 1970s, before the adoption in 1971 of the first code regarding reinforced concrete buildings (Legge 5/11/1971 n 1086) and

A. D'Ambrisi • M. De Stefano (✉) • M. Tanganelli • S. Viti
Dipartimento di Costruzioni e Restauro, Facoltà di Architettura, Università degli Studi di Firenze, Piazza Brunelleschi 6, 50121 Florence, Italy
e-mail: angelo.dambrisi@unifi.it; mario.destefano@unifi.it; marco.tanganelli@unifi.it

O. Lavan and M. De Stefano (eds.), *Seismic Behaviour and Design of Irregular and Complex Civil Structures*, Geotechnical, Geological and Earthquake Engineering 24, DOI 10.1007/978-94-007-5377-8_4,

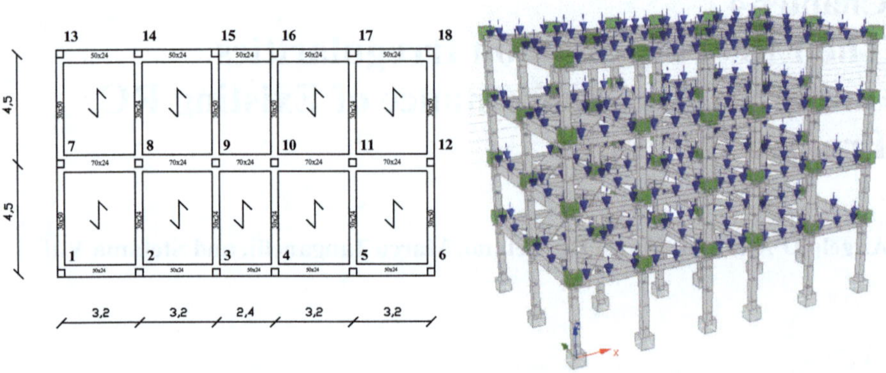

Fig. 4.1 Plan and 3D view of the considered sample structure

in 1974 of the first seismic code (Legge 2/2/1974 n. 64 1974). The evaluation of the seismic performance level of these buildings is therefore very important.

The present work deals with the evaluation of the safety of the RC buildings constructed in the 1960s and 1970s and tries to quantify the reduction in their seismic performance due to some of the typical loss of regularity.

One of the most common problems affecting such buildings is the noncoincidence between the mass and the stiffness centers and the consequent torsional effects. Usually the torsion is not considered in the design; it can be related to an asymmetry of the plan, to an irregular distribution of internal walls or balconies, or to a different use of a portion of the building. The torsional effects induced by these irregularity factors can be amplified by a nonhomogeneous distribution of the mechanical properties of the concrete very frequent in these kind of structures, poorly controlled during their construction phase.

The effect of all these irregularities on the seismic performance of existing RC framed buildings is evaluated herein with reference to a 4-story 3D framed sample structure designed for vertical loads only. The structure represents a typical existing RC building realized in Italy in the 1960s, before the introduction of seismic codes. The performed study focuses on the effect of the noncoincidence between the mass and the stiffness centers. The concrete mechanical properties and their variability are calibrated on the results of an extensive survey performed by the authors on a large sample of RC framed buildings realized before the adoption of seismic codes.

4.2 Sample Structure

The sample structure is a 4-story 3D reinforced concrete frame with two bays of 4.5 m in the *y* direction and 5 bays in the *x* direction, four of 3.2 m and one of 2.4 m, as shown in Fig. 4.1. The building is symmetric along the *x* and the *y* direction

Table 4.1 Mechanical properties of material

Materials	Class	Strength	Model
Concrete	C25/30	25/30 MPa	Mander
Steel	FeB38k	380 MPa	Bilinear

Table 4.2 Beam and column cross sections and reinforcement

Elements	B (cm)	H (cm)	Reinforcement
Beams – x	50	24	3Φ16 – 3Φ16
Beams – x	70	24	4Φ16 – 4Φ16
Beams – y	30	50	2Φ16 – 2Φ16
Beams – y	30	24	2Φ16 – 2Φ16
Columns	30	30	2Φ16 – 2Φ16

Table 4.3 Periods and modal masses of the sample structure

Period	Time (s)	% Participant mass (dir x)	% Participant mass (dir y)
1	0.927	0.0	88.1
2	0.891	83.4	0.0
3	0.806	4.7	0.0
4	0.303	0.0	8.9
5	0.289	8.8	0.0

and regular along its height. A C20/25 concrete and a FeB38k steel are assumed as materials, since they have mechanical characteristics (Table 4.1) compatible with those of the materials used in the 1960s.

The building is designed for vertical loads only, ignoring seismic loads. Vertical loads consist in dead loads and in live loads equal to 2 KN/m^2. The beam and column dimensions and reinforcement are summarized in Table 4.2.

The sample structure has been modeled with a fiber model using the computer code SeismoStruct (http://seismosoft.com) (Seismostruct v5.2.1 released 2011). The constitutive models used for the materials are that of Mander et al. (Mander et al. 1988) for the confined concrete, the three-linear model for the unconfined concrete, and the bilinear model for the reinforcing steel. Each structural element is modeled using four subelements, two external and two internal; the external subelements have a length equal to 1/15 of the total span. The contribution of the floors is computed introducing a rigid diaphragm at each floor bay. A preliminary modal analysis is performed to calculate the periods and the participation mass of the sample structure (Table 4.3).

4.3 Considered Irregularities

Some of the most common irregularities have been introduced in the sample structure to evaluate how much a not-computed torsional behavior can affect the seismic response of the building. The irregularities considered in the analysis are an asymmetric plan, an irregular distribution of balconies, a change of destination, and

Table 4.4 Values characterizing the assumed domain of f_c

Nominal value, f_{ck}	25.00 MPa
Mean value, f_{cm}	24.40 MPa
Reduced mean value, f'_{cm}	17.25 MPa
5% percentile value, $f_{ck,05}$	8.35 MPa

Table 4.5 Eccentricity of the "irregular" structure at each story

Level	Eccentricity (%)
0	1.90
1	4.90
2	5.09
3	5.09
4	0.31

a nonhomogeneity of the concrete mechanical characteristics. To obtain a building with an asymmetric plan, a further span having the same length of the others has been introduced along the *x* direction. An irregular distribution of balconies is obtained introducing a balcony on the right-hand side of the building. The changes of destination in old buildings are quite common. In the present case, it is assumed that part of the first story is changed from a residential destination to an office destination, with a consequent variation in the live loads.

In structures built in the 1960s, concrete can present mechanical properties significantly lower than the design properties due to the viscous phenomena, the degradation, but mainly to the low-quality level of the constructions realized in those years (Cristofaro 2009; D'Ambrisi et al. 2010). As a result, it is common to observe a strong nonhomogeneity in the distribution of concrete strength in the different parts of a building, with *coefficients of variation* up to 50% (Cristofaro 2009) and, consequently, an irregular distribution of the stiffness, different from that assumed in the analysis.

In the present work, a *coefficient of variation* of 40% has been assumed for the strength domain of concrete, while its mean value has been evaluated as the difference between the *mean value* and the *standard deviation* (Manfredi et al. 2007; American Society of Civil Engineering 2000). The nominal characteristic value of strength (f_{ck}), the average value including viscous effects (f_{cm}), the reduced value proposed by Manfredi et al. (f'_{cm}) (Manfredi et al. 2007), and the value corresponding to a 5% percentile of the distribution having f'_{cm} as mean value are reported in Table 4.4.

The effect of the irregularity in the distribution of the mechanical properties of concrete has been evaluated in a simplified way assuming that in the lower story the three columns belonging to the right column line have a strength equal to $f_{ck,\,05}$, while all the other columns have a strength equal to f'_{cm}.

All the above-described irregularities have been introduced in the sample building to obtain an irregular building whose seismic response can be compared with that of the sample structure. The eccentricities of each story of the irregular structure are summarized in Table 4.5.

4.4 Seismic Response

4.4.1 *Incremental Dynamic Analysis*

An incremental dynamic analysis has been performed on both the regular and the irregular structure with the computer code SeismoStruct (2011) considering as response parameters the top story displacement (*TD*) and the interstory drift (*ID*). The analysis has been performed considering a set of ground motions scaled to represent different seismic intensities, equal to 0.10, 0.15, 0.20, 0.25, 0.30, 0.35, and 0.40 g, respectively. For each value of PGA, the response of the structures has been defined elaborating the response domains from the assumed ensemble of ground motions.

4.4.2 *The Seismic Input*

The considered seismic input is constituted by an ensemble of seven ground motions extracted by a database of strong motions recorded in Italy (Itaca 2008) considering a PGA equal to 0.25 g, a nominal life of the structure of 50 years, and a magnitude ranging between 5.5 and 6.5. The data relative to the utilized ground motions are reported in Table 4.6. The selected ground motions have an average spectrum that approaches very well the one provided by Eurocode 8 (2005) for a soil of type B, as it is evident from Fig. 4.2.

4.4.3 *Assumed Limit States*

The maximum *ID* obtained for both the regular and the irregular sample structure has been compared with the standard values provided by FEMA 356 (American Society of Civil Engineering 2000) for the limit states *immediate occupancy* (*IO*), *life safety* (*LS*), and *collapse prevention* (*CP*) reported in Table 4.7.

Table 4.6 Ground motions data

Name	Location	Date (gg/mm/yyyy)	PGA (g)	Magnitude (L-S-W)	Duration (s)
Irpinia	Sturno	23/11/1980	0.225	6.5 – 6.8 – 6.9	70.75
Irpinia	Calitri	23/11/1980	0.174	6.5 – 6.8 – 6.9	85.99
L'Aquila	Colle Grilli	06/04/2009	0.446	5.8 –···– 6.3	100.00
L'Aquila	Aquil Park Ing	06/04/2009	0.353	5.8 –···– 6.3	100.00
L'Aquila	Aquil Park Ing	06/04/2009	0.330	5.8 –···– 6.3	100.00
L'Aquila	Centro Valle	06/04/2009	0.545	5.8 –···– 6.3	100.00
L'Aquila	Centro Valle	06/04/2009	0.657	5.8 –···– 6.3	100.00

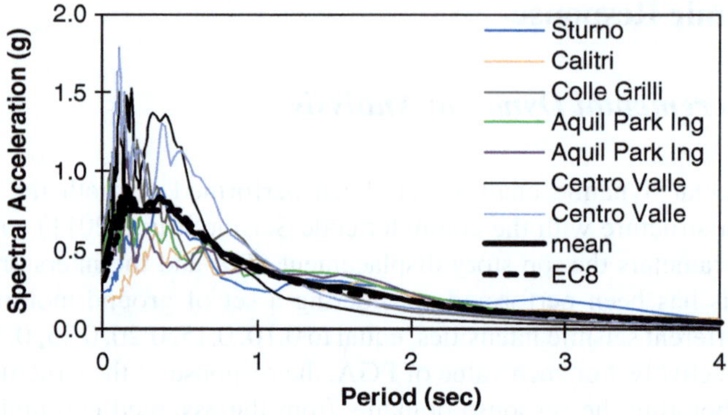

Fig. 4.2 Elastic response spectra of the considered ground motions

Table 4.7 Considered limit states

Limit states	Drifts (%)
Immediate occupancy	1
Life safety	2
Collapse prevention	4

4.5 Obtained Results

Figure 4.3 shows the results in terms of top story displacements (*TD*) obtained from the performed analyses for both the regular and the irregular structure. Each response domain refers to the values of the maximum *TD* obtained from the dynamic analysis in each column line, perpendicularly to the considered seismic excitation. Indeed, due to the introduced eccentricities, the response of the irregular structure varies along the longitudinal axis of the building.

As it can be observed from Fig. 4.3, the *TD* of the irregular structure at the column line 2, that is, the flexible side of the structure, is consistently larger than the corresponding value of the regular structure.

To better evidence the difference in the seismic response of the two considered structures, the values of the *TD* at each column line have been normalized with respect to the *TD* of the center of mass of the building. The obtained normalized displacements are shown in Fig. 4.4. As it can be observed from the figure, both at the stiff and at the flexible side of the structure, there is a difference of about 10% in the *TD* as a consequence of the introduced irregularities.

The increase in the *TD* due to the considered eccentricity is larger for low and medium values of PGA, for which the structure evidences an elastic response. With the increase of the PGA, indeed, the inelastic involvement of the structure increases and the sensitivity of its response to the torsional effects in terms of *TD* decreases (De Stefano and Pintucchi 2010).

The maximum values of the interstory drift obtained at each story are compared with the limit values provided by FEMA (Fig. 4.5) (American Society of Civil

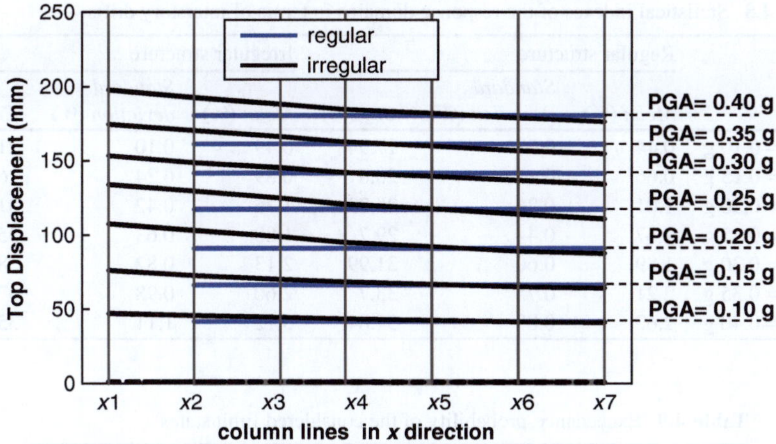

Fig. 4.3 Top story displacement at each column line

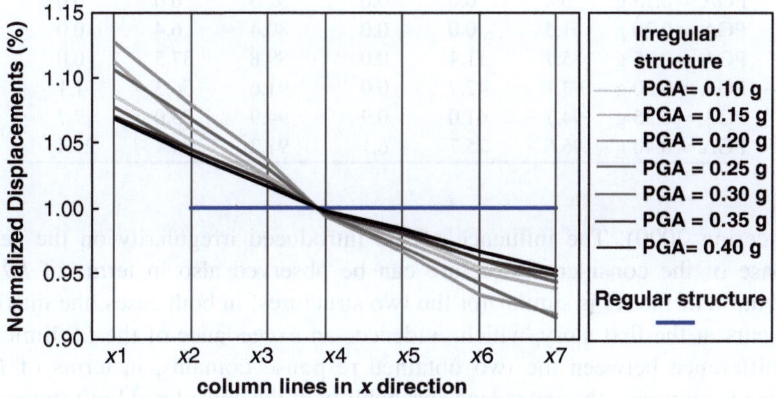

Fig. 4.4 Normalized top story displacement at each column line

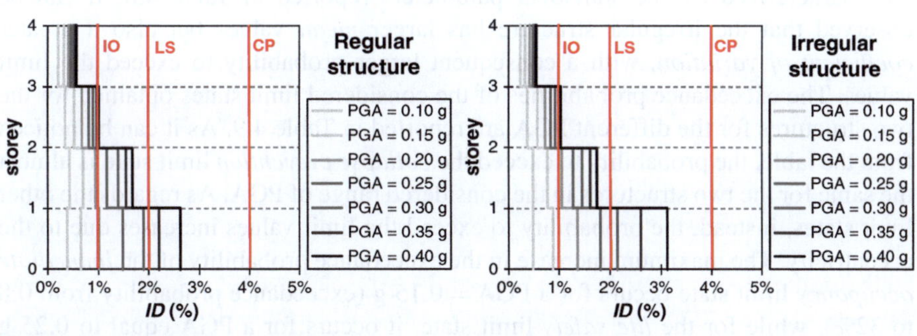

Fig. 4.5 Maximum interstory drift (*ID*)

Table 4.8 Statistical indexes of the response domains in terms of interstory drifts

	Regular structure			Irregular structure		
	Mean (%)	*Standard deviation* (%)	*Cov* (%)	*Mean* (%)	*Standard deviation* (%)	*Cov* (%)
PGA = 0.10 g	0.44	0.08	17.94	0.49	0.10	21.28
PGA = 0.15 g	0.66	0.14	20.8	0.89	0.24	26.9
PGA = 0.20 g	1.01	0.26	25.73	1.36	0.42	31.29
PGA = 0.25 g	1.47	0.44	29.7	1.80	0.63	35.2
PGA = 0.30 g	1.89	0.60	31.99	2.13	0.82	38.70
PGA = 0.35 g	2.21	0.75	33.7	2.60	0.98	37.7
PGA = 0.40 g	2.62	0.89	34.04	3.12	1.11	35.64

Table 4.9 Exceedance probability of the considered limit states

	Regular structure			Irregular structure		
	IO (%)	LS (%)	CP (%)	IO (%)	LS (%)	CP (%)
PGA = 0.10 g	0.0	0.0	0.0	0.0	0.0	0.0
PGA = 0.15 g	0.8	0.0	0.0	32.3	0.0	0.0
PGA = 0.20 g	51.5	0.0	0.0	80.4	6.4	0.0
PGA = 0.25 g	85.7	11.4	0.0	89.8	37.5	0.0
PGA = 0.30 g	93.1	42.7	0.0	91.6	56.3	1.1
PGA = 0.35 g	94.7	61.0	0.9	94.9	73.0	7.7
PGA = 0.40 g	96.6	75.7	6.1	97.2	84.4	21.4

Engineering 2000). The influence of the introduced irregularity on the seismic response of the considered structure can be observed also in terms of *ID*. The distribution of the *ID* is similar for the two structures; in both cases, the maximum *ID* occurs at the first story, which evidences an exceedance of the *LS* limit state. The difference between the two obtained response domains, in terms of *ID*, is evaluated comparing the exceedance probability of the considered limit states of the two structures. The exceedance probability of the limit values has been determined assuming a normal distribution for the two response domains; the two distributions are characterized by the statistical parameters reported in Table 4.8. It can be observed that the irregular structure has larger *mean* values but also a greater *coefficient of variation*, with a consequent larger probability to exceed the limit values. The exceedance probabilities of the considered limit states obtained for the two structures for the different PGA are reported in Table 4.9. As it can be noticed from the table, the probability to exceed the *collapse prevention* limit state is almost the same for the two structures in the considered range of PGA. As regards the other limit states, instead, the probability to exceed the limit values increases due to the eccentricity. The maximum increase in the exceedance probability of the *immediate occupancy* limit state occurs for a PGA = 0.15 g (exceedance probability from 0.8 to 32%), while for the *life safety* limit state, it occurs for a PGA equal to 0.25 g (exceedance probability from 11 to 37%).

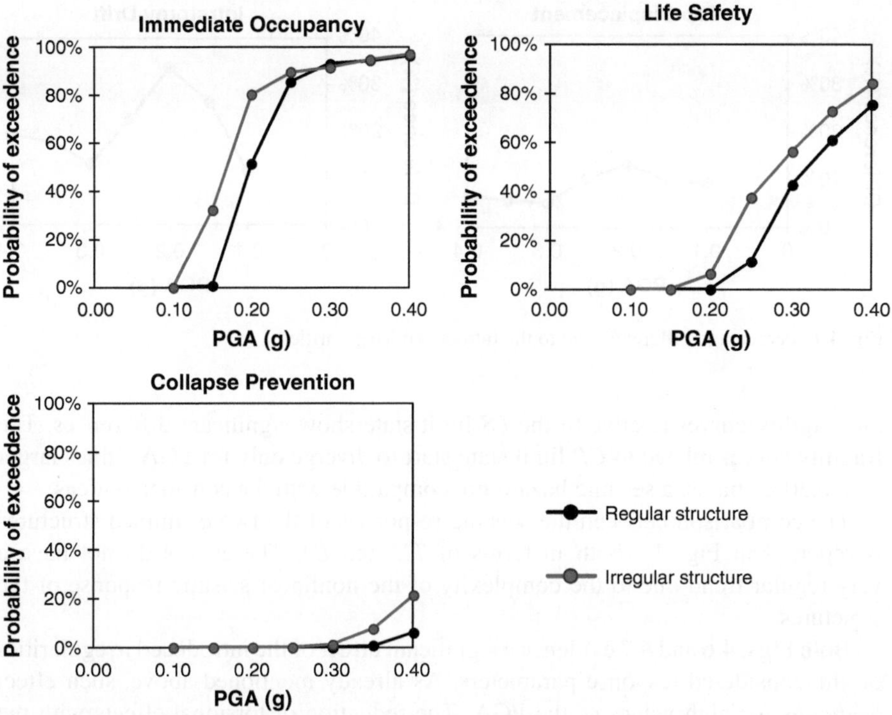

Fig. 4.6 Linearized fragility curves

The high sensitivity of the structure to the introduced eccentricity can be related to the low quality of the concrete and, therefore, to the reduced strength and stiffness of the structure.

In Fig. 4.6, the probabilities of the structures to exceed the considered limit states are plotted versus the PGA. The curves represent, in a linearized form, the fragility curves of the two structures with regard to the considered limit states. Each point of the curves represents the probability (P) of the response parameters (r) of the structure to exceed the limit value (lv) corresponding to the assumed performance level under a seismic input of given intensity (PGA) (Barron Corvera 2000; Reinhorn and Barron 1999), according to the following expression:

$$\text{Exceedance probability curve} = p\left[r > \frac{\text{lv}}{\text{PGA}}\right] \tag{4.1}$$

It can be noticed that the introduced irregularities induce a significant increase of the exceedance probability of the considered limit states. This increase depends on the range of considered PGA. In the range of interest for the considered buildings, that is, medium-low seismicity with a PGA ranging between 0.10 g and 0.20 g, the fragility curves diverge for the *IO* limit state only. For PGA larger than 0.20 g,

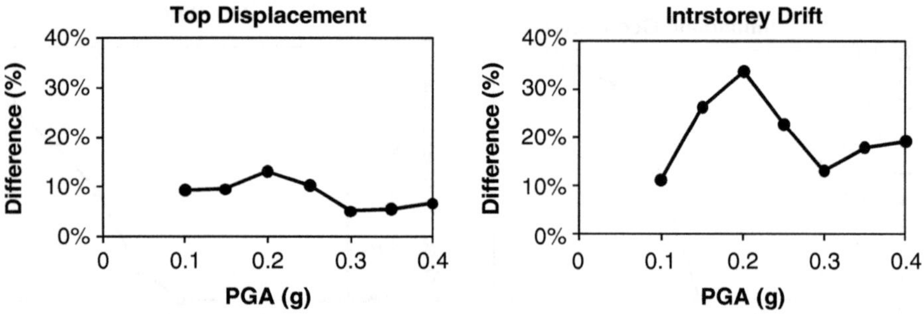

Fig. 4.7 Percentage difference due to the introduced irregularities

the fragility curves relative to the *LS* limit state show significant differences. The fragility curves related to *CP* limit state start to diverge only for PGA values larger than 0.30 g, that is, a seismic hazard not compatible with the considered areas.

The comparison between the seismic responses of the two examined structures is reported in Fig. 4.7 both in terms of *TD* and *ID*. The curves do not have a very regular trend due to the complexity of the nonlinear seismic response of the structures.

Both Figs. 4.6 and 4.7 evidence a significant effect of the introduced irregularities on the considered response parameters. As already mentioned above, such effect decreases for high values of the PGA. The reduction of torsional effects with the increase of the PGA is related to the inelastic involvement of the structure. In the case of strong earthquakes, the seismic response of the structure depends on the ductility available at local level; therefore it is important to check both the global and the local response parameters of the performed analysis.

To check the local ductility demand of the two sample structures, the values of the rotation at the chord, ϑ, at each beam end have been detected. Such values have been compared with the ultimate value of ϑ_u provided by Eurocode 8 (2005) and other national codes (NTC 2008). The exceedance of ϑ_u is suggested by EC8 as an acceptance conditions; therefore, if such limit is exceeded even in only one element, the structure does not satisfy the safety requirement. To check the inelastic involvement of the structures, a damage index is introduced, measuring the percentage of elements in which ϑ_u is exceeded. Damage index versus the PGA curves of the regular and irregular structures is reported in Fig. 4.8. The regular structure remains under the limit value of rotation in all cases, while in the irregular structure, a number of elements, increasing with the increase of the PGA (PGA larger than 0.20 g), exceed the limit. From these results, it is possible to assume that the irregular structure for high values of PGA experiences a large inelastic involvement, evidencing a distribution of damage that exceeds the acceptance level defined by the EC8.

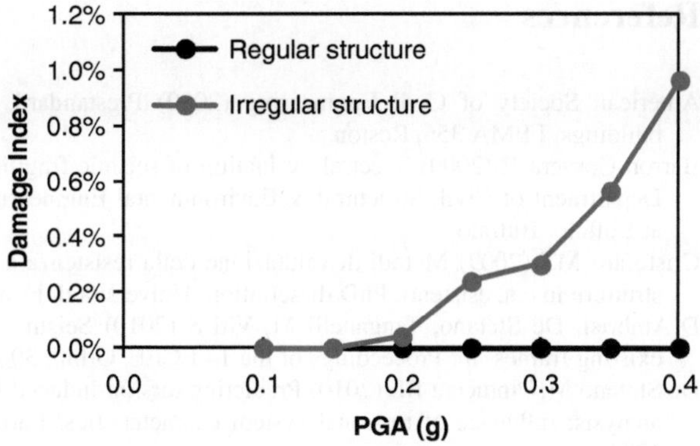

Fig. 4.8 Percentage of elements exceeding ϑ_u

4.6 Conclusions

In this chapter, the seismic performance of a typical RC framed structure built in Italy in the 1960s, designed for horizontal loads only, has been studied. The effect of common irregularities such as asymmetric plan, irregular distribution of balconies, different live loads, and nonhomogeneous mechanical properties of the concrete has been investigated, by introducing at each storey the equivalent eccentricity, that resulted to be around 5%. A seismic input consistent with the seismic hazard of the area has been considered assuming a set of seven ground motions, spectrum compatible with the EC8 provisions.

The introduced eccentricity, despite being very low, leads to non-negligible torsional effects. An increase larger than 10% has been found for the top story displacement, while the increase in the first-story interstory drift ranges between 7 and 25% depending on the considered PGA.

The evaluation of the exceedance probability of the assumed limit states has evidenced a significant sensitivity of the structure to the introduced irregularities. As regards the *life safety* limit state, an increase of the exceedance probability from 11 to 37% has been found (PGA = 0.25%), while for *immediate occupancy* limit state, the exceeding probability goes from 0.8 to 32% (PGA = 0.15 g).

The seismic performance level of the two structures has been studied comparing their fragility curves for each considered limit state. The comparison has evidenced a significant sensitivity of the seismic response to the introduced irregularities, depending, for each limit state, on the intensity of the seismic input.

The obtained results show that common irregularities significantly affect the seismic response of RC buildings realized with concrete having poor mechanical properties. To obtain more general results, further analyses should be performed on sample structures characterized by different geometrical and mechanical features.

Acknowledgments This work has been supported by the project DPC-ReLUIS 2010/2013 Task 1.1.2. "Strutture in cemento armato ordinarie e prefabricate."

References

American Society of Civil Engineering (2000) Prestandard for the seismic rehabilitation of buildings. FEMA 356, Reston

Barron Corvera R (2000) Spectral evaluation of seismic fragility of structures. PhD dissertation, Department of Civil, Structural & Environmental Engineering, State University of New York at Buffalo, Buffalo

Cristofaro MT (2009) Metodi di valutazione della resistenza a compressione del calcestruzzo di strutture in c.a. esistenti. PhD dissertation, Università di Firenze (in Italian)

D'Ambrisi, De Stefano, Tanganelli M, Viti S (2010) Seismic performance of irregular 3D RC existing frames. In: Proceedings of the 14 ECEE, Orhid, 30 Aug–3 Sept 2010

De Stefano M, Pintucchi BL (2010) Predicting torsion-induced lateral displacements for pushover analysis: influence of torsional system characteristics. Earthq Eng Struct Dyn 39(12):1369–1394

EC 8-3 (2005) Design of structures for earthquake resistance, part 3: strengthening and repair of buildings, European standard EN 1998-3. European Committee for Standardization (CEN), Brussels

Itaca (2008) Data base of the Italian strong motions data. http://itaca.mi.ingv.it

Legge 5/11/1971 n. 1086 (1971) Norme per la disciplina delle opere in conglomerato cementizio armato, normale e precompresso ed a struttura metallica. G.U. 321 del 21/12/1971 (in Italian)

Legge 2/2/1974 n. 64 (1974) Provvedimenti per le costruzioni con particolari prescrizioni per le zone sismiche. G.U. del 21/04/1974 (in Italian)

Mander JB, Priestley MJN, Park R (1988) Theoretical stress-strain model for confined concrete. J Struct Eng 114(8):1804–1826

Manfredi G, Masi A, Pinho R, Verderame G, Vona M (2007) Valutazione degli edifici esistenti in Cemento Armato. IUSS PRESS, Fondazione EUCENTRE, Pavia, ISBN: 978-88-6198-013-6

NTC (2008) D.M. Infrastrutture 14 gennaio 2008. Nuove Norme Tecniche per le Costruzioni, G.U. n: 29 del 4 febbraio 2009. (in Italian)

Reinhorn AM, Barron R (1999) Spectral evaluation of fragility of structures. In: Proceedings of first Romanian-American workshop in structural engineering, 1st RAWSE, Technical University of Iasi, Romania, 25–30 June 1999

Seismostruct v5.2.1 released (2011) Computer code, Finite Element package. (http://seismosoft.com)

Chapter 5
Influence of the Variability of Concrete Mechanical Properties on the Seismic Response of Existing RC Framed Structures

Angelo D'Ambrisi, Mario De Stefano, Marco Tanganelli, and Stefania Viti

Abstract In Italy, many RC framed buildings have been realized in the 1960s and the 1970s before the adoption of seismic codes. These buildings were designed for vertical loads only and were realized with concrete having poor and nonhomogeneous mechanical properties. This last condition is usually neglected despite it significantly affects the seismic response of the structures. A sample structure is considered to represent a typical existing RC building realized in Italy before the introduction of seismic codes. The structure is a four-story RC framed building designed for vertical loads only. The variability of concrete mechanical properties is evaluated on the base of an extensive survey developed by the authors on a large sample of RC framed buildings realized before the adoption of seismic codes. The seismic response of the sample structure is evaluated performing a nonlinear static analysis. The seismic demand is defined following the EC8 prescriptions. In the analysis the concrete strength has been described through a probabilistic domain, and a different value of strength, consistent with the assumed domain, has been assigned to each storey of the sample structure. Therefore, an irregular distribution of stiffness and strength is assumed along the height of the building. The seismic response of the structure is expressed in terms of maximum interstory drift. The increase of the interstory drift due to the considered irregularity is analyzed following the instructions of EC8 and FEMA 356. The obtained results show that the EC8 approach can result not

A. D'Ambrisi • M. De Stefano (✉) • M. Tanganelli • S. Viti
Dipartimento di Costruzioni e Restauro, Facoltà di Architettura, Università degli Studi di Firenze, Piazza Brunelleschi 6, 50121 Florence, Italy
e-mail: angelo.dambrisi@unifi.it; mario.destefano@unifi.it; marco.tanganelli@unifi.it; viti@unifi.it

O. Lavan and M. De Stefano (eds.), *Seismic Behaviour and Design of Irregular and Complex Civil Structures*, Geotechnical, Geological and Earthquake Engineering 24, DOI 10.1007/978-94-007-5377-8_5,

adequate to describe an irregular distribution of the concrete strength along the height of the building. The approach suggested by FEMA 356 provides a better representation of the seismic behavior of the sample structure, despite being not conservative.

5.1 Introduction

Many reinforced concrete buildings in Italy have been realized before the adoption of seismic codes; therefore, they are designed for vertical loads only and result not adequate with respect to seismic actions. In Italy, the first code regarding reinforced concrete buildings was adopted in 1971 (Legge 5/11/1971 n 1086), while the first seismic code was introduced in 1974 (Legge 2/2/1974 n. 64 1974), that is, after the intense constructive activity experienced by the country in the 1960s. Consequently, the Italian building heritage is constituted by a significant number of RC buildings realized with concrete having poor mechanical properties, whose seismic performance is well below the requirements of seismic codes.

The poor mechanical properties of concrete are worsened by the effect of the deterioration phenomena due to the age, as the viscous phenomena, the different sun exposure, and the maintenance of each part of the building, which can induce not negligible differences in the structural elements belonging to a building. Such intrinsic variability of mechanical properties of concrete is usually neglected in the seismic analysis of existing buildings, despite having a not negligible role. Seismic codes (EC 8-3 2005; American Society of Civil Engineering 2000) provide instructions for the mechanical characterization of the concrete in existing buildings, but they seem to underestimate the effect of its variability in the seismic response (Franchin et al. 2010; Jalameh et al. 2011).

In this chapter, the influence of the variability of concrete mechanical properties on the seismic response of RC buildings realized in Italy before the adoption of seismic codes is studied. A variation in the concrete strength along the building height is considered, while a constant value of the concrete strength is assumed in all the columns of each story. The concrete strength is evaluated on the base of a statistical study performed on a large number of experimental data on the concrete strength of buildings realized in the 1950s, the 1960s, and the 1970s in a central region of Italy (Cristofaro 2009). The seismic response, expressed in terms of interstory drift, is calculated performing a spectral analysis with the elastic spectrum provided by EC8 (2005). The results are then compared with those obtained applying the EC8 (2005) and the FEMA 356 (American Society of Civil Engineering 2000) provisions. This comparison evidences the important role played by the variability of concrete mechanical properties, which is not adequately taken into account by the considered codes (EC 8-3 2005; American Society of Civil Engineering 2000).

5.2 Concrete Mechanical Properties

5.2.1 Code Provisions

To simulate the seismic response of existing RC buildings, it is important to correctly define the concrete mechanical properties. Codes (EC 8-3 2005; American Society of Civil Engineering 2000) suggest to relate the concrete strength f_c to both *in situ* tests and the knowledge level of the structure.

Eurocode 8 (EC8) (2005) defines the knowledge level of the building through the confidence factor (CF) that is utilized to correct the mean value, f_{cm}, of the concrete strength obtained from *in situ* tests. The confidence factor is introduced as a function of the knowledge level of the material only, which is described on the base of the number and the type of performed tests (Table 5.1). It has to be underlined that, according to EC8 provisions, the variability of the strength distribution obtained from *in situ* tests does not affect the confidence factor. Therefore, EC8 seems to assume that the variability in concrete mechanical properties is not an intrinsic factor of the building, but it instead depends on the lack of information. If a satisfactory number of data is collected, numerical analyses can be performed adopting a concrete strength equal to the mean value obtained from the sample test (CF = 1, $f_c = f_{c,m}$).

FEMA 356 (American Society of Civil Engineering 2000) defines f_c as a function of the *coefficient of variation* (*cov*) of the domain obtained from the *in situ* tests.

Table 5.1 Knowledge levels and confidence factors according to Eurocode 8

Knowledge level	KL1	KL2	KL3
Geometry	From original outline construction drawings with sample visual survey or from full survey		
Details	Simulated design according to relevant practice and limited in situ inspection	From incomplete original detailed constructions drawings with limited in situ inspections or from extended in situ inspections	From original detailed construction drawings with limited in situ inspection or from comprehensive in situ inspection
Materials	Default value according to standards of the time and limited in situ testing	From original design specifications with limited in situ testing or from extended in situ testing	From original test reports with limited in situ testing or from comprehensive in situ testing
Analysis	Lateral Force Proc. modal response spectrum analysis	All	All
CF	1.35 (Italian annex)	1.2	1.0

Table 5.2 Design values of f_c domain (FEMA 356)

Design value of f_c	cov
$f_c = f_{c,m}$	$cov < 14\%$
$f_c = f_{c,m} - st\ dv$	$cov > 14\%$

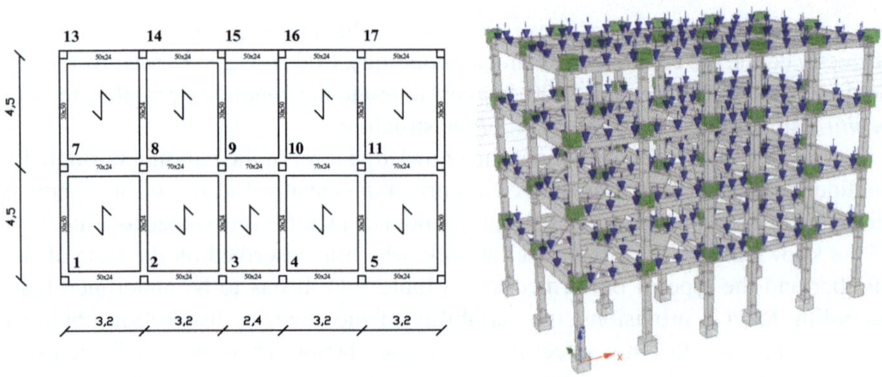

Fig. 5.1 Plan and 3D view of sample building

The value of f_c to use in the analysis, indeed, is assumed equal to the *mean* value of the sample only in the case of $cov < 14\%$. When the cov is larger than 14%, the knowledge level of the building is assumed to be not satisfactory, and further *in situ* tests are recommended. An intrinsic variability of the concrete mechanical properties due to the poor quality of the material is also considered. If despite the increase of the number of sample tests the cov remains larger than 14%, a reduced value of f_c is assumed, defined as the difference between the *mean* value and the *standard deviation* of the sample (Table 5.2).

5.2.2 *Sample Structure*

5.2.2.1 Geometrical and Mechanical Characteristics

The sample structure is a 4-story 3D reinforced concrete frame with two bays of 4.5 m in the y direction and five bays in the x direction, four of 3.2 m and one of 2.4 m, as shown in Fig. 5.1. The building is symmetric along the x and the y direction (D'Ambrisi et al. 2010).

Taking into account the results of the statistical analysis performed in Sect. 5.2, a concrete strength $f_c = 19.35$ N/mm^2 has been assumed, while a FeB38k steel is assumed as reinforcement since it has mechanical characteristics compatible with those of the materials utilized in the 1970s (Table 5.3). The building is designed for vertical loads only, ignoring seismic loads. Vertical loads consist in dead loads and in live loads equal to 2 KN/m^2. The beam and column dimensions and reinforcement are summarized in Table 5.3.

Table 5.3 Mechanical properties of materials and beam and column cross section dimensions and reinforcement

			Element	B (cm)	H (cm)	Reinf (top–bottom)
	f_c	Model	*x* beams	50	24	3Φ16–3Φ16
Concrete	19.35 MPa	Mander et al.	*x* beams	70	24	4Φ16–4Φ16
			y beams	30	50	2Φ16–2Φ16
	f_y	Model	*y* beams	30	24	2Φ16–2Φ16
Steel	380 MPa	Bilinear	Columns	30	30	2Φ16–2Φ16

	Number of columns	f_c
mean–1.5 st.dev(min)	18 (1 storey)	10.65 MPa
mean(mean)	36 (2 storeys)	19.35 MPa
mean+1.5 st.dev(max)	18 (1 storey)	28.05 MPa

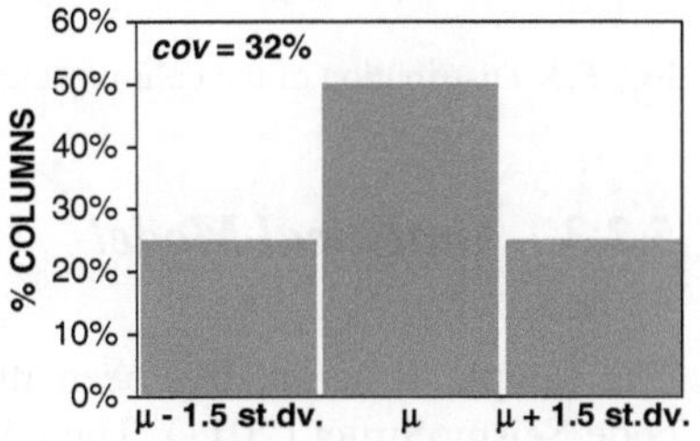

Fig. 5.2 Assumed distribution of f_c in the columns of the sample structure ($cov = 30\%$)

5.2.2.2 Distribution of the Concrete Strength in the Columns

The performed analysis is aimed to evaluate the effects of the strength variability along the height of the building on its seismic response. A constant value of f_c has been assumed at each story of the building to simplify the problem and to represent a most probable situation, since the columns of each story are usually realized at the same time and they usually present similar properties. The strength variability of the concrete of the columns significantly affects both their resistance (structural capacity) and the elastic stiffness of the building (seismic demand). The strength variability of the concrete of the beams does not seem, instead, to significantly affect the seismic response of the structure; therefore, a constant concrete strength value equal to the mean value has been assumed for the beams.

The variability of f_c has been simulated by three values only, the minimum (*min*), the mean (*mean*), and the maximum (*max*) value, respectively, obtained combining the mean value and the *standard deviation* of the assumed distribution, as can be noticed in Fig. 5.2. The mean value has been assigned to all columns of two stories, while the minimum and the maximum values have been assigned to the columns of one story only. In this way the *cov* evaluated on all the columns is equal to 32%, very close to the *cov* assumed for the concrete strength.

The assumed values of f_c together with their distribution are shown in Fig. 5.2. The distribution of the assumed values of f_c along the building height in the 12 cases which constitute the considered sample is reported in Fig. 5.3.

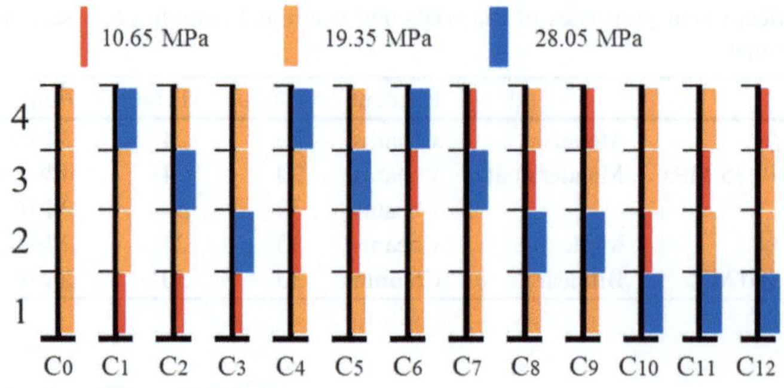

Fig. 5.3 Distribution of the column strength in elevation

5.2.3 *Analytical Model*

The sample structure has been modeled with a fiber model using the computer code SeismoStruct (2011). The constitutive models used for the materials are that of Mander et al. (1988) for the confined concrete, the three-linear model for the unconfined concrete, and the bilinear model for the reinforcing steel. Each structural element is modeled using four subelements, two external and two internal ones; the external subelements have a length equal to 1/15 of the total span. The contribution of the floors is computed introducing a rigid diaphragm at each floor bay.

5.3 Seismic Input

As seismic input it has been assumed the elastic spectrum provided by Eurocode 8 (type 1) for the soil type B. Different seismic intensities compatible with the seismic hazard of the area have been considered, corresponding to PGA values ranging between 0.15 and 0.30 g, with a step of 0.05 g. The assumed seismic spectra expressed as a function of T and of the spectral displacement are shown in Fig. 5.4.

5.4 Seismic Analysis

The seismic response of the structure is evaluated performing a spectral analysis. Both the structural capacity and the seismic demand are represented in the spectral coordinates with reference to a SDOF system. A bilinear diagram is assumed to represent the structural capacity of the system according to the N2 approach (Fajfar 1999).

In this section the results of the seismic analysis of the sample structure are compared with those obtained following the EC8 (2005) and the FEMA 356 provisions (American Society of Civil Engineering 2000).

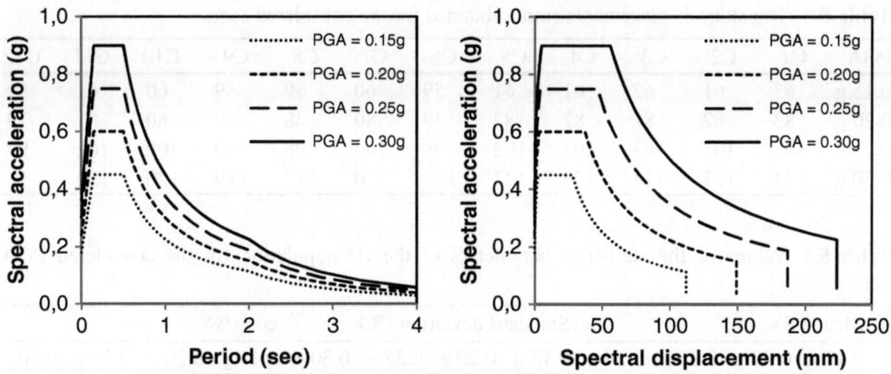

Fig. 5.4 Elastic spectra obtained by the EC8 provisions

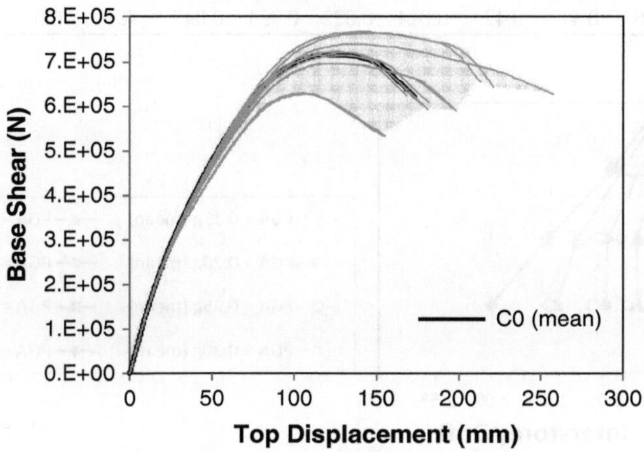

Fig. 5.5 Capacity curves for the considered cases

5.4.1 Response Domain of the Sample Structure

The seismic response of the building in both the considered cases is determined intersecting the capacity curve with the spectral demand. The capacity curves are defined assuming a horizontal force pattern proportional to the first vibration mode. The ultimate point of each curve corresponds to a value of the shear equal to the 85% of its maximum value. Each curve is linearized equalizing the areas delimited by the intersection of the two diagrams, to represent the capacity of the equivalent SDOF system. The capacity curves obtained for the considered cases (C1–C12) are shown in Fig. 5.5.

The elastic response of both models is determined intersecting the elastic branch of the capacity curve with the elastic spectral demand. The intersection point leads to the determination of the ductility factor of each case and, therefore, to the target displacement. The obtained top story displacement (*TD*) is shown in Table 5.4.

Table 5.4 Top story displacements (mm) obtained for the considered cases

PGA	C1	C2	C3	C4	C5	C6	C7	C8	C9	C10	C11	C12
0.15 g	63	61	62	61	61	59	60	59	59	60	61	59
0.20 g	83	82	82	82	82	79	80	78	79	80	81	79
0.25 g	103	102	103	102	102	98	100	98	99	100	102	99
0.30 g	124	123	124	123	122	118	120	117	119	120	122	118

Table 5.5 Values of the statistical parameters of the *ID* population for the considered PGA values

	Mean (%)				Standard deviation (%)				cov (%)			
	0.15 g	0.20 g	0.25 g	0.30 g	0.15 g	0.20 g	0.25 g	0.30 g	0.15 g	0.20 g	0.25 g	0.30 g
1	0.56	0.79	1.09	1.49	0.078	0.120	0.205	0.320	13.9	15.3	18.8	21.5
2	0.63	0.84	1.03	1.19	0.045	0.072	0.118	0.195	7.2	8.6	11.4	16.5
3	0.47	0.60	0.70	0.74	0.029	0.048	0.081	0.114	6.2	7.9	11.6	15.3
4	0.29	0.36	0.41	0.42	0.014	0.022	0.034	0.047	4.8	6.0	8.5	11.0

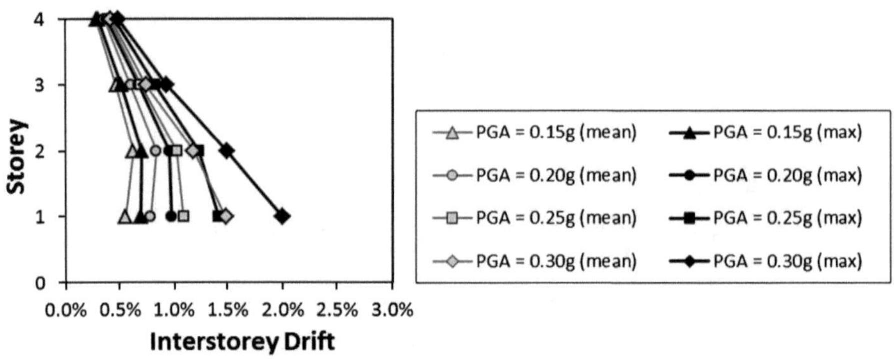

Fig. 5.6 Comparison between the *ID* calculated as mean and maximum value for each considered PGA

The distribution of *ID* along the building height is determined on the base of the first modal shape corresponding to the proper load step.

The values of the statistical parameters of the *ID* population for the considered PGA values are reported in Table 5.5. It can be noticed that the maximum value of *ID* is achieved at the first story; at this story the *cov* is in all cases over the 10%, reaching the 20% for the higher PGA.

The *ID* calculated as the mean value and the maximum value of each response distribution is shown in Fig. 5.6. As it can be seen in Fig. 5.6, due to the variability in the response distribution, the *ID* at the first story calculated as the maximum value is much larger than that calculated as the mean value for a PGA = 0.30 g. The increase in the *ID* due to the introduced variability, measured comparing the distribution relative to the maximum value with that relative to the mean value, is shown in Fig. 5.7.

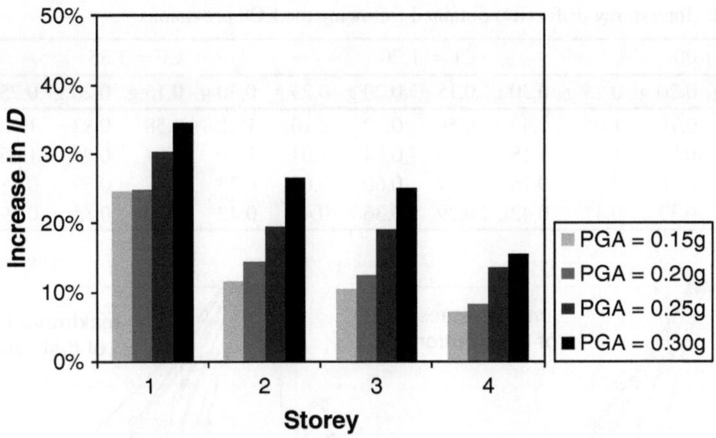

Fig. 5.7 Ratio between the *ID* calculated as mean and maximum value at each story

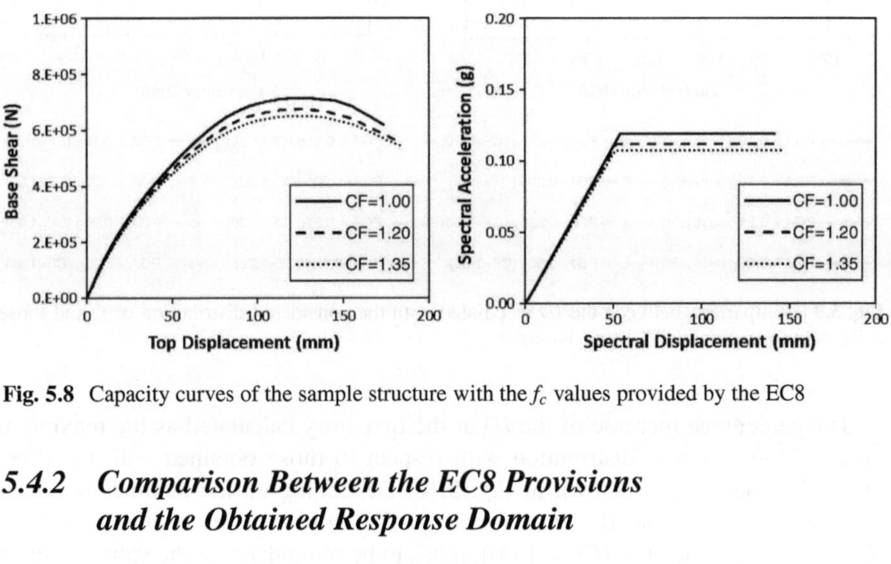

Fig. 5.8 Capacity curves of the sample structure with the f_c values provided by the EC8

5.4.2 Comparison Between the EC8 Provisions and the Obtained Response Domain

The seismic response of the sample structure is simulated according to the EC8 provisions for existing RC buildings. The variability of f_c is neglected, and its mean value is reduced through the confidence factor (CF). The capacity curves obtained with the three considered values of CF are shown in Fig. 5.8, while in the *ID* values obtained with the first modal shape are reported in Table 5.6.

In Fig. 5.9, the results obtained using the three reduced values of f_c are compared with those obtained using for f_c the mean and maximum values of the considered response distribution. While the results obtained using for f_c the mean value of the response domain do not differ from those obtained following the EC8 provisions, the *ID* obtained using the maximum value is much larger than those obtained following the EC8 previsions, particularly at the first story.

Table 5.6 Interstorey drifts (ID) obtained following the EC8 provisions

	CF = 1.00				CF = 1.20				CF = 1.35			
	0.15 g	0.20 g	0.25 g	0.30 g	0.15 g	0.20 g	0.25 g	0.30 g	0.15 g	0.20 g	0.25 g	0.30 g
1	0.53	0.76	1.05	1.42	0.56	0.79	1.10	1.42	0.58	0.82	1.12	1.55
2	0.62	0.83	1.02	1.18	0.63	0.84	1.04	1.20	0.64	0.85	1.05	1.22
3	0.47	0.60	0.70	0.75	0.47	0.60	0.68	0.73	0.47	0.59	0.68	0.73
4	0.29	0.37	0.41	0.42	0.29	0.36	0.40	0.42	0.29	0.35	0.39	0.41

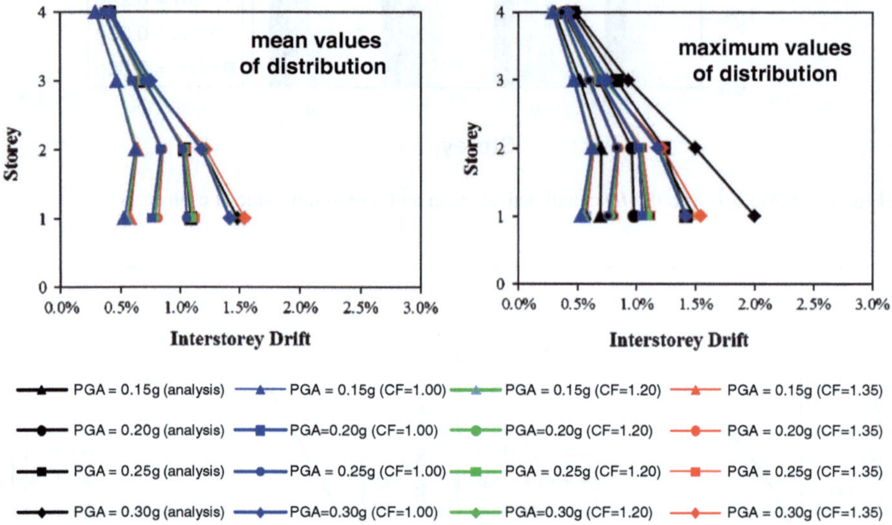

Fig. 5.9 Comparison between the *ID* calculated with the considered distribution of f_c and those calculated following the EC8 provisions

The percentage increase of the *ID* at the first story calculated as the maximum value of the response distribution with respect to those obtained with the three reduced values of f_c is shown in Fig. 5.10. Depending on the assumed value of CF, the increase in the *ID*, even for low PGA values, varies between the 20% (CF = 1.35) and the 30% (CF = 1.00). It has to be reminded that the value to select for CF does not depend on the variability of f_c.

5.4.3 *Comparison Between the FEMA Provisions and the Obtained Response Domain*

According to the FEMA 356 provisions, when the $cov > 14\%$, the number of tests on the concrete strength has to be increased until a $cov < 14\ \%$ is achieved, to use in the analysis the mean value of the obtained test results. If the $cov > 14\ \%$,

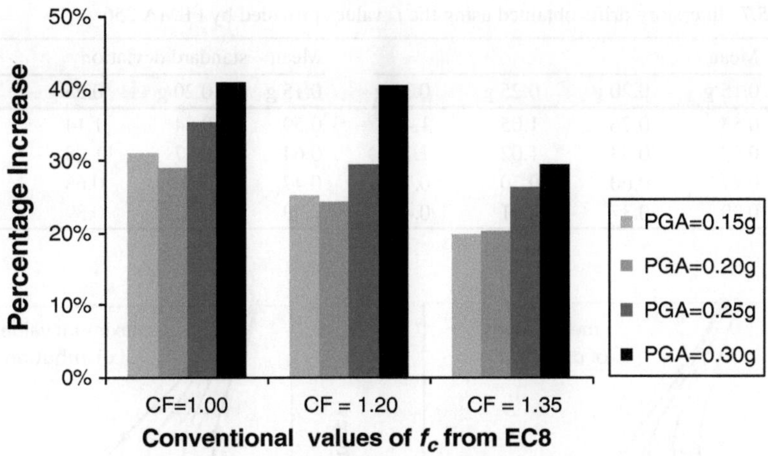

Fig. 5.10 Percentage increase of the *ID* at the first story considering the f_c variability

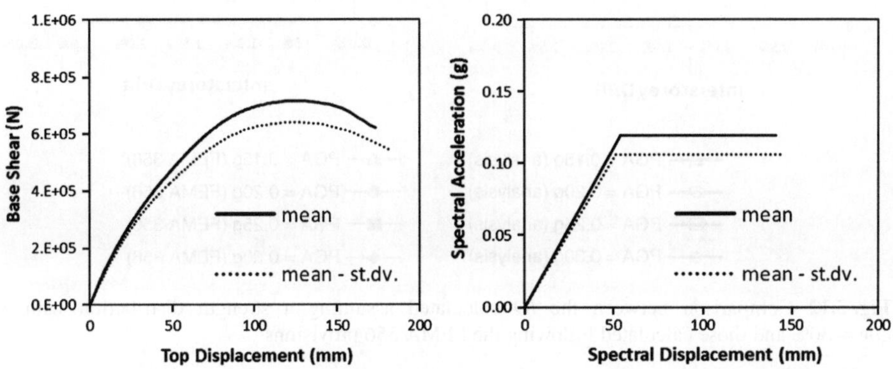

Fig. 5.11 Capacity curves of the sample structure obtained using the f_c values provided by FEMA 356

the value to assign to f_c is equal to the difference between the mean value and the *standard deviation* of the assumed distribution. To verify the suitability of the FEMA 356 provisions, two different analyses are performed. The first analysis refers to the assumed f_c distribution ($cov = 30\%$), whose results are compared with those obtained using a reduced value of f_c according to the FEMA provisions. The second analysis compares the results obtained from a distribution having a $cov = 14\%$ and those obtained using the mean value of f_c. This last analysis is aimed to evaluate the amount of effects of variability that the FEMA 356 approach considers to be negligible.

The capacity curves of the sample structure obtained following the FEMA 356 provisions are reported in Fig. 5.11, while the corresponding *ID* for each story and each PGA value is summarized in Table 5.7.

Table 5.7 Interstory drifts obtained using the f_c values provided by FEMA 356

	Mean				Mean – standard deviation			
St	0.15 g	0.20 g	0.25 g	0.30 g	0.15 g	0.20 g	0.25 g	0.30 g
1	0.53	0.76	1.05	1.42	0.59	0.84	1.14	1.53
2	0.62	0.83	1.02	1.18	0.64	0.87	1.08	1.23
3	0.47	0.60	0.70	0.75	0.47	0.60	0.68	0.73
4	0.29	0.37	0.41	0.42	0.29	0.35	0.39	0.41

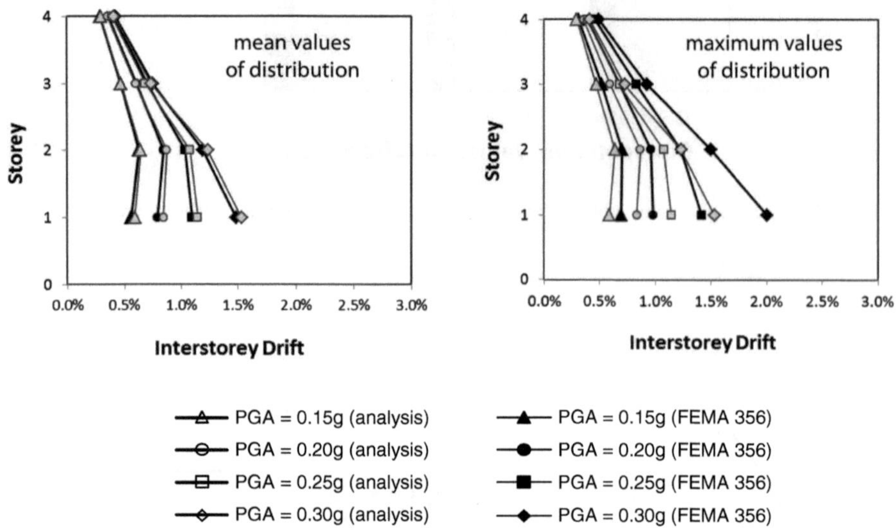

Fig. 5.12 Comparison between the *ID* calculated assuming a strength distribution with a $cov = 30\%$ and those calculated following the FEMA 356 provisions

The comparison between the *ID* obtained considering the mean and the maximum value of the assumed strength distribution and those obtained using the reduced value of f_c provided by FEMA 356 is shown in Fig. 5.12. Also in this case, as in the case of EC8, the code provisions lead to an underestimation of the maximum *ID*; however, the FEMA 356 provisions result more conservative than those provided by EC8.

To evaluate the suitability of the FEMA 356 provisions for a $cov < 14\%$, a distribution similar to the one already utilized in the above performed analyses but having a $cov = 14\%$ is considered. The same assumptions regarding the distribution of the strength along the height of the building are kept. The values of the statistical parameters of the *ID* distribution at each story are reported in Table 5.8, while the comparison between the *ID* obtained considering the maximum values of the strength distribution and those obtained following the FEMA 356 provisions is shown in Fig. 5.12.

Table 5.8 Values of the statistical parameters of the *ID* distribution with a *cov* = 14%

	Mean (%)				Standard deviation (%)				*cov* (%)			
St	0.15 g	0.20 g	0.25 g	0.30 g	0.15 g	0.20 g	0.25 g	0.30 g	0.15 g	0.20 g	0.25 g	0.30 g
1	0.54	0.76	1.05	1.43	0.025	0.044	0.081	0.117	4.5	5.8	7.7	8.2
2	0.62	0-83	1.02	1.18	0.015	0.028	0.051	0.090	2.4	3.3	5.0	7.6
3	0.47	0.60	0.65	0.75	0.012	0.019	0.187	0.051	2.5	3.1	28.7	6.9
4	0.29	0.36	0.41	0.43	0.006	0.010	0.021	0.020	2.2	2.7	5.1	4.7

5.5 Conclusions

The influence of the variability of concrete mechanical properties on the seismic response of RC buildings realized in Italy before the adoption of seismic codes has been studied.

The seismic response of the considered sample structure, expressed in terms of interstory drift, has been calculated performing a spectral analysis with the elastic spectrum provided by EC8. The results have been compared with those obtained applying the EC8 and the FEMA 356 provisions. The comparison has evidenced that the codes provisions are not adequate to provide a conservative prediction of the maximum interstory drift. Both examined codes, indeed, suggest to increase the number of the *in situ* tests to reduce the uncertainties in the concrete strength determination. According to the available experimental data, instead, the variability of the concrete mechanical properties is related to an intrinsic non-homogeneity of the material, due to the low quality of concrete itself, to a not adequate control during the construction process and to a not optimal preservation of the structure.

EC8 suggests to use in the analysis a concrete strength equal to the mean value of the results of the *in situ* tests, when a satisfactory knowledge level of the structure is achieved. As regards the considered case study, the response peaks obtained in terms of interstory drift considering the concrete strength variability exceed those obtained following the EC8 provisions of 30% for low values of PGA (0.15, 0.20 g) and 40% for higher seismic excitation (PGA = 0.30 g).

FEMA 356 takes into account the variability of concrete mechanical characteristics since it provides a value of f_c depending on the *cov* of the *in situ* tests results. Also in this case, however, the proposed analytical approach seems not to be fully adequate for large values of the *cov*. Indeed, while for *cov* < 14% the estimated peaks of *ID* underestimate only of 10% the results obtained with the assumed f_c distribution, when the *cov* = 30% the underestimation ranges between 18% for low PGA values and 30% for PGA = 0.30 g.

The performed analysis has underlined the important role played by a correct characterization of the concrete strength when existing RC buildings are investigated. When such buildings, realized with concrete having poor mechanical characteristics and with low quality standards, are studied, it is suitable to adopt an analytical approach more conservative than those provided by codes, probably set on buildings realized with concrete having better mechanical properties. In particular, the effects of a nonhomogeneous distribution of concrete mechanical properties should be considered in the analysis, with a special attention to the peaks in the seismic response due to the irregularity in elevation.

Acknowledgments This work has been supported by the project DPC-ReLUIS 2010/2013 Task 1.1.2. "Strutture in cemento armato ordinarie e prefabbricate."

References

American Society of Civil Engineering (2000) Prestandard for the seismic rehabilitation of buildings. FEMA 356, Reston

Cristofaro MT (2009) Metodi di valutazione della resistenza a compressione del calcestruzzo di strutture in c.a. esistenti. PhD dissertation, Università di Firenze

D'Ambrisi A, De Stefano, M, Tanganelli M, Viti S (2010) Seismic performance of irregular 3D RC existing frames. In: Proceedings of the 14th ECEE, Ohrid, 30 Aug–3 Sept 2010

EC 8-3 (2005) Design of structures for earthquake resistance, part 3: strengthening and repair of buildings, European standard EN 1998-3. European Committee for Standardization (CEN), Brussels

Fajfar P (1999) Capacity spectrum method based on inelastic demand spectra. Earthq Eng Struct Dyn 28:979–993

Franchin P, Pinto PE, Rajeev P (2010) Confidence factor? J Earthq Eng 14(7):989–1007

Jalameh F, Elafante L, Iervolino I, Manfredi G (2011) Knowledge-based performance assessment of existing RC buildings. J Earthq Eng 15(3):362–389

Legge 5/11/1971 n. 1086 (1971) Norme per la disciplina delle opere in conglomerato cementizio armato, normale e precompresso ed a struttura metallica. G.U. 321 del 21/12/1971

Legge 2/2/1974 n. 64 (1974) Provvedimenti per le costruzioni con particolari prescrizioni per le zone sismiche. G.U. del 21/04/1974

Mander JB, Priestley MJN, Park R (1988) Theoretical stress-strain model for confined concrete. J Struct Eng 114(8):1804–1826

Seismostruct v5.2.1 released (2011) Computer code, finite element package. http://seismosoft.com

Chapter 6
Seismic Response of Irregular Industrial Steel Buildings

Julien Richard, Sanda Koboevic, and Robert Tremblay

Abstract The suitability of the current Canadian procedures for seismic design of irregular industrial buildings is studied through the example of a titanium refinery located near Montreal, Canada. The building is braced by low-ductility concentrically braced steel frames and characterized by an irregular geometry, as well as an irregular mass and stiffness distributions. Three-dimensional elastic dynamic time history analyses were carried out for selected acceleration records compatible with the design spectra at the site. The results are compared to those obtained from the equivalent static force procedure and the response spectrum analysis method to evaluate how well these methods can predict deformations and forces in such highly irregular building. The study shows that the equivalent static method can adequately predict the displacements but may underestimate column and brace axial forces. In general, response spectrum analysis method provides appropriate estimates of the seismic response parameters given that a sufficient number of modes is considered in the analysis.

6.1 Introduction

In Canada, specific design procedures for seismic design of industrial buildings are not available. Instead, seismic design is done following requirements prescribed in the National Building Code of Canada (NRCC/NBCC 2005) that were mainly

---

J. Richard (✉)
Hatch and Associates, 5, Place Ville-Marie, bureau 200, Montreal, QC H3B 2G2, Canada
e-mail: jrichard@hatch.ca

S. Koboevic • R. Tremblay
École Polytechnique de Montréal, 2900 boul. Édouard-Montpetit, Campus de l'Université de Montréal, 2500 chemin de Polytechnique, Montréal (Québec) H3T 1J4, Canada
e-mail: sanda.koboevic@polymtl.ca; robert.tremblay@polymtl.ca

O. Lavan and M. De Stefano (eds.), *Seismic Behaviour and Design of Irregular and Complex Civil Structures*, Geotechnical, Geological and Earthquake Engineering 24, DOI 10.1007/978-94-007-5377-8_6,

developed for conventional office or residential buildings framed by regular, continuous and well-defined seismic force-resisting systems. In heavy industrial building, the structural elements are usually positioned to support heavy equipment and machinery needed for operation. For these reasons, such buildings most often have complex and irregular geometries, uneven distributions of mass and/or stiffness and dynamic characteristics that differ from those found in regular structures. The application of current design provisions to industrial buildings is not straightforward for practicing engineers and may result in inadequate designs (Daali 2004; Rolfes and MacCrimmon 2007). In particular, there exist uncertainties regarding the amplitude and distribution of the seismic force and deformation demand on industrial structures.

The global seismic response of steel-framed industrial buildings in past earthquakes was in general satisfactory. However, the reported structural damage to individual components or connections has resulted in disruption of operations in most of the industrial structures examined (Richard et al. 2009). Downtime periods may have detrimental social and economical consequences, including loss of income or unemployment, shortage of goods, electrical power and communications. Appropriate seismic provisions for design of industrial buildings are therefore needed to achieve adequate performance. Current data available in the literature regarding the seismic demand imposed on industrial steel buildings by strong ground motions are limited, and it is not known to what extent the seismic force and deformation demand on industrial structures can be adequately predicted using methods prescribed in current building design codes.

In this study, the seismic response of an irregular heavy industrial building is examined, with objective to validate if the analysis methods proposed in codes can be used to estimate deformations and forces under seismic loads. The study is conducted for an existing building near Montreal, Canada, that houses a titanium refinement process. Elastic three-dimensional time history analyses are carried out using a suite of ground motion records compatible with the design spectrum. The results for the seismic base shears, deformations and axial loads in braces and columns are compared to the predictions from equivalent static force and response spectrum analysis methods. Attention is directed to the appropriate number of modes to use in response spectrum analysis and to the possible impact of the direction of seismic loading on member force demand.

6.2 Building Studied and 3D Analysis Model

6.2.1 Seismic Action

A three-dimensional view and the layout of the columns at the base of the building studied are shown in Fig. 6.1a and b, respectively. The main portion of the building is 37 m × 59 m in plan and is 43 m high. An 8-m-tall penthouse is located in the

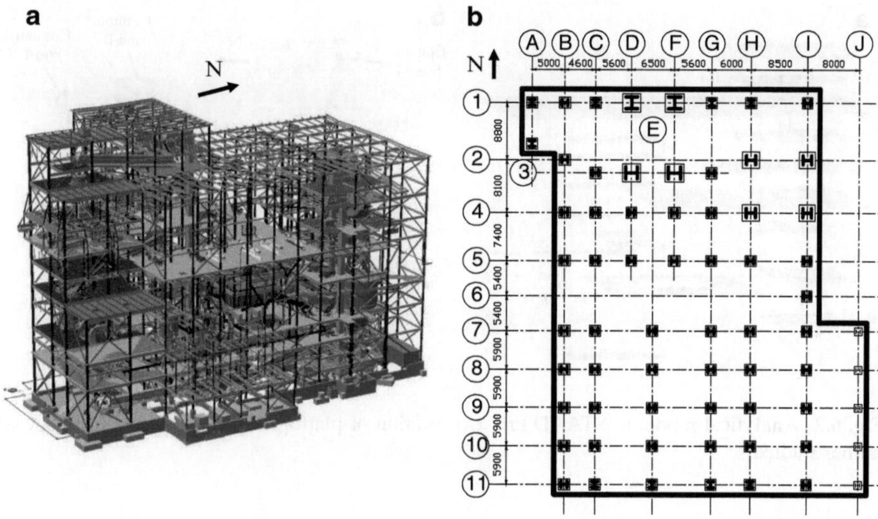

Fig. 6.1 Irregular building studied: (**a**) three-dimensional view, (**b**) column layout at the base

south-west part of the structure. The building also includes two extensions, one in the north-west corner (5 m × 6 m, 18 m high) and another one in the south-east corner (24 m × 8 m, 6 m high). A more recent extension located south of column line 11 is also shown in Fig. 6.1, and this addition was not considered in this study and is not shown in the other figures. With the exception of the roof level, the building does not have floors that extend through the whole building but rather a series of platforms that provide support and give access to different pieces of equipment. The platforms occupy mainly southern and northern portions of the building, thereby defining a large open space in the centre. Two large-capacity silos (750 and 1,200 t) are located between lines 1 and 4, and other smaller pieces of equipment are placed throughout the structure. Lateral loads are resisted by four low-ductility concentrically braced steel frames located on the perimeter of the building. In moderate seismic zones, such as Montreal, according to NBCC (2005), higher ductility system should have been selected, as the building height exceeds the height limits of 15 m permitted for low-ductility systems. Such systems, however, do not appear to be the most suitable for industrial building. In order to achieve good inelastic performance and high-energy dissipation capacity, they require well-tailored lateral load path with balanced strength hierarchy, a condition that can hardly be maintained when mass or structural modifications are imposed over the years due to evolving production processes. Knowing that low-ductility system would have been the preferred choice of designers for its simplicity and flexibility, the low-ductility braced frames are selected in this study to obtain data that could be further used to evaluate the relevance of current code restrictions on height. In addition to four principal braced frames on perimeter, other braced frames are provided in the vicinity of the heavy equipment and large openings. The building has several irregularities as described

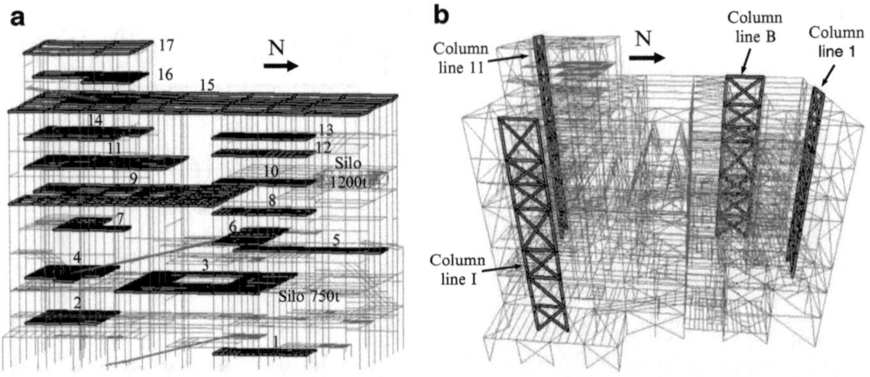

Fig. 6.2 Analytical model in STAAD.Pro: (**a**) location of platforms, (**b**) elevation of the braced frames studied

in the 2005 National Building Code of Canada, namely, mass irregularity, in-plane discontinuity of vertical lateral load-carrying elements, out-of-plane offsets and torsional sensitivity.

To facilitate the analysis and obtain more general results that could be representative of the behaviour of similar buildings, the geometry and the secondary framing arrangements were simplified. Figure 6.2 shows the final three-dimensional numerical model built in the program STAAD.Pro (Bentley and Inc. 2008). The model includes the columns, beams, braced frames and horizontal bracing as well as members supporting the main equipment. All the platforms were modelled, but only those with concrete slabs were assigned rigid diaphragm properties.

The total seismic weight of the structure ($W = 74{,}200$ kN) was determined as per NBCC (2005) and consisted of the weight of the structural members, 25% of the roof snow load and the weight of the main equipment in the fully loaded condition. In view of their importance and non-uniform distribution, the masses of the major equipments were assigned to additional nodes positioned at the centre of gravity of the equipments and linked to the rest of the structure with very stiff members. In this manner, local overturning moments due to the horizontal inertia forces acting above the base of the equipments could be incorporated in the analyses. The masses of the remaining equipment were assigned to the supporting columns at each floor in proportion to their tributary areas. For simplicity, the effects of accidental torsion, gravity loads and P-Delta effects were not considered in the analysis.

6.2.2 Selection and Scaling of Ground Motions

Elastic dynamic time history analyses were carried out for an ensemble of 14 simulated ground motion records described in Tremblay and Atkinson (2001) and compatible with the NBCC design spectrum for a Class C site in Montreal.

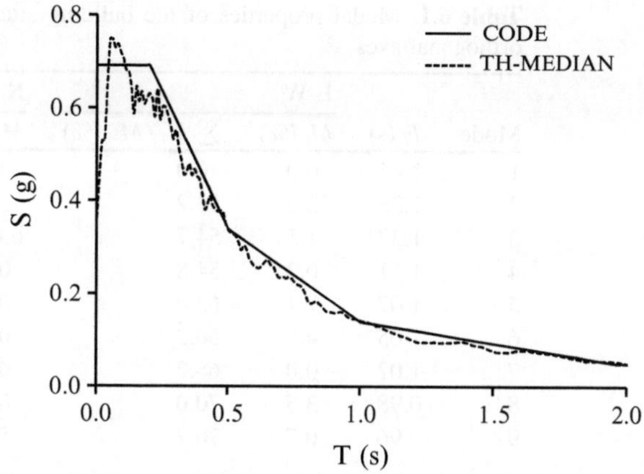

Fig. 6.3 Target design response spectrum (NBCC 2005) and median 5% damped elastic acceleration spectra for the ensemble of scaled ground motion

Simulated records were used due to the lack of historical ground motion records rich in high frequencies, which are typical at eastern North-American sites. The initial selection was done on the basis of the magnitude-distance scenarios that contribute the most significantly to the seismic hazard at the design location. The records were then calibrated to achieve compatibility between the acceleration spectra of individual records and the target design spectrum over the range of periods determined on the basis of the best visual fit between the two spectra as described in Richard (2009). The median acceleration response spectrum of scaled records and the target design spectrum are shown in Fig. 6.3.

6.3 Assessment of Analytical Methods Used in Design

6.3.1 Equivalent Static Force Method and Response Spectrum Method

Forces and deformations induced by seismic loads were first determined using the code-equivalent static force procedure and response spectrum analysis method. Selected response parameters were then compared to values obtained from elastic time history analysis to validate the application of current seismic design procedures to irregular industrial buildings. The seismic loading was not reduced to account for the ductility and the overstrength of the system so that direct comparison with the results of the elastic time history analyses could be made. According to NBCC (2005), dynamic analysis, either spectral or time history, is mandatory for such a highly irregular building. Nevertheless, the equivalent static method was also considered in this study as it is much simpler to use at the preliminarily design phase and is commonly employed by practicing engineers.

The fundamental periods were initially calculated using the NBCC empirical formula for braced frames, $T_a = 0.025\ h_n$, where h_n is the building height.

Table 6.1 Modal properties of the building studied along the building orthogonal axes

		E-W		N-S	
Mode	T_i (s)	M_i (%)	$\sum M_i/M$ (%)	M_i (%)	$\sum M_i/M$ (%)
1	1.46	0.0	0.0	0.2	0.2
2	1.38	53.2	53.2	2.1	2.1
3	1.32	1.5	54.7	63.4	65.7
4	1.11	0.2	54.8	0.2	65.9
5	1.07	7.1	62.0	0.0	65.9
6	1.05	4.2	66.2	0.2	66.1
7	1.02	0.0	66.2	0.0	66.1
8	0.98	3.8	70.0	0.7	66.9
9	0.96	0.7	70.7	5.1	72.0
...	...	...	...	...	...
70	0.38	0.1	90.0	0.1	86.2
...	...	...	...	...	...
95	0.33	0.5	92.9	0.4	90.3
...	...	...	...	...	...
126	0.26	0.3	97.1	0.1	96.1

Taking $h_n = 51$ m for the frame height at the penthouse location and $h_n = 43$ m elsewhere in the building, periods equal to 1.28 and 1.08 s were obtained, respectively. The building periods were also determined from thorough modal analysis. The results are summarized in Table 6.1. Although the building principal modes were aligned with the two orthogonal directions of the building, strong torsional components were present. Each of the two principal modes were associated with about 54% of the total mass, while the individual contribution of higher modes was relatively small and varied between 0.1 and 7% (see Table 6.1). Therefore, a large number of modes had to be considered (70 for the E-W direction and 95 for the N-S direction) to obtain the minimum 90% mass participation required by codes.

Rayleigh method gave periods equal to 1.29 and 1.28 s in the E-W and N-S direction, respectively. Values obtained from all three methods agreed well. These results indicate that the code empirical formulae and the Rayleigh method can provide reasonable estimates of fundamental periods for such irregular building.

The periods calculated with the Rayleigh method were used to determine the elastic base shear forces, V_e, according to NBCC's equivalent static force method. Following values were obtained: $V_e = 8{,}900$ kN along the E-W direction and $V_e = 9{,}000$ kN along the N-S direction. The force demand from the equivalent static method was obtained from two seismic load profiles in order to assess the approach for inclusion of higher modes: (1) the linear distribution with concentrated force at the top of the structure prescribed in NBCC (2005) and (2) the parabolic distribution as defined in ASCE 7-05 (ASCE 2005).

In the response spectrum analysis, modal contributions were combined using the CQC rule assuming 5% damping in all modes. Although in a real structure lower damping can be anticipated (2–3%), 5% damping was used in this study to maintain consistency with the level of damping considered in the NBCC design spectrum. As previously mentioned, large number of modes had to be included to obtain the minimum 90% mass participation required by codes: 70 and 95 modes for the E-W and N-S directions, respectively. The base shear forces from the response spectrum analysis, V_t, were equal to 5,690 and 6,460 kN representing only 64 and 72% of the seismic base shear forces obtained from the equivalent static method in corresponding directions. When including up to 126 modes, the combined mass participation increases to 97 and 96% in each of the two directions (see Table 6.1) and the two associated base shear forces increased to 71 and 79% V_e, respectively. Further increase in the number of modes in the response spectrum analyses did not lead to any further augmentation of the base shears. The difference between V_t and V_e is explained by the inherent inadequacy of the equivalent static force method to represent the dynamic response of complex structures.

For irregular building structures, NBCC (2005) requires that the resulting base shear obtained by response spectrum analysis corresponds to at least the value of the base shear obtained by an equivalent static method. The relevance of this requirement is questionable when the number of modes used in the spectral analysis is sufficient to reach convergence. However, more data are needed on this subject before more general conclusions can be drawn. In this study, the results of the response spectrum analysis conducted for the number of modes engaging 90% participating mass were scaled by the V_e/V_t ratio, as required in NBCC (2005), and then compared with other methods.

6.3.2 3-D Dynamic Time History Analysis

The modal superposition routine available in the program STAAD.Pro was used to carry out a three-dimensional dynamic time history analysis. Damping equal to 5% of the critical value was assigned to all modes. Preliminary analyses were performed to determine the appropriate number of modes required to adequately predict the seismic base shears, storey displacements and maximum brace forces. The number of modes was gradually increased to achieve 90% (70 modes and 95 modes in the E-W and N-S direction, respectively), 97% (126 modes) and nearly 100% (500 modes) participating mass. Increasing the participating mass from 90 to 97% resulted in peak base shear forces increasing by as much as 25%; however, no further base shear increase was observed beyond 126 modes. Similarly, brace axial loads changed significantly until 126 modes were included but remained fairly constant when additional modes were considered. Storey displacements were the least sensitive to the number of modes selected; however, for specific storeys, it was necessary to engage 97% of the mass to avoid changes in the results. Based on these three observations, 126 modes were included in the analysis. In view of the number of selected accelerograms, median results from time history analysis can be considered as representative for comparison with other methods, but 84th percentile values were also tracked to illustrate dispersion. The analyses were

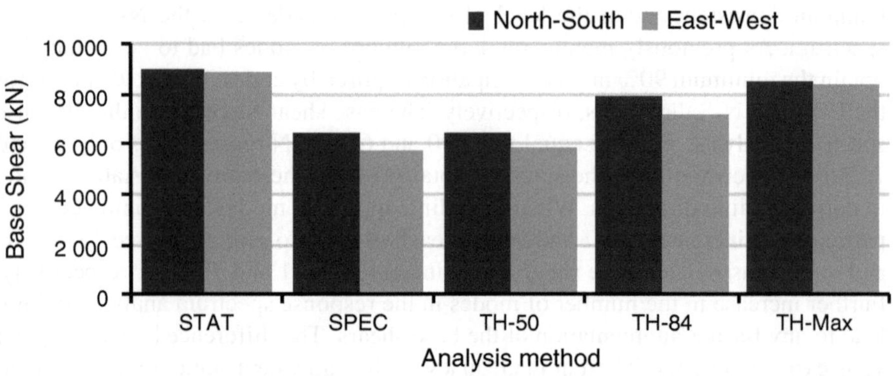

Fig. 6.4 Peak base shear from equivalent static force (STAT), response spectrum (SPEC) non-scaled and time history (TH-med and TH-84th) analysis methods

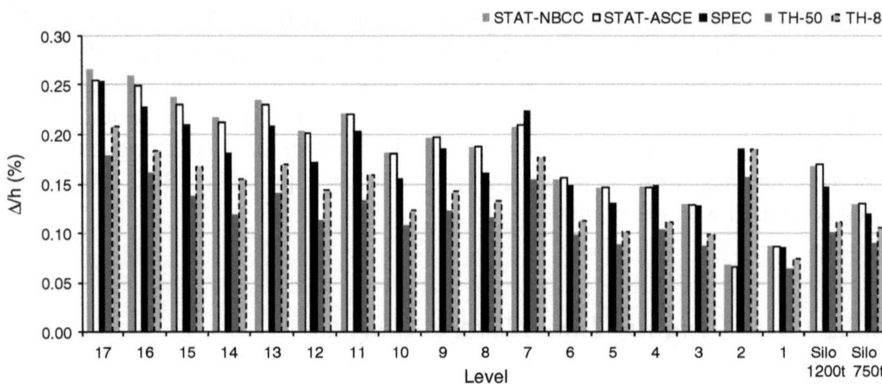

Fig. 6.5 Peak drift from equivalent static force (STAT), response spectrum (SPEC) scaled and time history (TH-med and TH-84th) analysis methods and for earthquake action along the N-S direction

conducted for two sets of orthogonal axes: the principal axes of the building and the set of axes oriented at 45° angle with respect to the principal axes. The latter set was selected to evaluate the impact that the direction of the analysis may have on the response.

6.3.3 Results and Discussion

The comparison between the three methods was done by observing the following response parameters: the seismic base shear, storey displacements and the axial forces in columns and braces of the braced frames. The results are shown in Figs. 6.4, 6.5 and 6.6.

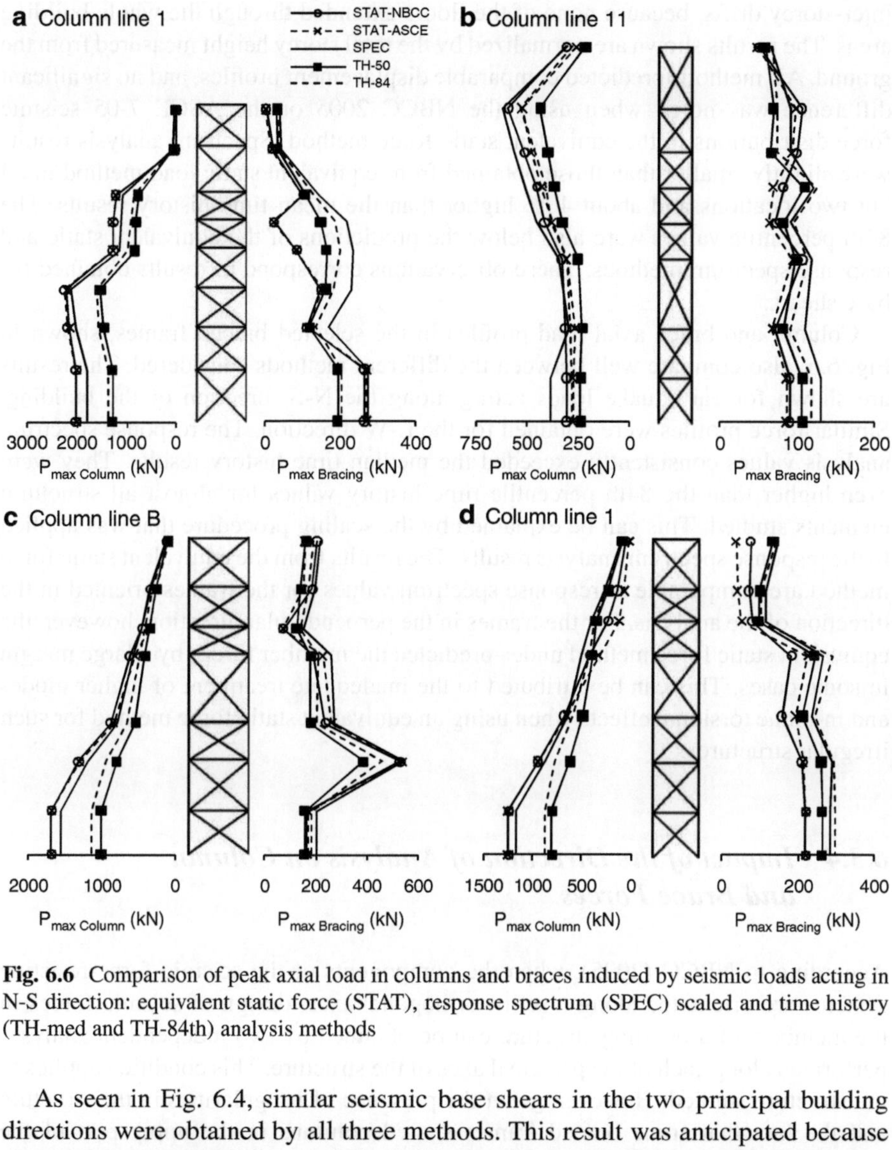

Fig. 6.6 Comparison of peak axial load in columns and braces induced by seismic loads acting in N-S direction: equivalent static force (STAT), response spectrum (SPEC) scaled and time history (TH-med and TH-84th) analysis methods

As seen in Fig. 6.4, similar seismic base shears in the two principal building directions were obtained by all three methods. This result was anticipated because the corresponding building periods in the two directions were comparable. Elastic base shears calculated using the equivalent static force method exceeded the median time history results by approximately 40% and were even slightly higher than the maximum peak values. Median results agree well with the values obtained from response spectrum analysis when no scaling of base shear is applied, confirming that the base shear estimates obtained from the equivalent static force method are probably conservative.

The calculated peak displacements under loads and ground motions applied in the N-S direction are shown in Fig. 6.5. Displacements are observed instead of inter-storey drifts, because none of the floors extended through the whole building areas. The results shown are normalized by the total storey height measured from the ground. All methods predicted comparable displacement profiles, and no significant difference was noted when using the NBCC 2005 or the ASCE 7-05 seismic force distributions in the equivalent static force method. Spectrum analysis results were slightly smaller than those obtained from equivalent static load method in all but two locations and about 40% higher than the mean time history results. The 84th percentile values were also below the predictions of the equivalent static and response spectrum methods. These observations correspond to results obtained for base shears.

Column and brace axial load profiles in the selected braced frames, shown in Fig. 6.6, also compare well between the different methods considered. The results are shown for earthquake loads acting along the N-S direction of the building. Similar force profiles were obtained for the E-W direction. The response spectrum analysis values consistently exceeded the median time history results. They were even higher than the 84th percentile time history values for almost all structural elements studied. This can be explained by the scaling procedure that was applied to the response spectrum analysis results. The results from the equivalent static force method are comparable to response spectrum values for the frames oriented in the direction of the analysis. For the frames in the perpendicular direction, however, the equivalent static force method under-predicted the member forces by a large margin in some cases. This can be attributed to the inadequate treatment of higher modes and in-plane torsional effects when using an equivalent static force method for such irregular structures.

6.3.4 Impact of the Direction of Analysis on Column and Brace Forces

According to NBCC (2005), when the components of the seismic force-resisting systems are orientated along a set of orthogonal axes, design of seismic forces in the members of a building structure can be obtained by two independent analyses performed along each of the principal axes of the structure. This condition applies to the structure studied. However, given the presence of irregularities in the structure and the fact that some braced frames shared columns in the upper part of the building, it was of interest to determine if the direction of the application of the loads could have impact on the member forces.

Figure 6.7 summarizes the results obtained from time history analysis for four directions of load application: two analyses along the principal axes of the building and two analyses along orthogonal axes oriented at 45° angle with respect to the building principal directions. Force envelopes obtained from the response spectrum analysis are also shown for comparison. Note that column lines B and I run in the N-S direction, while column lines 1 and 11 are in the E-W direction. As expected, the highest axial loads in diagonals of the braced frame were developed when

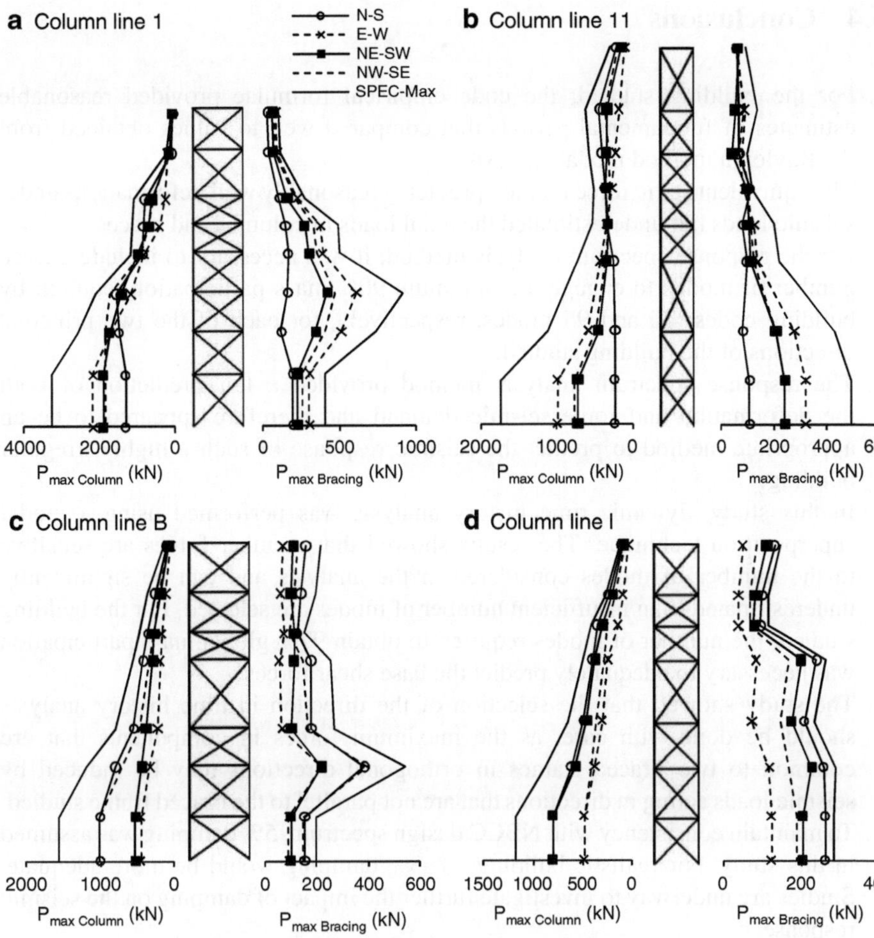

Fig. 6.7 Impact of direction of the analysis on peak axial load in columns and braces

seismic loads were acting in the direction of the braced frame. For the columns of braced frames in the N-S direction, the highest loads were induced by the same loading. However, in the upper columns of the frames along the E-W direction, up to 30% higher forces were recorded under the ground motions acting in directions other than the direction parallel to the frames, including loading at 45°. Such a result is explained by the fact that these columns were also part of another braced frame placed in the perpendicular direction to the studied frame. This type of framing arrangements is quite common in heavy industrial buildings, and caution should therefore be exercised when selecting the direction of the loading for seismic analysis. In all cases studied, member forces obtained from the response spectrum analysis were the highest and would therefore provide conservative estimates of design forces.

6.4 Conclusions

1. For the building studied, the code empirical formulae provided reasonable estimates of fundamental periods that compared well to values obtained from the Rayleigh method modal analysis.
2. The equivalent static force method predicted reasonably well deformations under seismic loads but underestimated the axial loads in columns and braces.
3. For the response spectrum analysis method, it was necessary to include a large number of modes to engage the minimum 90% mass participation required by building codes: 70 and 95 modes, respectively, for each of the two principal directions of the building studied.
4. The response spectrum analysis method provided a fair prediction of both the deformation and force seismic demand and therefore appeared to be an appropriate method to predict the seismic response of such a highly irregular building.
5. In this study, dynamic time history analysis was performed using a modal superposition technique. The results showed that member forces are sensitive to the number of modes considered in the analysis and can be significantly underestimated if an insufficient number of modes are selected. For the building studied, the number of modes required to obtain 97% global mass participation was necessary to adequately predict the base shear forces.
6. The study showed that the selection of the direction in time history analysis should be done with care, as the maximum forces in components that are common to two braced frames in orthogonal directions may be induced by seismic loads acting in directions that are not parallel to the braced frame studied.
7. To maintain consistency with NBCC design spectrum, 5% damping was assumed in this study. For realistic buildings, 2–3% damping would be more adequate. Studies are underway to investigate further the impact of damping on the seismic response.
8. The study was limited to a single building with specific dynamic characteristics, and the results cannot be generalized before similar study is conducted on additional structures. However, the results provide insight into possible limitations of current seismic design procedures and possible directions for future investigation.

Acknowledgements The authors would like to thank Mr. Robert A. MacCrimmon from Hatch Energy, Niagara Falls, ON, Canada, as well as Mr. Marco Campanelli, Mr. Terry Baumber and Mr. Jean-François Marchand from Hatch and Associates, Montreal, QC, Canada, for providing valuable data and comments related to research work presented. The financial support by the Natural Science and Engineering Research Council of Canada (NSERC) and Hatch and Associates, Montreal, PQ, Canada, through the NSERC Industrial Scholarship Program is gratefully acknowledged.

References

ASCE (2005), ASCE/SEI 7-05, including supplement no. 1, minimum design loads for buildings and other structures. American Society of Civil Engineers, Bacon Raton

Bentley Systems, Inc. (2008) STAAD.Pro. Version 20.07.03.16. Bentley Systems, Inc., Yorba Linda

Daali M (2004) Industrial facilities and earthquake engineering. In: Proceedings of the 13th World conference on earthquake engineering, Vancouver, BC, Canada, Paper No. 330

NRCC/NBCC (2005) National building code of Canada. National Research Council of Canada, Ottawa

Richard J (2009) A study of seismic behaviour of industrial buildings braced with low-ductility steel framing systems. MSc thesis, École Polytechnique, Montréal, QC, Canada (in French)

Richard J, Tremblay R, Koboevic S, MacCrimmon RA (2009) Seismic analysis and design approaches for crane-supporting steel structures. In: Proc., steel structures in seismic areas (STESSA), Philadelphia, USA, Paper No. 0182

Rolfes JA, MacCrimmon RA (2007) Industrial building design – seismic issues. Iron Steel Technol 4(5):282–298

Tremblay R, Atkinson GM (2001) Comparative study of the inelastic seismic demand of Eastern and Western Canadian sites. Earthq Spectra 17(2):333–358

Chapter 7
The Inter-storey Pounding Effect on the Seismic Behaviour of Infilled and Pilotis RC Structures

Maria J. Favvata and Chris G. Karayannis

Abstract The influence of the inter-storey structural pounding on the seismic behaviour of adjacent multistorey reinforced concrete structures with unequal storey heights is studied taking into account the local response of the infill panels. Results of more than 100 dynamic analyses indicate that the most important issue in the interaction between structures is the local response of the column of the tall structure that suffers the hit of the upper floor slab of the adjacent shorter and stiffer structure. This column appears to be in most of the times in a critical condition due to shear action. In this paper, the influence of the infill panels on the pounding problem of adjacent structures is studied. Two types of masonry-infilled structures are considered: (a) infilled frame and (b) infilled frame without infills at the base storey (pilotis frame). Results in terms of inter-storey drifts, shear requirements, ductility requirements and infill local seismic response are presented. The influence of the infill panels on the seismic performance of the critical column that suffers the hit led in all examined cases to an increase of the demands for shear and ductility, when compared to the corresponding values that are developed in the cases that the RC structure is studied without considering the infills. Thus, it can be demonstrated that in all examined inter-storey pounding cases, the presence of the infills was not enough for the amelioration of the excessive demands for shear and ductility of the column that suffers the impact. Nevertheless, the presence of the masonry infill panels has been proved as an important parameter for the safety of the building. Non-linear dynamic step-by-step analyses and special purpose elements are employed for the needs of this study.

M.J. Favvata (✉) • C.G. Karayannis
Division of Structural Engineering, Department of Civil Engineering, Democritus University of Thrace, 12, V. Sofias street, 67100 Xanthi, Greece
e-mail: mfavvata@civil.duth.gr; karayan@civil.duth.gr

O. Lavan and M. De Stefano (eds.), *Seismic Behaviour and Design of Irregular and Complex Civil Structures*, Geotechnical, Geological and Earthquake Engineering 24, DOI 10.1007/978-94-007-5377-8_7,

7.1 Introduction

An important cause of structural damage that under certain conditions can lead to collapse initiation is that of pounding between adjacent structures. Many analytical works on structural pounding have been reported during the last two to three decades (e.g. Anagnostopoulos and Spiliopoulos 1992; Chau et al. 2003; Maison and Kasai 1990, 1992; Liolios et al. 1991; Jankowski 2007, 2009). Nevertheless, most of the studies have yielded conclusions not directly applicable to the design of multistorey buildings potentially under pounding.

The influence of structural pounding on the seismic response of multistorey structures based on inelastic time history (dynamic) has recently been studied (Karayannis and Fotopoulou 1998; Karayannis and Favvata 2005a, b; Anagnostopoulos and Karamaneas 2008; Favvata et al. 2009; Abdel-Mooty et al. 2009). Moreover, results based on non-linear static analyses for the study of the interaction problem have also been reported in the literature (Favvata et al. 2008a).

Karayannis and Favvata (2005a) for the first time, examined the influence of the structural pounding problem on the ductility requirements and the overall seismic response of reinforced concrete structures with unequal heights designed according to the Eurocodes 2(EC2) (Eurocode 2 2002) and 8 (EC8) (Eurocode 8 2003). Results of 72 pounding cases between structures with equal inter-storey heights and each one for two different seismic excitations were presented and used in order to quantify the pounding effect. Moreover, initial results for the case of pounding between adjacent structures where the slabs of the one structure hit the columns of the other one were included. This phenomenon is referred by the authors as inter-storey pounding.

Furthermore, Karayannis and Favvata (2005b) present an extensive investigation on the inter-storey pounding problem between adjacent multistorey reinforced concrete frames with unequal total heights and different storey heights designed according to the codes EC2 and EC8. Fifty-two pounding cases each one for two different seismic excitations were examined. In these cases, the slabs of the short stiffer structure hit the columns of the multistorey structure at a point within the deformable height, and this phenomenon was referred to as inter-storey pounding. The effect of the number of stories on the response of the multistorey frame structures that suffers the pounding effect was investigated. The results of this study yielded the conclusion that the most important problem in the case of inter-storey pounding of reinforced concrete structures is the developing critical shear state at the columns that suffer the hit, since in these cases the demands of flexural ductility can more or less be safely satisfied. It is noted that the local damage of the critical column that suffers the impact as a result of seismic pounding had not been investigated until then.

In 2008, Anagnostopoulos and Karamaneas (2008) also underline the significant effect of the inter-storey pounding on the seismic response of the buildings in order to study the effectiveness of collision shear walls to help the colliding buildings avoid major damage due to interaction effect. However, only a particular case of adjacent 5-storey RC buildings with slightly different storey heights was studied, and the bearing response of the weaker vertical elements was under concern. The examined inter-storey pounding case was considered at the mid-height of the columns; thus, the flexural response and not the shear was critical for the response of the columns that suffer the hit. The two buildings were designed according to the Greek Earthquake Resistant Design and Reinforced Concrete Design codes.

The exterior reinforced concrete beam-column joints local inelastic response has also been reported in the literature as the key parameter for the study of the inter-storey pounding effect between adjacent structures (Favvata et al. 2009; Favvata 2006). Thus, special purpose joint element-model (Favvata et al. 2008b) was employed in the finite analysis mesh of the structural systems. The results of this investigation demonstrated that the possible local inelastic response of the exterior joints may be in some cases beneficial for the seismic behaviour of the critical column that suffers the impact. However, in all the examined cases, the developing demands for deformation of the exterior joints are substantially increased and severe damages can be observed due to the pounding effect.

Nevertheless, damages in reinforced concrete buildings during the recent earthquakes indicated that the interaction between masonry-infilled frames and bare frame can lead to unexpected effects on the seismic response of the RC building such as shear failure in columns, damage to joint region and soft-storey mechanism. Considering that brittle failures can occur by the concentration of high stresses transferred from the infills to the columns, the interaction between masonry infill and bare frame should also be considered in the seismic analysis of structures.

In view of the above, the role of the infill panels on the earthquake inter-storey pounding between adjacent structures is herein studied. Two types of masonry-infilled structures are considered: (a) infilled frame and (b) infilled frame without infills at the base storey (pilotis type). The case of bare frame structure is also examined for comparative reasons. Non-linear dynamic analyses and special purpose elements are employed for the needs of this.

7.2 Modelling Assumptions

7.2.1 Idealization of Inter-storey Structural Pounding

In this work the earthquake-induced interaction between adjacent structures with different total heights is studied, taking into account the local response of masonry-infilled panels. The actual condition and the model idealization of this interaction case are shown in Fig. 7.1.

In the case of the inter-storey pounding, the slabs of the diaphragms of each structure hit the columns of the other structure at a point within the deformable height. Contact points are taking into account at the levels of the floor slabs of the short structure, since it has been found (Karayannis and Favvata 2005a) that

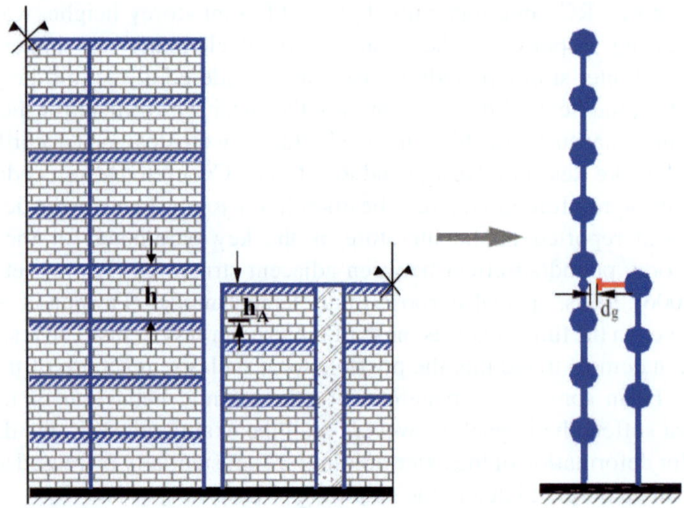

Fig. 7.1 Actual condition and model idealization of inter-storey pounding problem that includes infills panel local response

the response of the interacting structures is influenced only by the position and the characteristics of the contact point at the short structure's top floor. The influence of the other contact points on the results proved to be negligible for such types of structural pounding.

Collisions are simulated using special purpose contact elements that become active when the corresponding nodes come into contact. This idealization is consistent with the building model used and adequate for studying the effects of pounding on the overall structural response for the pounding cases under examination. In the case that the structures move one towards the other but the displacements are small and the existing gap is not covered, the contact element remains nonactive and the buildings continue to vibrate independently. In the case that the structures move one towards the other and the displacements bridge the existing gap or the structures are in contact from the beginning, then the contact element responds as a spring with almost infinite stiffness. More details on the characteristics and response of the contact element can be found in previous works by Karayannis and Favvata (2005a, b).

7.2.2 Simulation of the Infill Panels

For the simulation of the local response of the masonry infill panel, the equivalent diagonal strut model is used. For this purpose, two different types of elements have been examined by Karayannis et al. (2005). The first element was an inelastic truss element with bilinear response which carries loads only in compression.

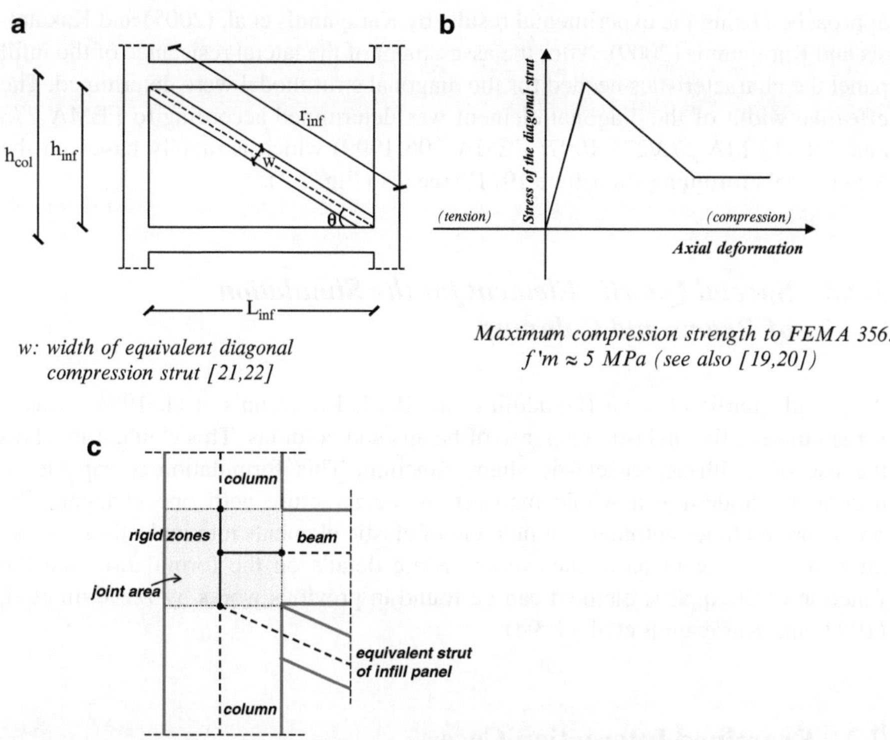

Fig. 7.2 Simulation of the infill panel based on the diagonal strut model (FEMA 273/274 1997; FEMA 306 1999): (**a**) characteristics of the equivalent diagonal strut; (**b**) response of the infill element; (**c**) equivalent diagonal struts are eccentrically connected within the frame

The response of the infill panel beyond the ultimate strength cannot be taken into account. Thus, in the analysis process when the infill panel reaches its ultimate strength it is assumed that the strength of the infilled frame decreases to the value of the strength of the bare frame at this point of loading. The second type of element which is used in this work is a special purpose beam-column element. This element accounts for more accurate definition of the response properties of infilled masonry than the previous one since it includes degrading branch (Fig. 7.2). Special attention has been given in the implementation of this element for the simulation of the infill panel in order to exhibit axial response only and not flexural one. Moreover, in the adopted implementation considering that the concentric struts cannot represent the forces imposed on the columns (and beams) of the frame by the infill, the equivalent diagonal struts are eccentrically connected within the RC frames (Fig. 7.2c, see also ASCE/SEI 41-06).

Further an important problem in modelling the infill panel is the determination of the response characteristics of the diagonal strut model, taking into account the actual conditions of the effective lateral confinement of the masonry by the reinforced concrete frame. The actual properties of the infill panel have been approached using the experimental results by Karayannis et al. (2005) and Kakaletsis and Karayannis (2009). After the assessment of the lateral resistance of the infill panel the characteristics needed for the diagonal strut model were determined. The effective width of the diagonal element was determined according to FEMA 273 and 306 (FEMA 273/274 1997; FEMA 306 1999) which is mainly based on the Mainstone's formula (Mainstone 1971) (see also Fig. 7.2).

7.2.3 *Special Quartic Element for the Simulation of Beams and Columns*

A special quartic element (Izzuddin et al. 1994; Karayannis et al. 1994) is used for simulating the inelastic response of beams and columns. This element involves the use of a 4th-degree elastic shape function. This formulation is capable of accurately modelling a whole member of the structure with one element. The procedure includes automatic refinement of elastic elements into inelastic elements after detection of material inelasticity. More details on the formulation and the function of the quartic element can be found in previous works by Izzuddin et al. (1994) and Karayannis et al. (1994).

7.3 Examined Interaction Cases

The most important issue in the inter-storey pounding cases is the local response of the external column of the tall building that suffers the impact from the upper floor slab of the adjacent shorter and stiffer structure. This impact takes place at a point of the deformable height of the column. The consequences of the impact can be very severe for the integrity of the column and may be a primary cause for the initiation of the collapse of the structure. This is the most critical case of interaction between adjacent buildings (Karayannis and Favvata 2005a, b).

In this respect, in the examined cases it is considered that the pounding takes place at points of the deformable height of the columns of the more flexible 8-storey frame structure. Each of the studied cases is examined for two different gap distances between the two structures and is analysed for the El Centro 1940 seismic excitation with maximum acceleration (α_{max}) scaled to be equal to the design acceleration of the examined structure ($\alpha_{max} = 0.3$ g). In these interaction cases, the total height of the 3-storey structure is greater than the total height of the 3rd floor and less than the total height of the 4th floor of the 8-storey frame, and thus the highest contact point of the two structures lies between the levels of the 3rd and the 4th floor of the 8-storey frame.

Further, a parametric investigation is performed considering two different positions of the contact point in order to point out the influence of the inter-storey pounding effect on the seismic behaviour of the critical column that suffers the hit. The examined positions are at 1/3 and 1/2 of the inter-storey height of the

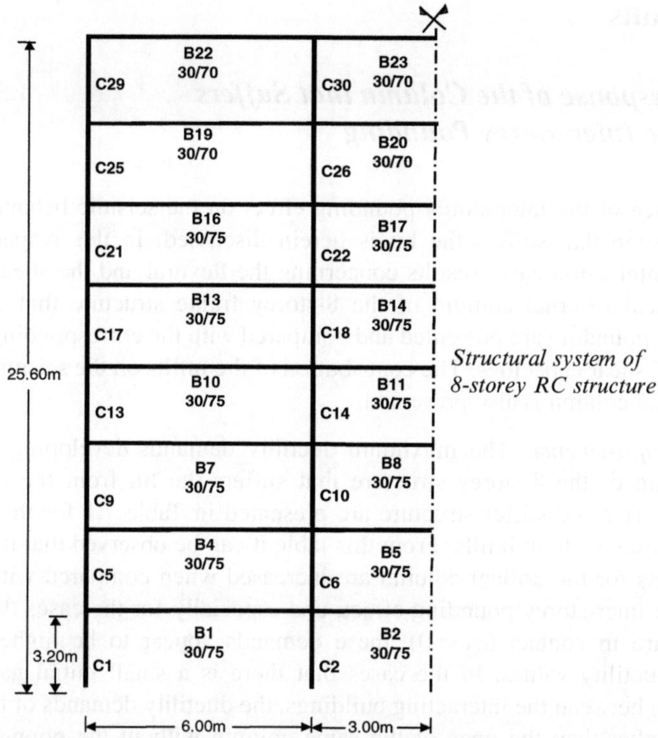

Fig. 7.3 Structural system of the examined 8-storey RC structure

column of 4th floor of the 8-storey frame. The influence of the gap size on the inter-storey pounding effect is also taken into account. Thus, all pounding cases are examined: (a) for structures in contact from the beginning ($d_g = 0$), (b) for initial gap distance between the two structures equal to $d_g = 2.0$ cm and (c) for the case that the structures vibrate independently without pounding effect.

Nevertheless, the main scope of this work is to incorporate the masonry infills as key parameter for the study of the inter-storey structural pounding. For this purpose, three types of multistorey structures that suffer the interaction effect are considered: (a) 8-storey fully infilled frame structure, (b) pilotis type 8-storey building (soft 1st storey) and (c) 8-storey bare frame structure (for comparison reasons). Moreover, the pounding cases between the 8-storey frame and 3-storey frame-wall structure are examined, considering that the highest contact point is located at the 1/3 of the inter-storey height of the 4th floor column of the 8-storey frame and that the structures are in contact from the beginning ($d_g = 0$). In these cases, the structural systems are subjected to a strong seismic action (Northridge, USA, 1994 with $\alpha_{max} = 0.45$ g).

Beams, columns and walls of all the examined structural systems were designed according to Eurocodes 2 and 8, meeting the Ductility Capacity Medium (DCM) criteria of the codes. The geometry of the 8-storey RC frame structure is shown in Fig. 7.3. The computer program used in this work is the well known nonlinear dynamic structural analysis program ADAPTIC (Izzuddin 1991).

7.4 Results

7.4.1 *Response of the Column that Suffers the Inter-storey Pounding*

The influence of the inter-storey pounding effect on the seismic behaviour of the critical column that suffers the hit is herein discussed. In this respect, for the examined interaction cases results concerning the flexural and the shear demands of the critical external column of the 8-storey frame structure that suffers the inter-storey pounding are presented and compared with the corresponding available flexural and shear capacities. The contribution of the infills on the seismic response of the critical column is also presented.

Ductility requirements. The maximum ductility demands developing in the 4th floor column of the 8-storey structure that suffers the hit from the slab of the adjacent shorter and stiffer structure are presented in Table 7.1 for the examined pounding cases without infills. From this table it can be observed that the ductility requirements for the critical column are increased when compared with the ones without the inter-storey pounding effect, and especially for the cases that the two buildings are in contact ($d_g = 0$), these demands appear to be higher than the available ductility values. In the cases that there is a small initial gap distance ($d_g = 2$ cm) between the interacting buildings, the ductility demands of the column are also higher than the ones of the same column without the pounding effect, but in most of the examined cases, they appear to be lower than the available ductility values.

However, the results of Table 7.1 indicate that changes of the position that the impact takes place within the deformable height of the column and changes of the initial gap distance between the adjacent structures have as a result a different influence on the response of the column.

In the examined pounding cases that include the local response of the masonry infill panels, an increase on the ductility demands of the critical column is observed (Table 7.2). In fact, as it can be observed in Fig. 7.4, during the seismic excitation

Table 7.1 Capacity demand of the external column that suffers the inter-storey pounding

	Ductility demands		Shear demands	
	$d_g = 0.0$ cm	$d_g = 2.0$ cm	$d_g = 0.0$ cm	$d_g = 2.0$ cm
Pounding at $h_A = 1/3$ h	7.64 (2.21)[a]	2.56 (0.74)	2.88[b] (250)[c]	2.72 (133)
Pounding at $h_A = 1/2$ h	8.74 (2.53)	2.60 (0.75)	1.75 (150)	1.57 (70)
Without pounding	Elastic	Elastic	Elastic	Elastic

Pounding cases between 8-storey frame and 3-storey structure (seismic excitation of El Centro 1940 – $\alpha_{max} = 0.3$ g)

[a]Ratio of the ductility demand to the available one

[b]Ratio of the maximum shear demand to the available shear strength

[c]Times of exceeding the available shear strength during the analysis

Table 7.2 Including infills local response; capacity demand of the external column that suffers the inter-storey pounding

8-storey frame:	Fundamental periods	Ductility demands	Shear demands
Bare	$T_{1,bare}=0.841$	2.27 (0.66)[a]	1.95[b] (31)[c]
Infilled	$T_{1,infilled}=0.641$	3.62 (1.05)	2.54 (49)
Pilotis	$T_{1,pilotis}=0.674$	2.58 (0.75)	2.48 (78)
Without pounding (*frame, infilled, pilotis*)		Elastic	Elastic

Pounding case between 8-storey frame and 3-storey structure at $h_A = 1/3$ h and $d_g = 0.0$ cm (seismic excitation of Northridge 1994 – $\alpha_{max} = 0.45$ g)
[a]Ratio of the ductility demand to the available one
[b]Ratio of the maximum shear demand to the available shear strength
[c]Times of exceeding the available shear strength during the analysis

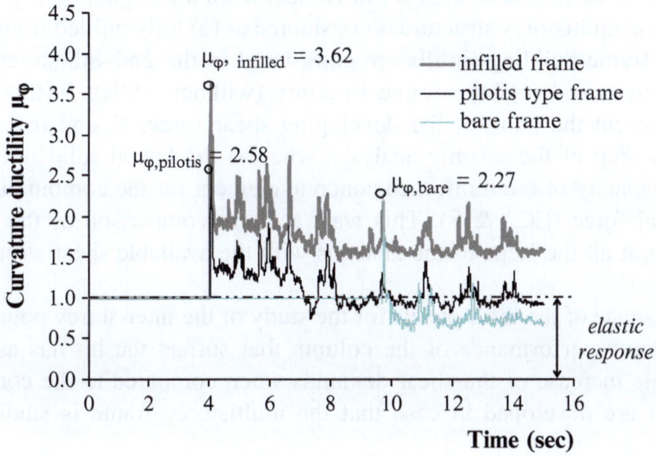

Fig. 7.4 Inter-storey pounding effect for the examined cases that include masonry infill panels (interaction at the point $h_A = 1/3$ h with $d_g = 0.0$). Time history ductility requirements of the external column of the 4th storey that suffers the hit (seismic excitation Northridge 1994 – $\alpha_{max} = 0.45$ g)

the ductility requirements of the critical column are substantially increased due to the presence of the infills. Moreover, comparison of the results of the seismic performance of the column that suffers the hit between the fully infilled frame and the pilotis type frame indicate that the inter-storey pounding effect is more intense in case of fully infilled frame than in case of pilotis.

Shear requirements of the critical column. The shear demands developing in the critical part of the column of the 8-storey structure that suffers the impact from the adjacent shorter and stiffer structure are presented in Tables 7.1 and 7.2 for all the examined pounding cases. These tables present the increase of the developing maximum shear forces due to pounding effect. Furthermore, the number of times during the analysis that the shear forces of the critical part of the column exceed its shear strength is also given in Tables 7.1 and 7.2.

For the cases that the infilled panels are not included, it can be observed that the inter-storey pounding effect induces a critical shear state for the seismic performance of the column since in all the examined cases, the developing shear forces exceed the capacity for shear strength of the member many times during the excitation (Table 7.1). Moreover, the pounding case where the highest contact point is considered at the 1/3 of the inter-storey height of the column of the 4th floor is proved to be the most critical interaction case for the safety of the 8-storey structure (Table 7.1). Similarly, in the cases that masonry infills are included in the analysis model, it can be observed that the maximum shear demands of the column that suffers the hit are increased due to the inter-storey pounding effect (Table 7.2).

Figure 7.5 presents the developing shear forces of the column of the 8-storey structure that suffers the hit at the point $h_A = 1/3$ h from the slab of the 3-storey structure, and the two structures are in contact from the beginning ($d_g = 0$) for the cases that the multistorey structure is considered as (a) fully infilled frame structure, (b) pilotis frame building (infills are considered in the 2nd–8th storeys and thus soft 1st storey) and (c) bare frame structure (without infills). In this figure, the points represent the pairs of the developing shear force, V, and the axial force, N, at every step of the seismic analysis, whereas the lateral solid lines show the available capacity of the reinforced concrete element for the combination of shear versus axial force (EC2 & 8). This way, a direct comparison of the developing shear force at all the steps of the analysis with the available shear strength can be obtained.

The presence of the infill panels for the study of the inter-storey pounding effect on the seismic performance of the column that suffers the hit has as a result a considerable increase of the shear demands when compared to the corresponding values that are developed in case that the multistorey frame is studied without infills.

7.4.2 *Local Response of Infills and Overall Drifts Due to the Inter-storey Pounding Effect*

The results presented herein are for the pounding cases between the infilled 8-storey frame structures and the 3-storey frame-wall building considering that the highest contact point is located at the 1/3 of the inter-storey height of the 4th floor column of the tall structure and that the structures are in contact from the beginning ($d_g = 0$).

In this aspect, Fig. 7.6 comprises the influence of the structural inter-storey pounding on the local response of the masonry infills. The results are for both examined types of infilled frames: (a) fully infilled and (b) pilotis type. Comparative results in terms of maximum developed deformations of the infill panels are presented between the cases that (a) the structural system suffers the inter-storey pounding effect and (b) the structures are vibrating independently (without interaction effect). In this figure, the capacity deformation at ultimate strength of the infills as well as the ultimate deformation at failure are shown. In this way, a

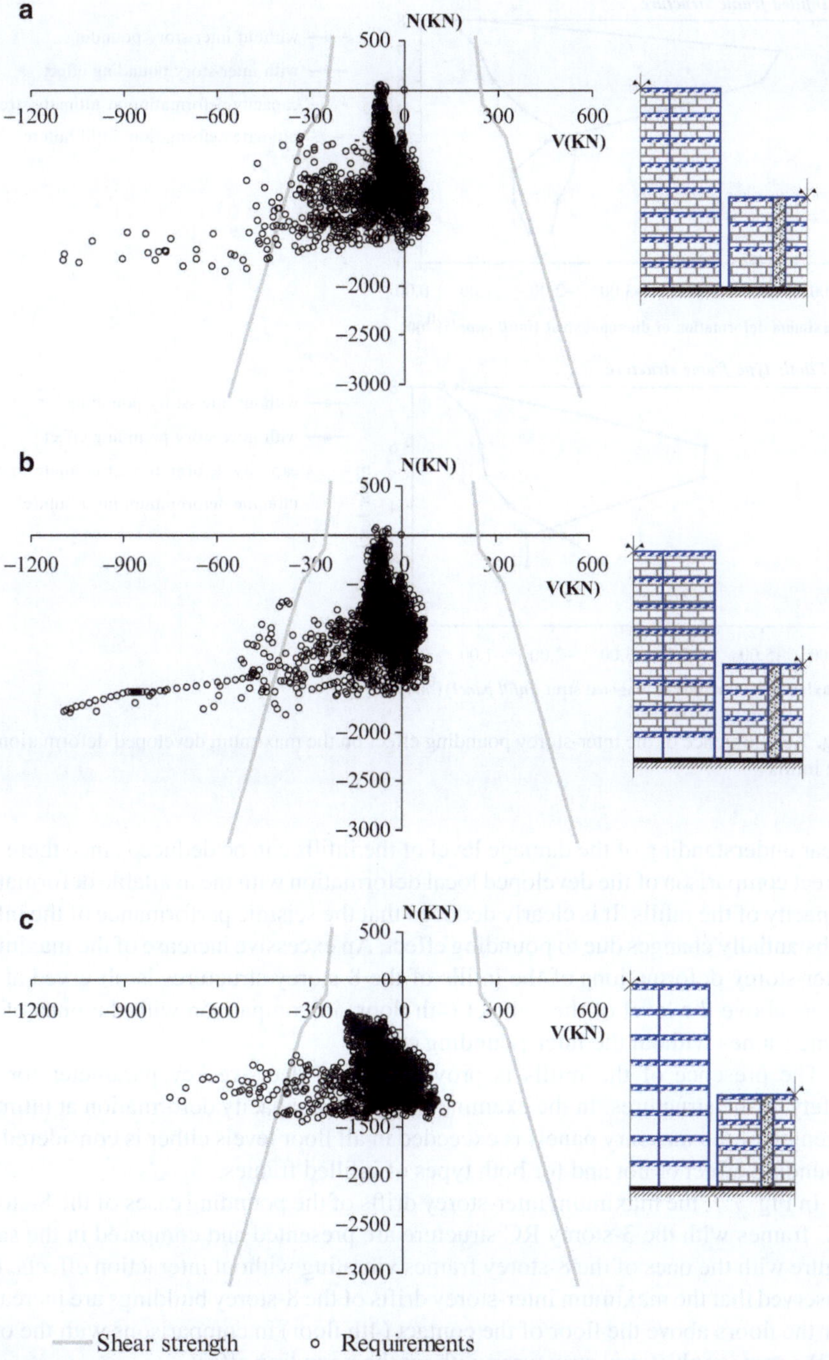

Fig. 7.5 Influence of the infill panels on the pounding problem of the 8-storey frame structure. Developing shear forces of the external column of the 4th storey that suffers the inter-storey pounding: (**a**) infilled frame, (**b**) pilotis type frame, (**c**) bare frame

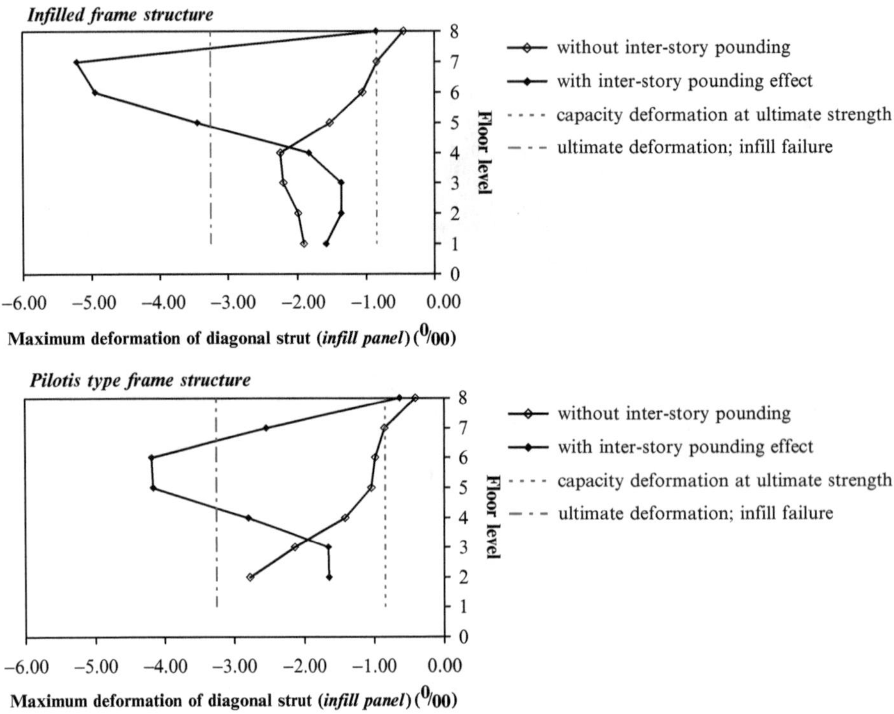

Fig. 7.6 Influence of the inter-storey pounding effect on the maximum developed deformations of the infills

clear understanding of the damage level of the infills can be deduced since there is a direct comparison of the developed local deformation with the available deformation capacity of the infills. It is clearly deduced that the seismic performance of the infills substantially changes due to pounding effect. An excessive increase of the maximum inter-storey deformations of the infills of the 8-storey structures is observed at the floors above the level of the contact (4th floor) in comparison with the ones of the same frames without the inter-pounding effect.

The presence of the infills is proved to be a decisive key parameter for the safety of the structures. In the examined cases, the capacity deformation at ultimate strength of the masonry panels is exceeded in all floor levels either is considered the pounding effect or not and for both types of infilled frames.

In Fig. 7.7, the maximum inter-storey drifts of the pounding cases of the 8-storey RC frames with the 3-storey RC structure are presented and compared in the same figure with the ones of the 8-storey frames vibrating without interaction effects. It is observed that the maximum inter-storey drifts of the 8-storey buildings are increased for the floors above the floor of the contact (4th floor) in comparisons with the ones of the same multistorey structures without the pounding effect.

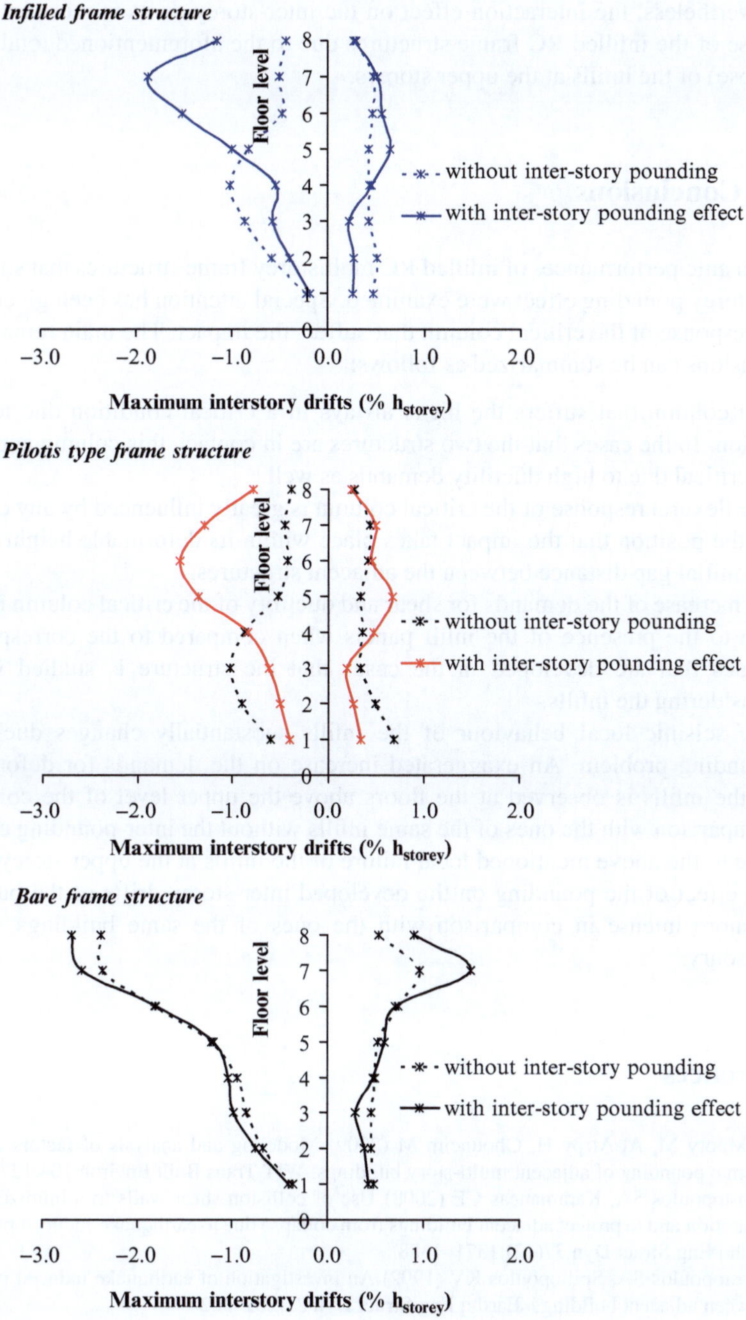

Fig. 7.7 Influence of the inter-storey pounding effect on the maximum developed drifts

Nevertheless, the interaction effect on the inter-storey drifts is more intense in the case of the infilled RC frame structures due to the aforementioned total failure (collapse) of the infills at the upper storeys.

7.5 Conclusions

The seismic performances of infilled RC multistorey frame structures that suffer the inter-storey pounding effect were examined. Special attention has been given to the local response of the critical column that suffers the impact. The main remarks and conclusions can be summarized as follows:

1. The column that suffers the hit is always in a critical condition due to shear action. In the cases that the two structures are in contact, this column appears to be critical due to high ductility demands as well.
2. The flexural response of the critical column is greatly influenced by any changes on the position that the impact takes place within its deformable height and on the initial gap distance between the adjacent structures.
3. An increase of the demands for shear and ductility of the critical column is noted due to the presence of the infill panels when compared to the corresponding values that are developed in the cases that the structure is studied without considering the infills.
4. The seismic local behaviour of the infills substantially changes due to the pounding problem. An exaggerated increase on the demands for deformation of the infills is observed at the floors above the upper level of the contact in comparison with the ones of the same infills without the inter-pounding effect.
5. Due to the above mentioned local failure of the infills at the upper storey levels, the effect of the pounding on the developed inter-storey drifts of the buildings is more intense in comparison with the ones of the same buildings without masonry.

References

Abdel-Mooty M, Al-Atrpy H, Ghouneim M (2009) Modeling and analysis of factors affecting seismic pounding of adjacent multi-story buildings. WIT Trans Built Environ 104:127–138

Anagnostopoulos SA, Karamaneas CE (2008) Use of collision shear walls to minimize seismic separation and to protect adjacent buildings from collapse due to earthquake-induced pounding. Earthq Eng Struct Dyn 37(12):1371–1388

Anagnostopoulos SA, Spiliopoulos KV (1992) An investigation of earthquake induced pounding between adjacent buildings. Earthq Eng Struct Dyn 21:289–302

Chau KT, Wei XX, Guo X, Shen CY (2003) Experimental and theoretical simulations of seismic poundings between two adjacent structures. Earthq Eng Struct Dyn 32:537–544

Eurocode 2 (2002) Design of concrete structures – part I: general rules and rules for building, CENprEN 1992-1-1, European Committee for Standardization, Brussels

Eurocode 8 (2003) Design of structures for earthquake resistance – part I: general rules, seismic actions and rules for building. Doc CEN/TC250/SC8/N335 prEN 1998-1-1, European Committee for Standardization, Brussels

Favvata MJ (2006) Investigation of the seismic response and performance of multistory reinforced concrete structures. Special simulation of joints – interaction of structures. Dissertation, Department of Civil Engineering, Democritus University of Thrace, Xanthi, Greece

Favvata MJ, Karayannis CG, Liolios AA (2008a) Non-linear static versus dynamic analysis for the interaction of structures. In: Proceedings of the 7th European Conference on Structural Dynamics, Southampton, 7–9 July 2008, p E318

Favvata MJ, Izzuddin BA, Karayannis CG (2008b) Modelling exterior beam-column joints for seismic analysis of RC frame structures. Earthq Eng Struct Dyn 37(13):1527–1548

Favvata MJ, Karayannis CG, Liolios AA (2009) Influence of exterior joint effect on the inter-story pounding interaction of structures. Struct Eng Mech 33(2):113–136

FEMA 273/274 (1997) NEHRP guidelines for seismic rehabilitation of buildings. Federal Emergency Management Agency, Washington, DC

FEMA 306 (1999) Evaluation of earthquake damaged concrete and masonry wall buildings. Infilled frames. Federal Emergency Management Agency, Washington, DC

Izzuddin BA (1991) Nonlinear dynamic analysis of framed structures. PhD thesis, Department of Civil Engineering, Imperial College, University of London

Izzuddin BA, Karayannis CG, Elnashai AS (1994) Advanced nonlinear formulation for reinforced concrete frames. J Struct Eng (ASCE) 120(10):2913–2935

Jankowski R (2007) Assessment of damage due to earthquake-induced pounding between the main building and the stairway tower. Key Eng Mater 347:339–344

Jankowski R (2009) Non-linear FEM analysis of earthquake-induced pounding between the main building and the stairway tower of the Olive View Hospital. Eng Struct 31(8):1851–1864

Kakaletsis DJ, Karayannis CG (2009) Experimental investigation of infilled reinforced concrete frames with openings. ACI Struct J 106(2):132–141

Karayannis CG, Favvata MJ (2005a) Earthquake-induced interaction between adjacent reinforced concrete structures with non-equal heights. Earthq Eng Struct Dyn 34:1–20

Karayannis CG, Favvata MJ (2005b) Inter-story pounding between multistory reinforced concrete structures. Struct Eng Mech 20(5):505–526

Karayannis CG, Fotopoulou MG (1998) Pounding of multistory RC structures designed to EC8 & EC2. In: Proceedings of the 11th European conference on earthquake engineering (in CD form), ISBN 90 5410 982 3. Balkema, Rotterdam

Karayannis CG, Izzuddin BA, Elnashai AS (1994) Application of adaptive analysis to reinforced concrete frames. J Struct Eng (ASCE) 120(10):2935–2957

Karayannis CG, Kakaletsis DJ, Favvata MJ (2005) Behaviour of bare and masonry infilled R/C frames under cyclic loading: experiments and analysis. In: Proceedings of the 5th conference on Earthquake Resistant Engineering Structures (ERES), Skiathos, Greece, May 2005

Liolios A, Anagnostides G, Vasiliadis L, Elenas A (1991) A numerical approach for a hemivariational inequality arising in the earthquake interaction between adjacent structures with different story levels. In: Proceedings of international conference on computational engineering science, Melbourne, pp 632–635

Mainstone RJ (1971) On the stiffness and strengths of infilled frames. In: Proceedings of Inst. of Civil Engineers (ICE), Supplement (IV): p 7360, pp 57–90

Maison BF, Kasai K (1990) Analysis for type of structural pounding. Struct Eng, ASCE 116(4):957–977

Maison BF, Kasai K (1992) Dynamics of pounding when two buildings collide. Earthq Eng Struct Dyn 21:771–786

Chapter 8
Modal Irregularity in Continuous Reinforced Concrete Bridges. Detection, Effect on the Simplified Seismic Performance Evaluation and Ways of Solution

Gustavo Ayala and Marco Antonio Escamilla

Abstract This chapter presents the preliminary results of an ongoing investigation on the modal irregularity in structures, particularly in reinforced concrete viaduct-like bridges, and the effect of this structural/demand characteristic on the seismic performance evaluation of such structures using simplified methods of analysis. The main objective of this chapter is to understand modal structural irregularity and its effects on the performance results obtained from approximate elastic analysis procedures prescribed by most codes or simplified nonlinear analysis methods using modal spectral analyses. It is shown that modal irregularity may be present in bridges with relatively close modes, and it may also occur when, in the inelastic range, the instantaneous modes of the bridge change their composition and that this irregularity may lead to erroneous results, particularly when a modal combination rule is involved. To overcome this problem, a new analysis method for the seismic performance evaluation of bridges exhibiting modal irregularity is presented. This method has as origin a simplified seismic evaluation method based on the performance of a reference single degree of freedom system derived from the capacity curve of the bridge, calculated using evolutive modal spectral analyses with their corresponding dissipated hysteretic energy correction. To show the application of this method, the seismic performance of two bridges, one considered regular and the other with evident modal irregularity, subjected to an earthquake record of two different intensities, one that keeps them within the elastic range of behaviour and the other that takes them into the inelastic range. Finally, the validity of this approximation is shown by comparing the "exact" seismic performance of the bridges, obtained by nonlinear step-by-step analysis, with the corresponding performances obtained using the simplified method. To show the validity of the

G. Ayala (✉) • M.A. Escamilla
Instituto de Ingeniería, Universidad Nacional Autónoma de México, Ciudad Universitaria, México DF C.P. 04510, Mexico
e-mail: GAyalaMilian@iingen.unam.mx; MEscamillaG@iingen.unam.mx

O. Lavan and M. De Stefano (eds.), *Seismic Behaviour and Design of Irregular and Complex Civil Structures*, Geotechnical, Geological and Earthquake Engineering 24, DOI 10.1007/978-94-007-5377-8_8,

procedure proposed for the construction of the capacity curve of a bridge, a comparison of results obtained from conventional force pushover analysis, evolutive modal spectral analysis with a correction for hysteretic energy dissipation and the incremental dynamic analysis; using as reference the last one and stressing the potential and limitations of the second.

8.1 Introduction

Current seismic design codes emphasize in their norms that structures should be designed to avoid collapse, by dissipating part of the input energy of an earthquake through the hysteresis at their damaged elements and by controlling the location and intensity of the accepted damage. Earthquake damage observations after the occurrence of events of significant intensity have shown, however, that many of the affected structures had not experienced behaviours consistent with these norms and the assumptions accepted in their definition. Among the different types of structures in this situation, bridges are one of the affected types of structures.

Bridges are apparently simple structures which behaviour under design seismic demands could be satisfactorily predicted with currently accepted methods; unfortunately this has not been the case, possibly because many of the assumptions used in their seismic analysis and design are not directly applicable to bridges as most of the research efforts on the seismic behaviour of structures, that has led to this assumptions, have been focused to buildings, ignoring the fact that building structures have a very different behaviour to bridges. This situation is normally accentuated when modal combination rules in modal spectral analysis, derived for the analysis of buildings, are used.

The generality of design codes is based on the use of the equal displacements rule (Veletsos and Newmark 1960), which postulates that the maximum inelastic displacement of a single degree of freedom, SDOF, oscillator may be approximated by the corresponding maximum elastic displacement. It has been shown, however, that this rule is not directly applicable to short period oscillators and to seismic demands characteristic of soft ground conditions and even less to complex structures where the damage caused by earthquake invalidates the assumption that each mode of vibration remains constant, ignoring the fact that behind the definition of each mode there is a shape that changes with damage, something that does not occur when each mode is assumed to be equivalent to a SDOF system. This fact has been indirectly brought to the attention of researchers involved in the seismic evaluation of bridges, who have experienced the risk of having wrong performance results from simplified analysis methods due to this "irregularity".

Current literature on the seismic evaluation and design of bridges has diverse proposals to detect the degree of irregularity through indexes, defined with equations in terms of structural characteristics that establish the conditions that guarantee the validity of the results of simplified analysis procedures. This outcome makes evident the need of a simplified method of analysis which gives valid results even when these conditions are not met.

This chapter proposes two new parameters to characterize regularity defined as modal regularity indexes and a seismic evaluation procedure for viaduct-type bridges which explicitly considers modal irregularity to give valid evaluation results. It is shown that modal irregularity in bridges may not only occur when modal frequencies are relatively close to each other but also when the modes that contribute most to the performance of the structure change in form and order of participation when passing from one damage state to another. Further to the problems and limitations of the variants of the equal displacements rule and their implementation in the seismic evaluation and design of bridges, this chapter finds that modal irregularity is another factor that may significantly affect the approximation of the results obtained from these methods and that this factor alone may lead to erroneous results, when compared with those obtained from nonlinear step-by-step analyses.

8.2 Bridge Regularity

Up till now, the notion of regularity in current seismic design codes is either not present or not clearly defined. Aware of this situation, some researchers have focused their studies on understanding the influence of irregularity in the evaluation of the seismic performance of bridges proposing regularity indexes for the classification of these structures which may help the analyst in deciding what approximate method to use in their evaluation or seismic design.

Even though the definition of regularity varies slightly from code to code, the factors which influence the performance of an irregular structure are not known. Both, codes and researchers, agree that, if a bridge is irregular, the results obtained from its evaluation or seismic design may not be correct.

According to the AASHTO LRFD code (2007), most common bridges are irregular; this classification is based primarily on the relationship of mechanical properties (pier stiffness and mass) and geometric characteristics (length of piers and spans). Another code that addresses the concept of regularity in the analysis of structures is the Eurocode for seismic design, EC8 (CEN 1994). This code, however, in its section devoted to bridges, gives a vague definition of regularity with no paragraph describing the degrees of irregularity or an index for its classification. According to this code, the degree of regularity depends on the ductility factor, q, i.e. the more ductile a bridge may be, the more irregular.

8.2.1 Elastic Regularity Index

These indexes are based on the hypothesis that a structure may have an irregular behaviour even when it does not incur in the inelastic range. Elastic regularity indexes are calculated considering only the elastic properties of the structure and are used to classify the structure only as regular or irregular and/or to help the analyst decide whether a particular method of analysis is applicable (Isakovic and Fischinger 2000; Maalek et al. 2009).

Calvi and Elnashai (1994) proposes an elastic regularity index, I_R, with the purpose of evaluating the use of seismic design methods for bridges, such as that proposed by the EC8 (CEN 1994), as well as to provide the analyst with a fast and simple to apply tool to classify the structure, as regular or irregular. This regularity index combines the mode shape of the deck with the mode shapes of the entire bridge. For its calculation, this index only considers elastic properties of the structure and is defined by Eq. 8.1.

$$I_R = \sqrt{\frac{\sum_{i=1}^{n}\left(\frac{\Phi_i^B}{\Phi_i^B M \Phi_i^B} M \frac{\Phi_i^D}{\Phi_i^D M \Phi_i^D}\right)^2}{n}} \tag{8.1}$$

where

Φ_i^B: modal shape of the entire bridge structure
Φ_i^D: modal shape of the deck
n: number of modes considered in the modal spectral analysis

The magnitude of the index I_R varies over a range of 0–1, with a regular bridge normally having an index close to unity, while an irregular bridge an index close to zero. The use of the regularity index proposed by (Calvi and Elnashai 1994) when applying the design approach of the EC8 CEN (1994) may yield designs with low safety levels for ductile structures.

Other authors (Isakovic and Fischinger 2000; Fischinger and Isakovic 2003; Maalek et al. 2009) have carried out extensive research on the influence of irregularities in the seismic behaviour of viaduct-type reinforced concrete bridges finding that the degree of regularity of a bridge varies according to parameters such as the stiffness ratio of the deck and the piers, the torsional sensitivity (Isakovic and Fischinger 2000), the type of supports and the relative location of the pier with the largest stiffnesses.

Isakovic and Fischinger (2000) consider a bridge as irregular if the performance obtained with a single mode analysis method differs from that obtained with a conventional modal analysis. To validate the use of single mode analysis methods, SM, for the seismic evaluation of bridges (Isakovic and Fischinger, 2000) propose an elastic regularity “index”. This “index” is based on a comparison of the normalized lateral deformed configuration, obtained by linear static analysis using two different lateral load patterns.

8.2.2 *Inelastic Regularity Index*

Inelastic regularity indexes are based on the assumption that a structure which responds in the elastic range cannot have an irregular behaviour; i.e. for a structure to show an irregular behaviour, it must enter the inelastic range. Their formulation considers the accumulated damage in the structure calculated using nonlinear analysis methods; however, the current definitions of inelastic regularity index do not consider the change in mode shapes and order of participation, from one damage state to another. The most representative inelastic regularity indexes are based on a simple comparison of the normalized lateral deformed shapes, considering different lateral load patterns (Isakovic and Fischinger 2000; Maalek et al. 2009).

The FRI index is based on the assumption that the degree of regularity of a viaduct-type bridge depends on the regularity of the lateral deformed configuration of the deck. For its definition, the normalized deformed configuration of the bridge under study is compared with a deformed configuration totally regular obtained from the nonlinear static analysis of the bridge deck. For the calculation of the FRI, Eq. 8.2 is used.

$$FRI^{UL,\ SM} = \frac{|S^{D} - S^{B}|}{|S^{D}|} \tag{8.2}$$

where
S^D: area under the normalized lateral deformed configuration from the analysis of the deck
S^B: area under the normalized lateral deformed configuration from the analysis of the bridge

Index SRI is defined in the same way as index FRI; however, to calculate the SRI index, the entire bridge is used to define both deformed configurations. For the first analysis, a pushover analysis is carried out using a load pattern based on the fundamental mode. To calculate the second deformed configuration, a load pattern obtained from a spectral modal analysis is used. To calculate the SRI index, Eq. 8.3 is used.

$$SRI = \frac{|S^{SM} - S^{MM}|}{|S^{SM}|} \tag{8.3}$$

where
S^{SM}: area of the normalized lateral deformed shape obtained from the single mode analysis
S^{MM}: area of the normalized lateral deformed shape obtained from the multimode analysis

Indexes FRI and SRI are also defined within a range from 0 to 1 and have the same interpretation than the elastic regularity "index" proposed by Isakovic and Fischinger (2000).

Isakovic and Fischinger (2000) propose and use an inelastic "index" with the purpose of validating the use of the nonlinear static method, N2, for the seismic evaluation of irregular bridges. This inelastic regularity "index" has the same interpretation than the elastic regularity "index", proposed by the same authors (Isakovic and Fischinger 2000); however, to calculate it nonlinear static analyses are required for defining the lateral deformed configurations. The first deformed configuration is generated using the load pattern of the N2 method. The load pattern used to calculate the second deformed configuration is defined from the results of the first analysis. This inelastic "index" is calculated with Eq. 8.4.

$$\text{index} = \frac{\sum_{i=1}^{n-1} \left| |\emptyset_i^{\mathrm{N2}}| - |\emptyset_i^1| \right|}{\sum_{i=1}^{n-1} |\emptyset_i^{\mathrm{N2}}|} \tag{8.4}$$

where
$\emptyset_i^{\mathrm{N2}}$: normalized lateral displacement of point i obtained by means of the N2 method
$\emptyset_i^1$: normalized lateral displacement of point i obtained from a second pushover analysis

8.3 Modal Regularity in the Simplified Methods of Analysis

Modal irregularity is manifested as a change in the shapes of the modes of vibration, from one damage state to another during the response of a structure and occurs only in the inelastic range. The degree of modal regularity can vary from a slight change in the mode shapes to even the inversion of these shapes and their corresponding participation factors.

Modal irregularity occurs in any type of structure responding in the inelastic range of behaviour; however, in the case of short viaduct-type bridges, changes of mode shapes may occur when a number of elements (piers) behave into the inelastic range. This change in modal shapes may even be shown immediately after the first yielding of an element, a phenomenon that rarely occurs in buildings, where modal irregularity manifests only when the structure presents considerable damage, and this explains why the building analysts are not normally aware and do not consider this effect on the application of simplified methods.

Although modal irregularity is not the only factor affecting performances calculated with simplified methods of analysis, different from those calculated with the "exact" method, this effect can lead to completely erroneous results if not considered in the analysis. Some of the factors which significantly affect the modal irregularity of viaduct-type bridges are:

1. The ratio of superstructure to the substructure stiffnesses
2. The ratios of stiffnesses between different piers
3. The damage model
4. The characteristics of the seismic demand

This chapter proposes versions of modal regularity index, I_{RM1} and I_{RM2}, aiming to provide the analyst with a tool to identify, with a simple formulation, the degree of irregularity of a structure and the elements to decide which method of analysis is acceptable to use.

Table 8.1 Classification of the investigated bridges using different regularity criteria

		Regularity index			
Bridge	AASHTO	I_R	Index	FRI	I_{MR}
V213P	Irregular	0.63	0.80	0.61	0.78
V232P	Irregular	1	0.92	0.97	1

The indexes I_{RM1} and I_{RM2}, defined by Eqs. 8.5 and 8.6, consider inelastic behaviour by using modal analyses of the structure in the elastic and inelastic ranges. They are based on the evolution of modal shapes during its inelastic response from a damage state to another. Table 8.1 presents and compares the results of applying different proposals for these definitions of irregularity for the two bridges analysed, one considered as "irregular", V213P, and the other as "regular", V232P.

$$I_{RM} = \sum_{j=1}^{n} \alpha \Phi_i^j \Phi_{i-1}^j \tag{8.5}$$

$$I_{RM} = \sum_{j=1}^{n} \alpha \left| \frac{\Phi_i^j}{\sum_{j=1}^{n} \left| \Phi_i^j \right|} - \frac{\Phi_{i-1}^j}{\sum_{j=1}^{n} \left| \Phi_{i-1}^j \right|} \right| \tag{8.6}$$

$$\alpha^j = \frac{\Gamma_i^j S_{ai}^j}{\sum_{j=1}^{n} \Gamma_i^j S_{ai}^j} \tag{8.7}$$

where Φ_i^j : normalized modal shape of mode j for event i Γ_i^j : modal participation factor of mode j for event i S_{ai}^j: spectral pseudo-acceleration corresponding to mode j and event i.

8.4 Proposed Procedure

The procedure proposed in this section is aimed to approximate in a direct way the nonlinear performance of a bridge-like structure subjected to a given seismic demand, without requiring the application of any other procedure for calculating the performance. The capacity curve of the structure is generated by a series of successive modal spectral analyses, i.e. evolutive modal spectral analyses.

The origin of this procedure is the simplified method of seismic evaluation proposed by Requena and Ayala (2000), based on a capacity curve calculated with equivalent seismic forces of increasing intensity and on the performance of a reference SDOF system. The method considers the stiffness degradation of the structure when behaving in the inelastic range and the contribution of higher modes of vibration, using the results of modal spectral analyses of models of the structure updated in accordance with each damage configuration during loading.

The procedure proposed preserves most characteristics of the original (Requena and Ayala 2000), also including other contributions, like replacing equivalent lateral loads by lateral displacements obtained from modal spectral analyses (Alba et al. 2005), and considering a correction for hysteretic energy dissipation under seismic demands through an equivalent damping ratio, ξ_{eq}, for each level of damage in the equivalent SDOF system characterized by the capacity curve (Mendoza and Ayala 2011; Cárdenas et al. 2011). This new method is aimed at the seismic assessment of irregular viaduct-type reinforced concrete bridges by calculating the performance of the structure directly from the independent modal capacity curves with nonlinear behaviour branches corresponding to comparable modal shape configurations, minimizing in such a way the ill effects of modal irregularities on the determination of the seismic performance of the structure obtained by adding the participation of all modes, with a modal combination rule.

8.4.1 Determination of the Seismic Performance of a Structure

The performance of a bridge is evaluated using the independent modal capacity curves constructed with the step-by-step procedure described in the previous section, i.e. calculating as many points in these curves as plastic hinges are sequentially formed in the bridge. Each point is defined from a modal spectral analysis of the structural model corresponding to each damage level. The procedure to calculate the capacity curve involves the following steps:

1. Definition of the seismic demand: The procedure proposed allows as seismic demands to the structure a design spectrum or a single earthquake record.
2. Definition of the capacity curves of the structural elements: The moment-curvature diagrams of the required sections are calculated using standard procedures.
3. Determination of the elastic branch of the capacity curve: The elastic branch of the capacity curve is defined by the yield point (D_f–V_f), corresponding to the intensity of seismic demand producing the first damage in the structure. The scaling factors, SF^1, for each bridge pier are calculated using Eq. 8.8 where M_Y is the yield moment, M_G the moment due to gravity loads and M_{DS} is the moment produced by the unscaled seismic demand. The SF^1 finally required in the analysis is the minimum of the scaling factors calculated for each pier.

$$S_f^1 = \frac{M_y - M_G}{M_{DS}} \tag{8.8}$$

4. Definition of the elastic range of the modal capacity curves: A particular curve is defined for each mode that significantly influences the response of the structure. The required number of modes is based on the degree of irregularity of the bridge.

5. Definition of a new model for the structure: The model of the structure is updated considering the accumulated damage of previous demand intensities.
6. Calculation of the next preliminary performance point in the corresponding inelastic branch of the capacity curve: Once the structure enters the range of inelastic behaviour, the next preliminary performance point of the capacity curve is defined by the application of a scaled seismic demand, enough to produce the next damage level. To calculate the increase in seismic demand, Eq. 8.9 is used, where Mac is the accumulated moment.

$$S_{\mathrm{f}}^{2} = \frac{M_{\mathrm{y}} - M_{\mathrm{ac}}}{M_{\mathrm{DS}}} \tag{8.9}$$

7. Correction of the preliminary performance point in the capacity curve: The correction of the preliminary performance point in the capacity curve to consider the hysteretic energy dissipation is carried out using the procedure of (Mendoza and Ayala 2011).
8. Calculation of the performance points in the modal capacity curves: To define these points in the modal capacity curves, it is required to consider as inelastic branches those corresponding to modal configurations similar to the elastic, minimizing in such a way the effects of modal irregularity.
9. Calculation of the first performance point in the inelastic range: This point of the capacity curve is calculated by applying a modal combination rule, Eq. 8.10.

$$r = \sqrt{\sum_{i=1}^{n} r_n^2} \tag{8.10}$$

10. Calculation of the subsequent performance points associated to other damage levels: To generate the points of subsequent performance levels in the capacity curve, the calculation process must be continued from step 5.

The method ends when a target displacement is reached or a collapse mechanism in the structure occurs. In Fig. 8.1, the above procedure is schematically illustrated.

8.5 Application Examples

To illustrate the influence of modal irregularity on simplified analyses and validate the procedure proposed, two examples of viaduct-like reinforced concrete bridges 200 m long, divided into four spans of 50 m each are considered. The first bridge, V232P, considered "regular" (Calvi and Elnashai 1994; Isakovic and Fischinger 2000; Maalek et al. 2009) and the second, V213P, considered "irregular" (Calvi and Elnashai 1994; Isakovic and Fischinger 2000; Maalek et al. 2009). The letter "V" indicates the type of bridge, viaduct; the letter "P" indicates the type of support

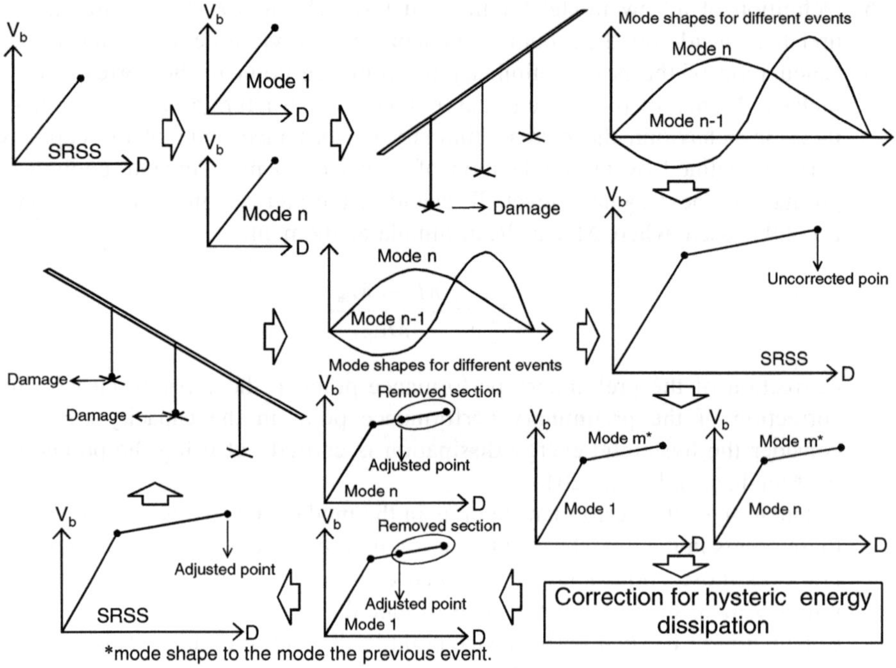

Fig. 8.1 Procedure of the new proposal to evaluate the performance of an existing bridge structure

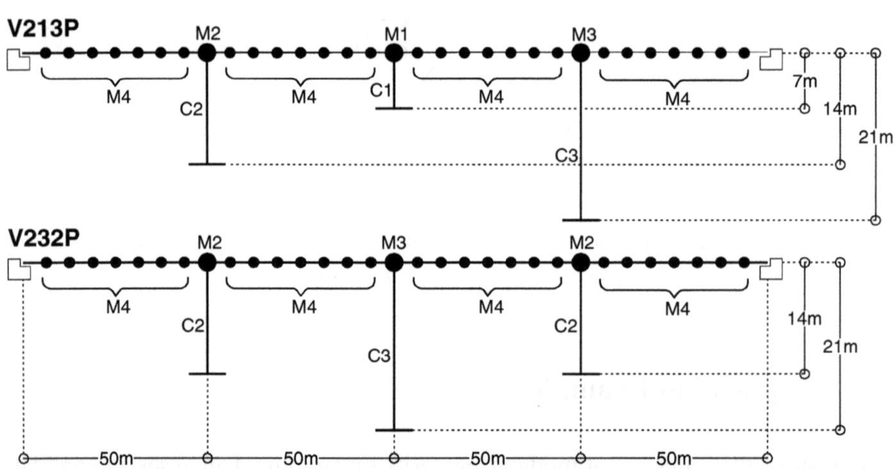

Fig. 8.2 Bridges V213P and V232P

on the bridge abutments, pinned support, and the intermediate numbers, VxxxP, indicate the relationship between pier heights (Fig. 8.2). Tables 8.2, 8.3, 8.4 show the mechanical properties of the materials and the geometrical and mechanical

Table 8.2 Mechanical properties of the materials

	f'_C (MPa)	E_C (MPa)	γ_C (kN/m^3)
Deck	27	25,000	25
Pier	27	20,000	25

Table 8.3 Geometric and mechanic characteristics of the decks, V213P y V213P

Deck	A (m^2)	I_X (m^4)	M1 (kN)	M2 (kN)	M3 (kN)	M4 (kN)
Box beam	6.97	88.46	1654.3	2018.3	2382.3	1290.3

Table 8.4 Geometric and mechanic characteristics of the substructures

				Transversal section	
Substructure	Pier	M_Y (kN-m)	M_Y (kN-m)	A (m^2)	I_X (m^4)
	C1-V213P	51,100	31,150		
Strong	C2-V213P	30,500	51,100	0.16	7.39
	C3-V213P	31,150	30,500		
	C1-V232P	–	–		
Weak	C2-V232P	75,200	51,100	0.16	2.21
	C3-V232P	39,160	30,500		

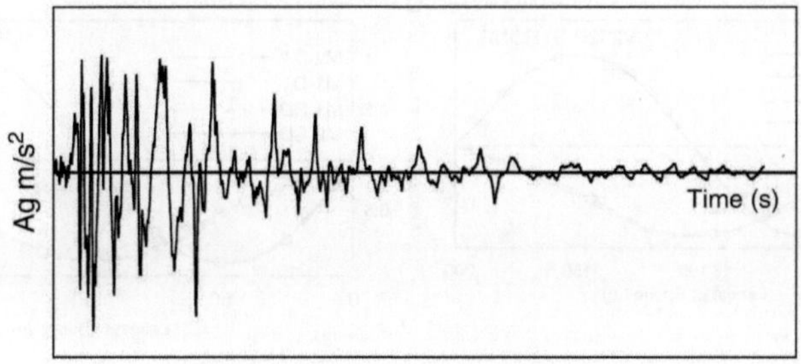

Fig. 8.3 Seismic record used in the evaluation of the bridges studied

characteristics of the decks and the piers of the bridges used as examples. The decks of both bridges are considered to behave elastically, even though the structures are subject to demands of considerable magnitude. Inelastic behaviour is considered to occur only in the piers.

8.5.1 Seismic Demand

The demand used in the evaluation of seismic performance of the examples is the EW component of the Takatori Station record of the 17 January 1995 earthquake in Kobe, Japan (see Fig. 8.3). To carry out the evolutive modal spectral analyses, the commercial structural analysis program SAP2000, (CSI 2000), was used, and for the incremental dynamic analyses, IDA, the program DRAIN 2DX (Prakash et al. 1992).

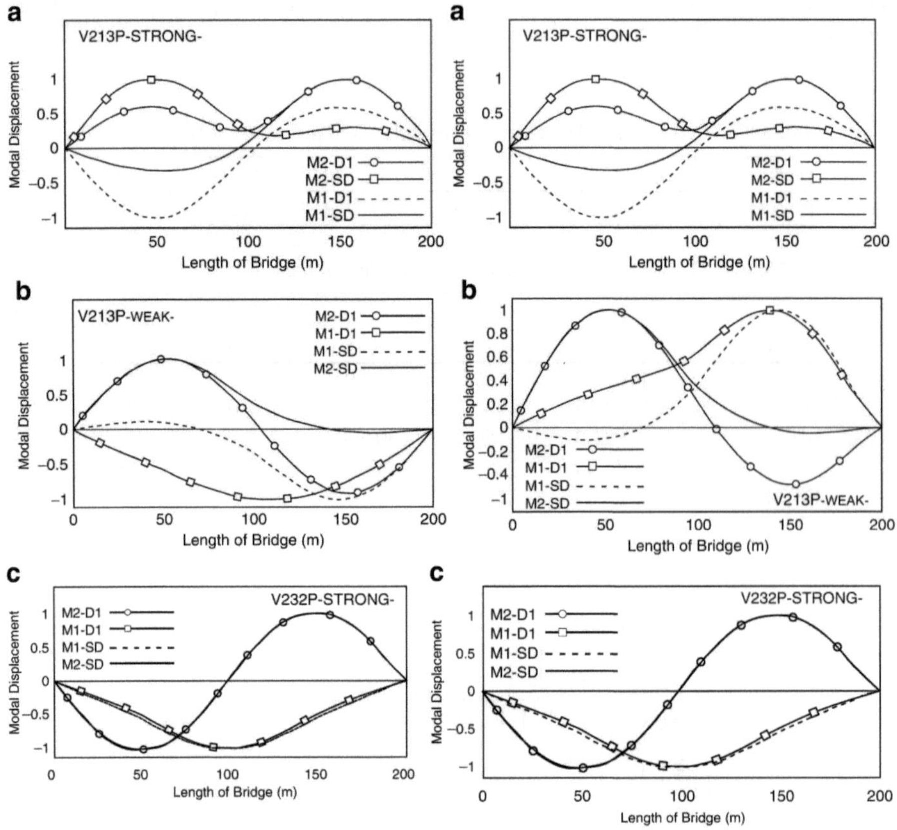

Fig. 8.4 (**a**)–(**c**) Modal shapes of the bridges

8.5.2 Detection of Modal Irregularity in Viaduct-Type Bridges

To show that modal irregularity can occur in viaducts with vibration periods close to each other, and when the modes that contribute most to the total response of the structure change their shape when passing from one damage state to another during their inelastic response, two reinforced concrete viaduct-type bridges are analysed calculating their degree of irregularity.

Figure 8.4a–c shows the results obtained from the modal analysis of the V232P and V213P bridges for different damage states. Two different damage models were considered, the first introducing plastic hinges at the bases of the piers of the substructure and the second introducing a stiffness degradation at the bases of the pier elements; modal analyses were also performed considering two different values of the lateral stiffness of the substructure.

Fig. 8.5 Capacity curve of the bridge, V232P, without correction for dissipation of hysteretic energy

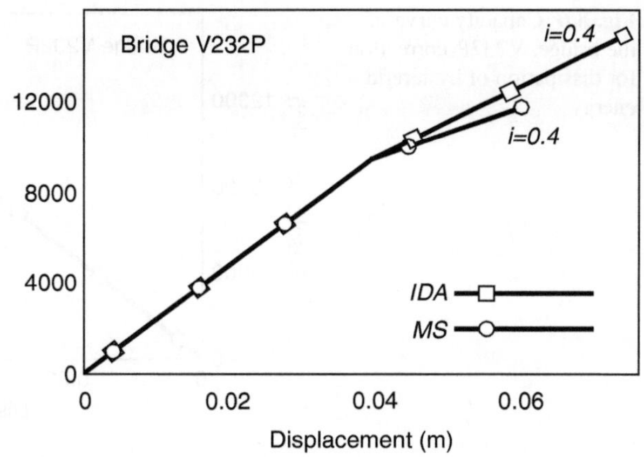

Figure 8.4a shows that using a simple model to represent damage (plastic hinges) in the assessment of structures using simplified methods can lead to structures with a high degree of irregularity, in particular for short viaducts. Figure 8.4b shows that decks of high stiffness when compared to the stiffness of the substructure produce unacceptable degrees of irregularity; if the pier with the highest contribution to the lateral stiffness is damaged, a structure with a high degree of irregularity may result, regardless of the relationship between the stiffnesses of the decks and piers. Figure 8.4c shows that viaducts regular in geometry, length and pier stiffnesses have a lower degree of irregularity than irregular viaducts.

8.5.3 Performance Evaluation of the Investigated Examples

To validate the procedure proposed study, the behaviour of one bridge, V232P, subjected to a given seismic demand is investigated. Their capacity curves were generated using the two options, one with correction for hysteretic energy dissipation (Mendoza and Ayala 2011) and the other without it and the IDA. To determine the effectiveness of the simplified methods of analysis, their capacity curves are compared with those obtained considered as "exact", calculated with the IDA.

Figure 8.5 shows a comparison of the capacity curves for the viaduct V232P, obtained without the correction for hysteretic energy dissipation, and the IDA. In this figure, it may be observed that the displacements of the characteristic point chosen to plot the capacity curve do not correspond, for the same intensity of seismic demand, to the IDA displacements; i.e. the capacity and the IDA curves are similar in form but the performances are different.

Figure 8.6 shows the capacity curves for the V232P bridge obtained by the simplified method but now considering a correction for hysteretic energy dissipation, in accordance with the procedure proposed (Mendoza and Ayala 2011), SM, and

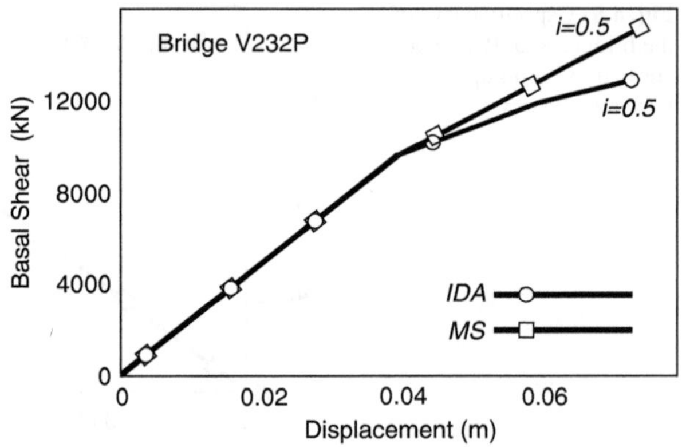

Fig. 8.6 Capacity curve of the bridge, V232P, correction for dissipation of hysteretic energy

the IDA. This figure shows similar magnitudes of displacements in correspondence with the same intensity of the seismic demand; i.e. the performances calculated by both methods are comparable.

8.6 Conclusions

This chapter presented the preliminary results of an ongoing investigation on the modal irregularity occurring in structures, particularly viaduct-type bridges, discussing its undesirable effects on the approximate methods of seismic evaluation and proposing an alternative approximate evaluation procedure which eliminates these effects. The procedure is based on basic approximations of structural dynamics concepts used to sequentially correct the capacity curve for dissipation of energy due to hysteresis by considering equivalent damping ratios. To illustrate the potential of the procedure proposed, the details of each step involved in its application are presented.

From the analysis of the contents of this chapter, the following conclusions may be extracted:

1. The regularity indexes proposed in this chapter may be used to support the analyst in deciding whether the simplified methods of analysis are suitable for the evaluation of one bridge in particular.
2. The geometric characteristics of a bridge are not always the only parameters, and rarely the most appropriate, needed to define the degree of modal irregularity of a viaduct-type bridge, as this irregularity also depends on other factors such as the characteristics of the damage under seismic demand intensities.
3. To determine the degree of modal irregularity, it is necessary that the structure under evaluation incurs the nonlinear range of behaviour. Considering that in the

linear range there is an exact correlation between performance and intensity of the seismic demand for a simplified method and the corresponding "exact" IDA, the degree of irregularity for bridges responding in the elastic range must be investigated from a different viewpoint to that followed in this chapter.
4. The use of the lateral deformed configuration as the most important characteristic to measure the degree of modal irregularity of a bridge structure may lead to unreliable results, as in general, this lateral deformed configuration is calculated using a conventional modal combination rule which approximation has been tested on the evaluation of the seismic performance of buildings and not as much on the evaluation of bridges.

Acknowledgements The constructive comments and the interest shown by Dr. Mauro Niño in the preparation of this chapter, the sponsorship of the National Council of Science and Technology of the project no. 82839, "Development of conceptual, theoretical models and simplified methods for the evaluation and seismic design of structures based on performance" and the scholarship of the second author during his graduate studies are acknowledged.

References

AASHTO, AASHTO LRFD (2007) Bridge design specifications, 4th edn. American Association of State Highway and Transportation Officials, Washington, DC
Alba F, Bento R, Ayala AG (2005) Seismic performance evaluation of plane frames vertically regular and irregular. In: Proceedings of the IV European workshop on the seismic behaviour of irregular and complex structures, EAEE, Thessaloniki, Greece
Calvi GM, Elnashai AS (1994) A. Pavese, influence of regularity on the seismic response of RC bridges. In: 2nd International workshop on seismic design and retrofitting of RC bridges, 8–13 August, New Zealand, pp 83–93
Cárdenas R, Ayala AG, Mendoza M (2011) Comparative study of simplified seismic analysis and incremental dynamic analysis in the calculation of the capacity curve (in Spanish). In: Proceedings of the XVII National Congress in earthquake engineering, SMIS paper V-32. Puebla, Mexico
CEN (1994) Eurocode 8: design of structures for earthquake resistance. Part 2: Bridges, Committee European of Normalisation, European Pre-standard, ENV1998-2, Brussels
CSI (2000) SAP2000, integrate finite element analysis and design of structures. Computer and Structures Inc., Berkeley
Fischinger M, Isakovic T (2003) Inelastic seismic analysis of reinforced concrete viaducts. Struct Eng Int 13(2):111–118
Isakovic T, Fischinger M (2000) Regularity indices for bridge structures. In: Proceedings of the XII World Conference on earthquake engineering, paper No 1725. Auckland, New Zealand
Maalek S, Akbari R, Maheri M (2009) The effect of higher modes on regularity of single-column-bent highway viaducts. Bridges Struct 5:29–43
Mendoza M, Ayala AG (2011) Performance based evaluation procedure for reinforced concrete buildings: development and validation (in Spanish, sent for review and publication to Revista de Ingeniería Sísmica)
Prakash V, Powell GH, Filippou F (1992) DRAIN-2DX: base program user guide. SEMM Report 92-29, University of California, Berkeley

Requena M, Ayala AG (2000) Evaluation of a simplified method for determination of the nonlinear seismic response of RC frames. In: Proceedings of the XII World Conference on earthquake engineering, paper No 2109. Auckland, New Zealand

Veletsos A, Newmark N (1960) Effect of inelastic behaviour on the response of simple systems to earthquake motions. In: Proceedings of the II World conference on earthquake engineering, vol 2, Tokyo, Japan, pp 895–912

Chapter 9
Influence of Infill Panels on the Seismic Response of Existing RC Buildings: A Case Study

Marco Tanganelli, Stefania Viti, Mario De Stefano, and Andrei M. Reinhorn

Abstract An analytical evaluation of the seismic response of irregular RC buildings with infill panels designed without considering modern seismic provisions is presented in this chapter. Many RC buildings have been built in the 1960s and 1970s in Italy (and many other countries), before the seismic code was issued. Such buildings are usually irregular, due to mass and stiffness distribution, and they exhibit torsional effects when subjected to horizontal loads. One of the factors contributing to the torsional effects is the infill panels built on the perimeter of structures. Usually there is no discontinuity between the structure and the infill, so that the stiffness of the structure is largely affected by the distribution of the infill walls that can be irregular both in plan and in elevation. This work presents the effect of masonry infill in the seismic response of precode RC buildings. A numerical analysis has been performed on a case study, simulating the seismic response of the structure, including masonry infill panels. The case study of an existing building, rectangular in plan, having a mild mass eccentricity and an irregular distribution of infill panels, and, therefore, a different eccentricity at each storey is selected for the evaluation. The seismic performance of the building has been evaluated comparing the seismic response with the limit conditions provided by Italian Technical Code, both in terms of interstorey drift and shear capacity. The seismic response of the case-study building is satisfactory in respect to deformation parameters (chord rotation, top displacement, interstorey drift) but exhibits an unacceptable amount

---

M. Tanganelli • S. Viti • M. De Stefano (✉)
Dipartimento di Costruzioni e Restauro, Facoltà di Architettura, Università degli Studi di Firenze, Piazza Brunelleschi 6, 50121 Florence, Italy
e-mail: marco.tanganelli@unifi.it; viti@unifi.it; mario.destefano@unifi.it

A.M. Reinhorn
Department of Civil, Structural and Environmental Engineering, University at Buffalo, The State University of New York, 135 Ketter Hall, Amherst Campus, Buffalo, NY 14260, USA
e-mail: reinhorn@buffalo.edu

O. Lavan and M. De Stefano (eds.), *Seismic Behaviour and Design of Irregular and Complex Civil Structures*, Geotechnical, Geological and Earthquake Engineering 24, DOI 10.1007/978-94-007-5377-8_9,

of shear in some of the elements, experiencing "short-column" failures. A possible solution for the problem is proposed, by removing some of the partial infill panels, in order to reduce the shear concentration demands and eccentricity.

9.1 Introduction

Irregular RC buildings are an important part of the Italian modern constructions. In fact, there has been a peak in the construction activity in the years 1960 and 1970, before the seismic technical code was issued (Legge 2/2/1974 n. 64 1974). Many existing buildings, therefore, have been constructed without considering horizontal earthquake loads at all. As a result, there are numerous buildings (D'Ambrisi et al. 2010) with irregularities both in plan and in elevation not considered in the design.

While demolition and reconstruction would be the best solution for the above-mentioned buildings, cost constrains prevent such solution. Retrofitting and upgrade of the buildings is not a convenient approach either, since it would be expensive without providing the advantages of new constructions (existing building is not adequate for equipment, energy saving, etc.). Therefore, the most convenient improvement of the seismic performance should be very simple and inexpensive, and, possibly, it should not require any interruption of buildings' functions.

In this work, the effect of the masonry infill on the seismic response of existing RC-framed buildings is evaluated. In these buildings, in fact, the infill panels made of traditional masonry are built without any discontinuity with the structural frame, affecting the seismic response of the entire structure. An irregular distribution of infill panels can induce, however, also torsional effects in otherwise regular structures.

A case study is considered, for an existing building in a hospital complex, having a high priority for the seismic protection. The structure has some mild irregularities due to a nonsymmetric distribution of columns in plan. The masonry infill panels built tight in the RC frames have an irregular distribution in plan of the building, inducing a different eccentricity between the stiffness and mass centers at each storey. Such plan irregularity is coupled to vertical irregularities at each floor due to short columns construction when using partial infill panels, that is, infill panels not covering the whole storey height. Indeed, this situation occurs quite often, since the short infill panels are a common way to introduce horizontal window spaces in the external walls without interrupting the curtain walls. Horizontal windows have been extensively used in the Italian architecture and other countries constructed in the 1970s.

Seismic performance of the building has been evaluated by comparing the seismic response of the building, neglecting and including the infill panels, with the limit conditions provided by Italian Technical Code (Norme Tecniche delle Costruzioni 2008; D.M. 14 Gennaio 2008) for *interstorey drift* and *shear capacity*. The analysis shows that, due to the partially infill panels, the limit shear capacity (required by the code) is exceeded in short columns even when the entire building system is still in the elastic range.

A possible way to improve the seismic performance of the structure is proposed, consisting of the replacement of some of the infill panels, in order to avoid short columns and shear concentration. The proposed modified feature of the panels has even the effect of reducing the eccentricity between mass and stiffness centers in plan.

9.2 Case Study

The case study, shown in Fig. 9.1, is a 3-storey RC-framed building built in 1978. Since the structural design was made in 1972, the applied technical codes (D. Pres. Rep. Del 25/011962 n 1684; Legge 5/11/1971 n 1086) were not adequate to provide seismic safety according to the current standards. The building is part of a hospital complex, so it is classified in class of use IV (strategic buildings).

The column's position is symmetric only in the *x*-direction, where there are two bays having equal spans (3.15 m). In *y*-direction, the two bays have different spans, of 3.35 and 6.35 m, respectively. Each interstorey height is 3.62 m.

The concrete has a strength class of R_{ck} 250 ($f_c = 20$ MPa), while the steel used for the reinforcement is FeB38k ($f_y = 380$ MPa). All the beams have the same geometry (30 × 60 cm), while the column cross sections, oriented along the *y*-direction, vary between 30 × 50 cm at the first storey and 30 × 35 at the third storey. The RC structure has been modeled by means of the fiber model implemented in Seismostruct (2011).

Three of the sides of the building have infill panels with different geometrical features. Panels are made of hollow brick masonry, with void percentage <45%, and their mechanical characteristics, taken from NTC2008, are presented in Table 9.1 along with their geometrical characteristics.

Figure 9.2 shows the computational model implemented in Seismostruct (2011) to determine the infill panel behavior.

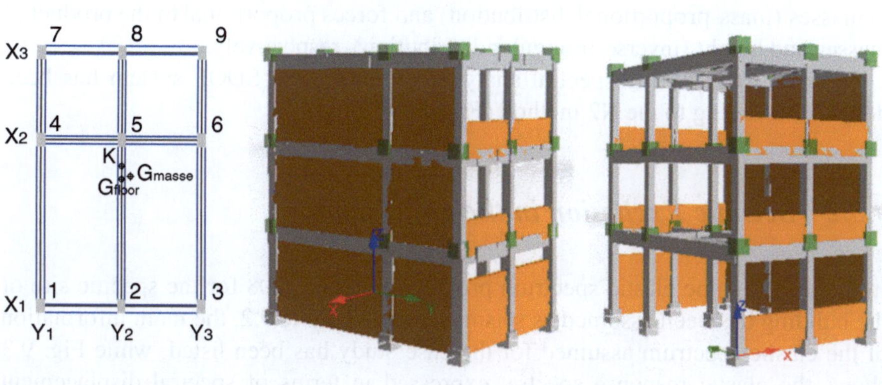

Fig. 9.1 Case study plan and 3D views

Table 9.1 Geometrical and mechanical properties of the infill panels

Material		Geometry of panels						
			x1 side		*x3 side*		*y3 side*	
Strength	4 MPa		*L*	*H*	*L*	*H*	*L*	*H*
Stiffness	3,600 MPa	PT	2.85	1.2	2.85	2.6	5.85–2.85	3.0
Thickness	30 cm	1P	2.85	1.2	2.85	1.2	5.95–2.95	3.0
Model	Strut and shear	2P	2.85	1.2	2.85	1.2	6.0–3.0	3.0

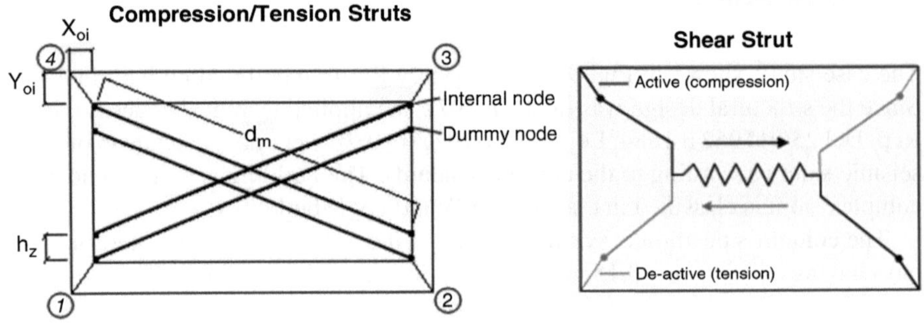

Fig. 9.2 Assumed model for the infill panels (By Seismostruct 2011)

9.3 Evaluation of the Seismic Performance of the Structure

9.3.1 Seismic Analysis

The seismic response of the structure has been determined by performing a nonlinear static analysis (Viti et al. 2006; Reinhorn 1997), based on spectral description of the seismic demand and capacity of the structure. The spectral capacity of the case study has been found by assuming two different horizontal patterns for the inertial forces, according with the NTC2008 provisions, that is, forces proportional to masses (mass-proportional distribution) and forces proportional to the product of masses and height (inverse-triangular distribution), respectively.

In order to apply the spectral analysis, an equivalent SDOF system has been defined, according to the N2 method (Fajfar 1999).

9.3.2 Seismic Excitation in the Evaluation

In the analysis, the elastic spectrum provided by NTC2008 for the specific site of the building has been assumed as seismic input. In Table 9.2, the main information of the elastic spectrum assumed for the case study has been listed, while Fig. 9.3 shows the elastic response spectra, expressed in terms of spectral displacement versus spectral acceleration, for the considered limit states.

Table 9.2 Data related to seismic input of the case study

			PGA			
Soil type	Class of use	Nominal life	SLO	SLD	SLV	SLC
B	IV	50 years	0.118 g	0.148 g	0.322 g	0.379 g

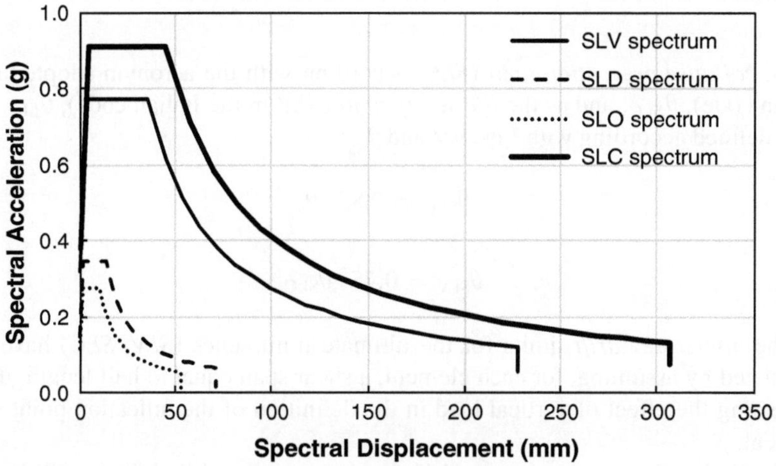

Fig. 9.3 Elastic composite response spectra used in evaluation

9.4 Limit States

The assumed response parameters are the *top displacement* (*TD*), that is, the *lateral displacement* of the target point at the top level, the *interstorey drift* (*ID*), and the *shear force* (*V*), measured at each column line of the structure.

The limit states described by the Italian Code (Circ. Min. 2009) have been considered in checking the seismic performance of this case study.

9.4.1 Limit States Assumed for Interstorey Drifts

The limit states assumed for the *interstorey drift* have been defined as a function of the chord rotation. NTC2008 quantifies the ultimate rotation, θ_u through Eq. 9.1:

$$\theta_u = \frac{1}{\gamma_{el}} 0.016 \cdot (0.3^{\nu}) \left[\frac{\max(0.01;\omega')}{\max(0.01;\omega)} f_c \right]^{0,225} \left(\frac{L_V}{h} \right)^{0,35} 25^{\left(\alpha\, \rho_{sx} \frac{f_{yw}}{f_c}\right)} \left(1.25^{100\rho_d}\right) \tag{9.1}$$

The limit rotations assumed to define the ultimate limit states have been quantified as a function of θ_u. In particular, the rotation values assumed, respectively,

Table 9.3 Limit values of *interstorey drift* (%) in columns

	SLO			SLD			SLV			SLC		
	1st st.	2nd st	3rd st.	1st st.	2nd st	3rd st.	1st st.	2nd st	3rd st.	1st st.	2nd st	3rd st.
x-direction	0.33	0.33	0.33	0.5	0.5	0.5	1.9	2.1	2.3	2.5	2.7	3.0
y-direction	0.33	0.33	0.33	0.5	0.5	0.5	2.2	2.3	2.4	2.9	3.0	3

as the *collapse prevention limit* (*SLC*, according with the acronym adopted in the Italian code), θ_{SLC}, and as the *life safety limit* (*SLV* in the Italian code), θ_{SLV}, have been defined according with Eqs. 9.2 and 9.3:

$$\theta_{\text{SLC}} = 0.85 \cdot \theta_u \tag{9.2}$$

$$\theta_{\text{SLV}} = 0.75 \cdot \theta_{\text{SLC}} \tag{9.3}$$

The *interstorey drift* limits for the ultimate limit states (*SLV, SLC*) have been calculated by assuming, for each element, a shear span equal to half length, that is, neglecting the effect of vertical load in the definition of the inflection point of the element.

The limit values of *interstorey drift* for the operational limit states (*SLO, SLD*) have been assumed according to the code provisions, that is, 0.5% for the *SLD* and 2/3 of such values for the *SLO*. The assumed values for *interstorey drifts* have been listed in Table 9.3.

9.4.2 Limit States Assumed for Shear Capacity

Shear limit states' values provided by NTC2008 for reinforced concrete elements have been assumed as limit conditions for both columns and beams of the case study. The shear limit is described by the following expressions:

$$V_{\text{Rd}} = \min\left(V_{\text{Rsd}}, V_{\text{Rcd}}\right) \tag{9.4}$$

$$V_{\text{Rsd}} = \frac{0.9d \cdot a_{\text{sw}}}{s \cdot f_{\text{yd}}} (\text{ctg}\,\alpha + \text{ctg}\,\theta) \sin\,\theta \tag{9.5}$$

$$V_{\text{Rcd}} = 0.9\,d \cdot b_w \cdot \alpha_c \cdot f'_{\text{cd}} \cdot \frac{(\text{ctg}\,\alpha + \text{ctg}\,\theta)}{(1 + \text{ctg}\,\theta)} \tag{9.6}$$

where α is the angle between the element axis and the shear reinforcement and θ is the angle between the element axis and the compressed truss. Depending on the amount of the angle θ assumed, ranging between 21.5° and 45°, the limit shear can

Table 9.4 Shear capacity limit (*N*) in the columns

Limit state	$\theta = 45°$ (cotg $\theta = 1.0$)			$\theta = 21.5°$ (cotg $\theta = 2.5$)		
Column level	1st st.	2nd st	3rd st.	1st st.	2nd st	3rd st.
x-direction	40,100	40,100	40,100	100,400	100,400	100,400
y-direction	69,900	55,000	47,600	174,700	137,500	119,000

Table 9.5 Shear capacity (N) limit in the beams

Beam specification	$\theta = 45°$ (cotg $\theta = 1.0$)	$\theta = 21.5°$ (cotg $\theta = 2.5$)
x-direction, inside	84,800	211,900
x-direction, side	84,800	211,900
y-direction, span = 6.35 m	84,800	211,900
y-direction, span = 3.15 m	84,800	211,900

vary substantially. Tables 9.4 and 9.5 show the values of shear limit states obtained, for columns and beams, respectively, for the two values of θ.

It has been noted that an inclination of 21.5° between the compressive truss and the element axis seems inadequate, especially for columns. Most part of *international codes* suggest to consider, as shear limit, the value corresponding to $\theta = 45°$, with the addition of the contribution of the tensile concrete. In this work, both of the above limit state values of θ provided by NTC2008 have been considered.

9.5 Results

9.5.1 Eccentricity in the Structural System

The eccentricity of each storey has been evaluated by comparing the mass and the stiffness centers in the two directions, for the bare frame (FR) and for the frame including the infill contributions to the stiffness (FR_IN).

Positions of (stiffness) rigidity centers in the building, at each floor level, have been found by applying an approximate procedure, as its rigorous definition for multistorey buildings cannot be achieved, except for some special cases. The capacity curves for the 2D structure's frame have been determined in the two directions, and from them, the lateral stiffness has been calculated at each level as the ratio between floor shear and the corresponding interstorey drift. Once values of lateral stiffness have been obtained at one floor, the location of center of rigidity (also known as center of stiffness) has been determined.

Two alternative horizontal force patterns, one mass-proportional and the other inverse-triangular, have been considered while determining the capacity curves. Table 9.6 lists the positions of the centers of rigidity obtained, together with the

Table 9.6 Position of mass and rigidity centers and their eccentricities

	Mass center		Rigidity (stiffness) center							
			FR				FR_IN			
			Mass-prop.		Static-prop		Mass-prop.		Static-prop	
	X_m	Y_m	X_K	Y_K	X_K	Y_K	X_K	Y_K	X_K	Y_K
1	3,509	4,963	3,274	5,129	3,267	5,126	3,714	7,147	3,709	7,125
2	3,515	4,957	3,208	5,146	3,227	5,182	3,400	5,007	3,485	4,993
3	3,302	4,918	3,144	5,161	3,149	5,213	3,360	4,998	3,354	5,001
	Global dimensions		Eccentricity							
			FR				FR_IN			
			Mass-prop.		Static-prop		Mass-prop.		Static-prop	
	L_X	L_Y	e_x (%)	e_y (%)	e_x (%)	e_y (%)	e_x (%)	e_y (%)	e_x (%)	e_y (%)
1	6,300	9,700	3.7	−1.7	3.8	−1.7	−3.3	−22.5	−3.2	−22.3
2	6,300	9,700	4.9	−1.9	4.6	−2.3	1.8	−0.5	0.5	−0.4
3	6,300	9,700	2.5	−2.5	2.4	−3.0	−0.9	−0.8	−0.8	−0.9

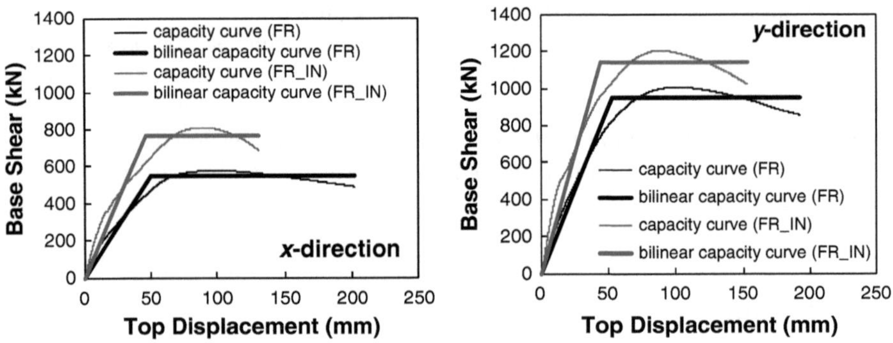

Fig. 9.4 Capacity curves for the mass-proportional horizontal pattern

corresponding eccentricities with respect to the mass centers as percentages of the global floors dimensions.

9.5.2 Capacity Curves and Seismic Response of the Structure

Figures 9.4 and 9.5 show the capacity curves obtained for the case study (with and without the infill panels contribution) in the main directions for the two horizontal load patterns.

The values of the target displacements obtained for the assumed limit states for the two horizontal patterns considered are shown in Tables 9.7 and 9.8 for the bare frame and the frame with infill panels.

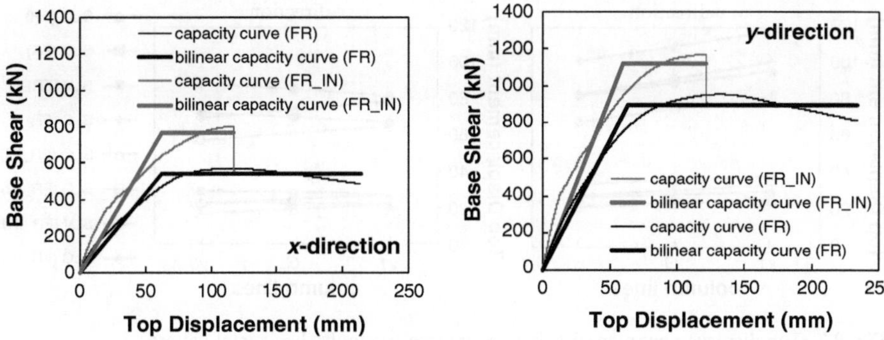

Fig. 9.5 Capacity curves for the inverse-triangular horizontal pattern

Table 9.7 Target top displacement (FR)

	Mass-prop pattern				Inverse-triangular pattern			
	SLO	SLD	SLV	SLC	SLO	SLD	SLV	SLC
x-direction	22.1	28.3	69.0	84.2	28.8	36.8	89.7	109.7
y-direction	17.2	21.9	53.6	65.5	22.5	28.8	70.3	85.6

Table 9.8 Target top displacement (FR_IN)

	Mass-prop pattern				Inverse-triangular pattern			
	SLO	SLD	SLV	SLC	SLO	SLD	SLV	SLC
x-direction	17.9	22.9	56.0	68.1	24.2	31.1	75.6	92.2
y-direction	14.4	18.5	44.9	55.0	19.4	25.0	61.1	74.4

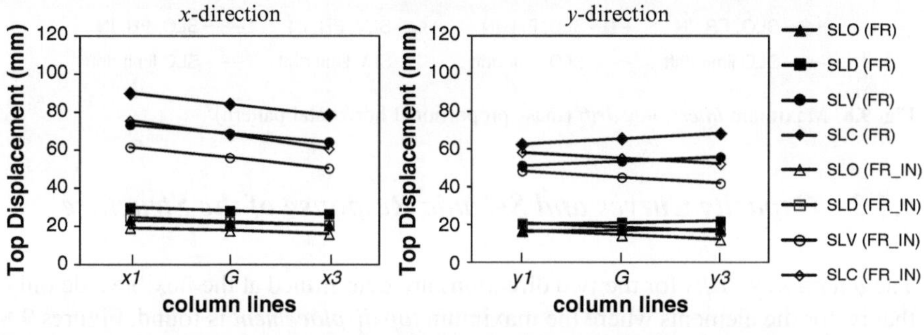

Fig. 9.6 Top displacement plan distribution (mass-proportional horizontal pattern)

Figures 9.6 and 9.7 show the plan-wise trend of the maximum top displacement along the two main directions of the structure. It has to be noted that, for both the assumed horizontal patterns, in the *y*-direction, the infill panels induce a change in the torsional behavior of the building. In fact, the rigid side moves from the *y1* column line when the panels are neglected to the *y3* column line when they are included in the analysis.

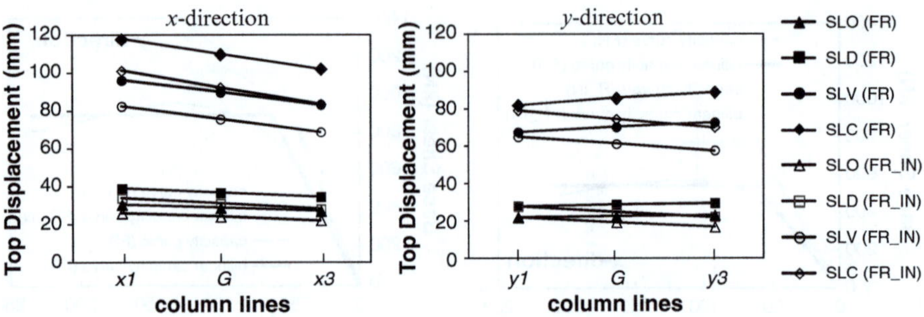

Fig. 9.7 Top displacement plan distribution (inverse-triangular horizontal pattern)

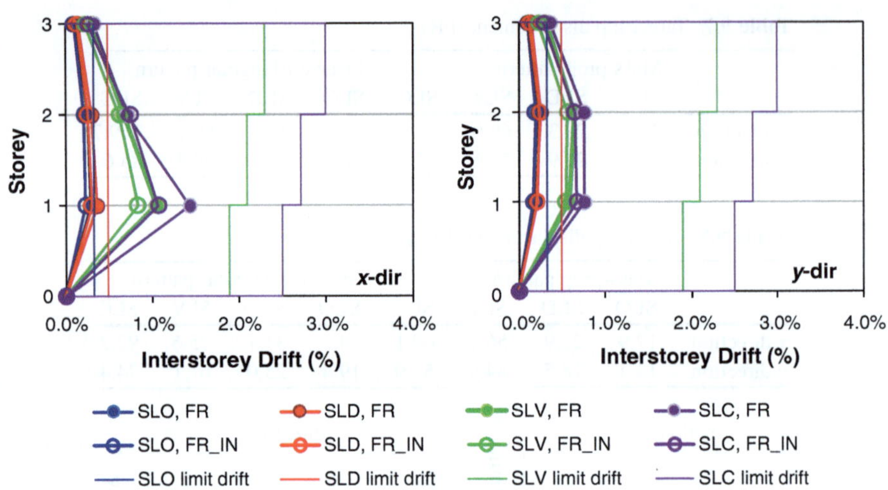

Fig. 9.8 Maximum *interstorey drift* (mass-proportional horizontal pattern)

9.5.3 Capacity Curves and Seismic Response of the Structure

The *interstorey drifts* for the two directions are determined at the flexible side only, that is, for the elements where the maximum *top displacement* is found. Figures 9.8 and 9.9 show the values of the maximum *ID* obtained in the two directions for the two considered horizontal patterns.

It can be seen that the structure achieves larger *interstorey drift* in *x*-direction. The drift of the *FR* model is larger than those of the *FR_IN* in all performed analyses. In fact, the infill panels are still in service when the target displacements are achieved for all the limit states considered.

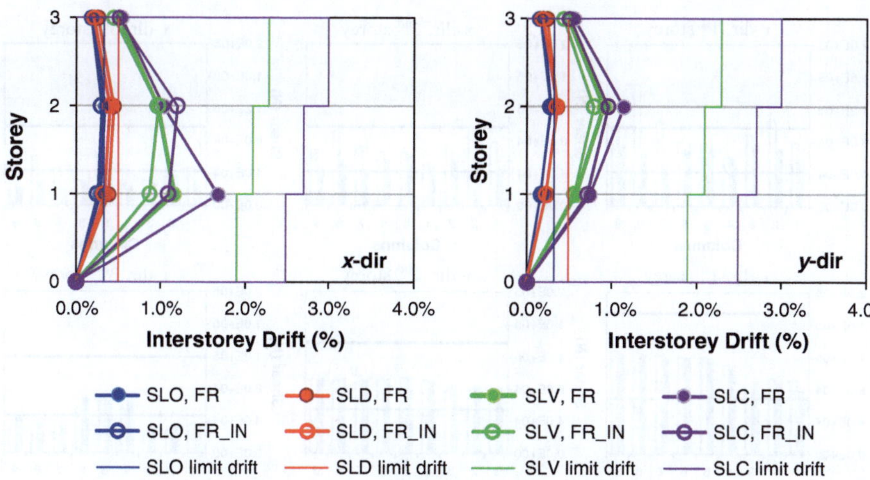

Fig. 9.9 Maximum *interstorey drift* (inverse-triangular horizontal pattern)

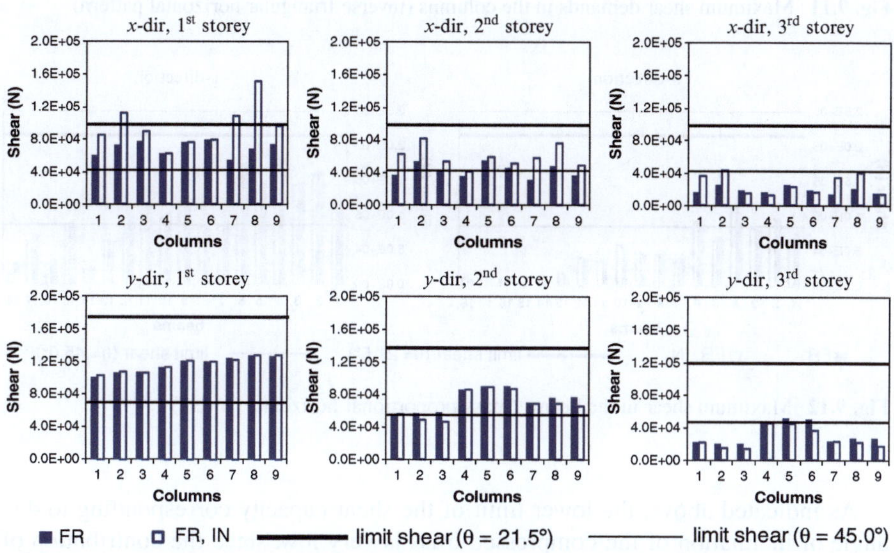

Fig. 9.10 Maximum shear demands in the columns (constant horizontal pattern)

9.5.4 Capacity Curves and Seismic Response of the Structure

Shear stress demands in the elements were checked at each step of the analysis performed. In Figures 9.10, 9.11, 9.12, and 9.13, the maximum shear demands in each element are compared with the correspondent limit as required by NTC2008.

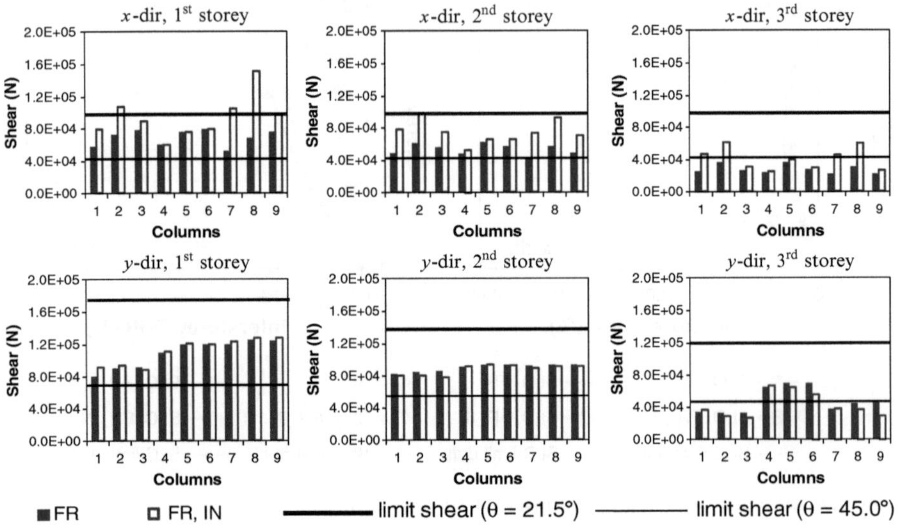

Fig. 9.11 Maximum shear demands in the columns (inverse-triangular horizontal pattern)

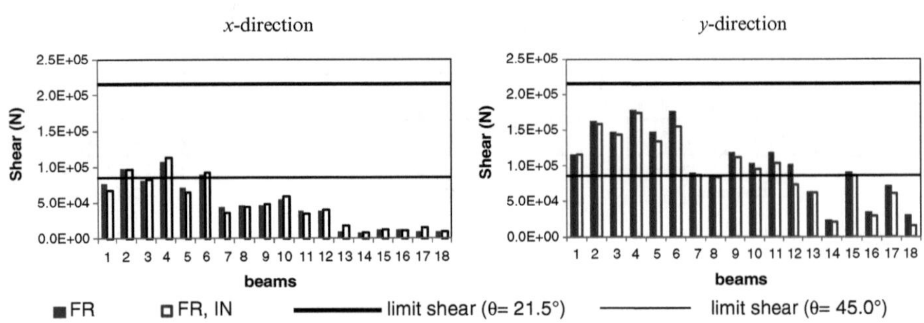

Fig. 9.12 Maximum shear in the beams (mass-proportional horizontal pattern)

As indicated above, the lower limit of the shear capacity corresponding to 45° angle of inclination of the compressed truss is very low, since the contribution of the tensile elements of the truss is completely neglected.

The higher value of the shear limit state is further considered as the effective limit. Figures 9.10 and 9.11 show that the shear demand in some columns exceeds also the effective limit state, when the infill panels are included in the analysis. It can be noted that the columns with larger shear demand are those located in the first storey of the frame $x3$, where the partial infill panels are placed. In fact, these columns have an effective shear span much shorter than their height, forcing them to behave like "short columns."

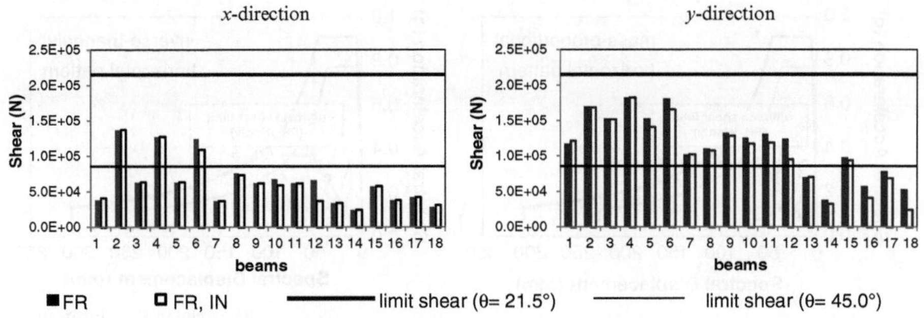

Fig. 9.13 Maximum shear in the beams (inverse-triangular horizontal pattern)

9.6 Suggested Improvement of Structure

A possible retrofit strategy is proposed for the case study, in order to improve its seismic performance, reducing the shear demand, which is the only parameter that does not fit the technical requirements due to partial infill panels along the short sides of the building.

The proposed variation consists in the replacement of the short curtain walls, placed along $x1$ and $x3$ column lines, with alternative panels made of the same material but spanning the entire height with a vertical space for windows or doors. The proposed variation still permits the introduction of windows in all spans of the frames, and it does not change the building architecturally and functionally.

The new infill panels have the same height of the storey and do not induce short-column problems. In order to further improve the seismic response of the building, the panels introduced in the short sides have a width smaller than frame spans, thus reducing also the mass eccentricity in x-direction.

The capacity curve of the proposed structure in the x-direction is shown in Figure 9.14, together with those of the original structure, with and without infill panels. The spectral demands and all the limit states are shown in the same figure, for the specific structures.

The seismic response of the structure, in terms of *interstorey drift*, is in all cases smaller than the limit states considered. It can be seen that the maximum shear demand for all the models considered exceeds the lower limit ($\theta = 45°$) required by NTC2008, while only the model with the short infill panels (FR_IN) exceeds even the higher limit ($\theta = 21.5°$). It has been observed that the maximum shear demand exceeds the ultimate shear limit in the FR_IN model well before the global structure yields.

The shear demand in the model representing the proposed retrofitted structure, FR_MOD, remains below the ultimate shear limit state, fulfilling all the requirements of the current code.

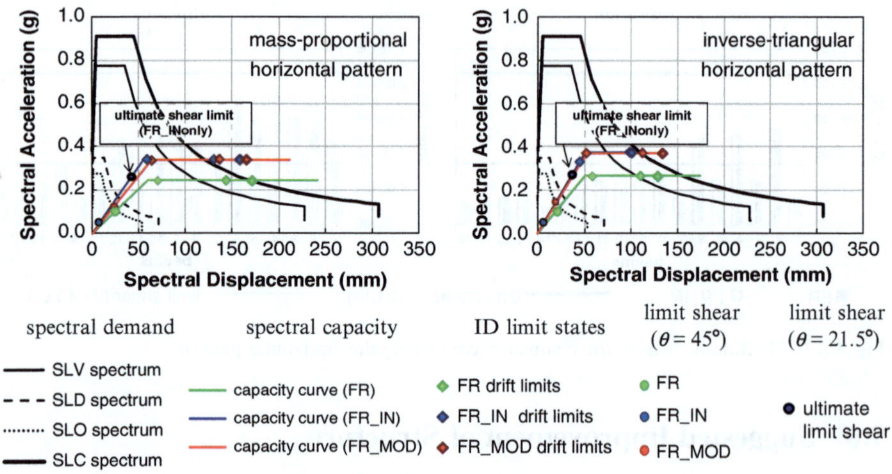

Fig. 9.14 Comparison between spectral capacity and limit states (*x*-direction)

9.7 Remarks and Conclusions

This chapter shows that the presence of infill panels and their distribution in plan of the building affect substantially the seismic performance in both lateral and torsional directions. In fact, the partially infilled walls may produce both mass and stiffness eccentricities. In particular, when "short-column" behavior occurs, the failure of columns reduces the lateral resistance and rigidity, often asymmetrically in the structure, leading to horizontal torsion.

In spite the torsional effects, the contribution of the infill of frames is extremely important since it reduces overall displacements and drift below the level of development of cracks in the main structural elements, as shown in the case study, for the strong seismic excitation.

As shown also in the case study, the increase of shear forces in the "short columns" near the partial infill bays of the frames leads to local damage. The increased shear in the "short columns" exceeds the limit capacity required by the modern Italian Technical Code.

In order to improve the seismic performance of such structures, the short infill panels causing the increased shear in the "short columns" can be replaced with alternative panels, having the same height of the storey, as shown in the case study. Additionally, since a proper geometrical distribution of the replacement infill panels can be used, the eccentricity of the structure in *x*-direction can be decreased, with further benefits in the seismic response of the structure.

Acknowledgments This work has been supported by the project DPC-ReLUIS 2010/2013 Task 1.1.2. "Strutture in cemento armato ordinarie e prefabricate."

References

Circ. Min. 2 febbraio 2009, n. 617 – Istruzioni per l'applicazione delle 'Nuove norme tecniche per le costruzioni' di cui al decreto ministeriale 14 gennaio 2008 (in Italian)

D. Pres. Rep. Del 25/011962 n (1684) Provvedimenti per l'edilizia, con particolari prescrizioni per le zone sismiche (in Italian)

D.M. 14 Gennaio (2008) Nuove norme tecniche per le costruzioni (NTC). G.U. n. 29 del 04/02/2008, Suppl. Ord. N. 30 (in Italian)

D'Ambrisi A, De Stefano M, Tanganelli M, Viti S (2010) Seismic performance of irregular existing RC framed structures. In: Proceedings of the 14th ECEE, Ohrid, 30 Aug–3 Sept 2010

Fajfar P (1999) Capacity spectrum method based on inelastic demand spectra. Earthq Eng Struct Dyn 28:979–993

Legge 5/11/1971 n. 1086 (1971) Norme per la disciplina delle opere in conglomerato cementizio armato, normale e precompresso ed a struttura metallica (in Italian)

Legge 2/2/1974 n. 64 (1974) Provvedimenti per le costruzioni con particolari prescrizioni per le zone sismiche. G.U. del 21/04/1974

Reinhorn A (1997) Inelastic analysis techniques in seismic evaluations. In: Krawinkler H, Fajfar P (eds) Seismic design methodologies for the next generation of codes. Balkema, Lisse, pp 277–287

Seismostruct v5.2.1 released (2011) Computer code, Finite Element package. (http://seismosoft.com)

Viti S, Cimellaro GP, Reinhorn AM (2006) Retrofit of a hospital through strength reduction and enhanced damping. Smart Struct Syst 2(4):339–355

Part III
Seismic Analysis of Irregular Structures

Chapter 10
Physically Based Prediction of the Maximum Corner Displacement of One-Storey Eccentric Systems

Tomaso Trombetti, Stefano Silvestri, Giada Gasparini, and Michele Palermo

Abstract This chapter gives a new insight into the dynamic behaviour of one-storey eccentric systems, with particular attention devoted to provide a comprehensive physically based formulation of the maximum corner displacement amplification, which involves three contributions (translational response, torsional response and their combination). It is shown that the largest amplifications, which mainly occur for the class of torsionally flexible systems, are due to the translational contribution through to the shift in the fundamental period of the eccentric system with respect to the one of the equivalent not-eccentric system. A simplified method for the estimation of the maximum corner displacement based on the physical properties of the system is finally obtained.

10.1 Introduction

Since the late 1970s, it is known that structures characterised by noncoincident centre of mass and centre of stiffness, commonly defined as eccentric (or asymmetric) systems, when subjected to dynamic excitation develop a coupled lateral-torsional response that may considerably increase their local peak response, such as the corner displacements (Kan and Chopra 1977a, b; Hejal and Chopra 1987; Lu and Hall 1992; Naeim and Kelly 1999; Tso 1990).

In order to effectively apply the performance-based design approach to seismic design, there is a growing need for code-oriented methodologies aimed at predicting deformation parameters. Thus, the estimation of the displacement demand at

---

T. Trombetti • S. Silvestri • G. Gasparini • M. Palermo (✉)
Dipartimento di Ingegneria Civile, Ambientale e dei, Materiali (DICAM), Università di Bologna, Via Risorgimento 2, 40136 Bologna, Italy
e-mail: tomaso.trombetti@unibo.it; stefano.silvestri@unibo.it; giada.gasparini4@unibo.it; michele.palermo7@unibo.it

O. Lavan and M. De Stefano (eds.), *Seismic Behaviour and Design of Irregular and Complex Civil Structures*, Geotechnical, Geological and Earthquake Engineering 24, DOI 10.1007/978-94-007-5377-8_10,

different locations, especially for eccentric structures, appears a fundamental issue. Furthermore, the ability to predict the torsional response of eccentric systems can be also useful to improve the capability of one of the most actually used seismic design approaches (i.e. push-over analysis (Perus and Fajfar 2005)).

Since the early 1990s, Nagarajaiah et al. (1993), investigating the torsional coupling behaviour of base-isolated structures, observed that, for the specific class of torsionally stiff asymmetric structures, the maximum centre mass displacement can be well approximated by the maximum displacement of the equivalent not-eccentric system.

In previous research works (Trombetti 1994; Trombetti and Conte 2005; Trombetti et al. 2008a, b, c), the authors identified a structural parameter, called "alpha", capable of measuring the attitude of one-storey asymmetric systems to develop rotational responses and proposed a simplified procedure, called "Alpha method", for the estimation of the maximum torsional response. In its original formulation, the "Alpha method" was based on the aforementioned assumption of equal maximum displacement response between the eccentric system and the equivalent not-eccentric system.

The objective of this chapter is to provide a more comprehensive investigation on the dynamic properties of one-storey eccentric systems, with specific focus on the class of the so-called torsionally flexible systems, which showed a greater attitude in developing consistent corner displacement amplifications (Trombetti et al. 2008b, c).

10.2 Problem Formulation

Let us consider the one-storey eccentric structure (i.e. a system characterised by noncoincident centre of mass, CM, and centre of stiffness, CK) displayed in Fig. 10.1 (the origin of the reference system is located at CM). It is assumed that the diaphragm is infinitely rigid in its own plane, and that the lateral-resisting elements (e.g. columns, shear walls) are massless and axially inextensible. The self-torsional stiffness (k_θ) of each lateral-resisting element is also neglected. Under this assumption, the following three degrees of freedom are assumed: (1) longitudinal centre mass displacement, $u_{y,\mathrm{CM}}$; (2) transversal centre mass displacement, $u_{x,\mathrm{CM}}$; and (3) centre mass rotation, $u_{\theta,\mathrm{CM}}$, which coincides with the floor rotation, u_θ. The system is subjected to a one-way dynamic excitation (e.g. free vibrations or seismic input) along the longitudinal direction (namely, the y-direction).

From simple trigonometric relationships, with reference to the plan view of the system given in Fig. 10.1, the longitudinal corner side displacement, i.e. the displacement of the flexible side of the system (e.g. point B, the farther from CK), $u_{y,B}$, at any generic instant t, is given by:

$$u_{y,B}(t) = u_{y,\mathrm{CM}}(t) - u_\theta(t) \cdot \frac{L}{2} \tag{10.1}$$

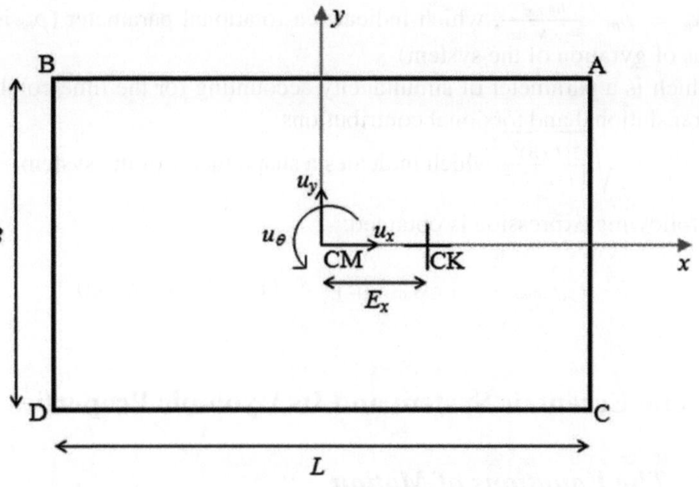

Fig. 10.1 Plan view of the in-plane eccentric system with the indication of the degrees of freedom

Estimating the corner displacement according to Eq. 10.1 requires the development of time-history analyses. Nevertheless, the practical engineer is interested in the absolute maximum value, $u_{y,B,\max}$, of the corner displacement response history. Thus, the main purpose of this research work is to provide a simple formula for the evaluation of $u_{y,B,\max}$, starting from:

$$u_{y,B,\max} = u_{y,\mathrm{CM},\max} \oplus u_{\theta,\max} \cdot \frac{L}{2} \tag{10.2}$$

which highlights that the maximum corner displacement depends on the following three contributions:

- Translational contribution, as given by the maximum absolute displacement response $u_{y,\mathrm{CM},\max}$ of the centre of mass (see Sect. 4)
- Torsional contribution, as given by the product of the maximum absolute rotational response $u_{\theta,\max}$ and the lever arm $L/2$ (see Sect. 5)
- Combination of the translational and torsional contributions of above, as indicated by symbol $\oplus$ (see Sect. 6)

Manipulation of Eq. 10.2 leads to:

$$u_{y,B,\max} = u_{y,\mathrm{CM},\max,\mathrm{N-E}} \cdot \frac{u_{y,\mathrm{CM},\max}}{u_{y,\mathrm{CM},\max,\mathrm{N-E}}} \left(1 \oplus \frac{u_{\theta,\max}}{u_{y,\mathrm{CM},\max}} \cdot \frac{L}{2}\right) \tag{10.3}$$

Introducing:

- $\delta = \frac{u_{y,\mathrm{CM},\max}}{u_{y,\mathrm{CM},\max,\mathrm{N-E}}}$, which indicates the centre mass displacement amplification with respect to the equivalent not-eccentric system (N-E)

- $A \cdot \alpha_u = \rho_m \cdot \frac{u_{\theta,\max}}{u_{y,\mathrm{CM},\max}}$, which indicates a rotational parameter (ρ_m is the mass radius of gyration of the system)
- B, which is a parameter of simultaneity accounting for the time combination of the translational and torsional contributions
- $\phi = \frac{L}{2\rho_m} = \sqrt{\frac{3(L/B)^2}{1+(L/B)^2}}$ which indicates a shape factor of the system

The following expression is obtained:

$$u_{y,B,\max} = u_{y,\mathrm{CM},\max,\mathrm{N-E}} \cdot \delta \cdot (1 + A \cdot B \cdot \alpha_u \cdot \phi) \quad (10.4)$$

10.3 The Eccentric System and Its Dynamic Properties

10.3.1 The Equations of Motion

Under the following additional assumptions:

- The total lateral stiffness k of the system is the same along the x- and the y-direction (i.e. $k = k_x = k_y$, where k_x and k_y are the translational stiffness along the x- and the y-direction, respectively).
- The rotational response u_θ developed under dynamic excitation is small enough to allow the approximation $u_\theta \cong \sin(u_\theta) \cong \tan(u_\theta)$.
- The longitudinal eccentricity is equal to zero (i.e. $E_y = 0$). This case, for any given transversal eccentricity Ex, maximises the rotational response of the system in free vibrations (Trombetti and Conte 2005).

The dynamic coupled lateral-torsional response of the system under consideration (Fig. 10.1) is governed by the following set of coupled differential equations of motion (Trombetti and Conte 2005), written in a reference system with origin located at CM:

$$m \begin{bmatrix} \ddot{u}_x(t) \\ \ddot{u}_y(t) \\ \rho_m \ddot{u}_\theta(t) \end{bmatrix} + [C] \begin{bmatrix} \dot{u}_x(t) \\ \dot{u}_y(t) \\ \rho_m \dot{u}_\theta(t) \end{bmatrix} + m\omega_L{}^2 \begin{bmatrix} 1 & 0 & 0 \\ 0 & 1 & e_x\sqrt{12} \\ 0 & e_x\sqrt{12} & \Omega_\theta^2 + 12e_x^2 \end{bmatrix} \begin{bmatrix} u_x(t) \\ u_y(t) \\ \rho_m u_\theta(t) \end{bmatrix} = \begin{bmatrix} p_x(t) \\ p_y(t) \\ \frac{p_\theta(t)}{\rho_m} \end{bmatrix} \quad (10.5)$$

In Eq. 10.5, m is the mass of the system, $e_x = E_x/D_e$ is the relative eccentricity (hereafter it will be simply indicated as e), D_e is the equivalent diagonal equal to $12\rho_m$, $\Omega_\theta = \omega_\theta/\omega_L$ is a dimensionless parameter that measures the torsional

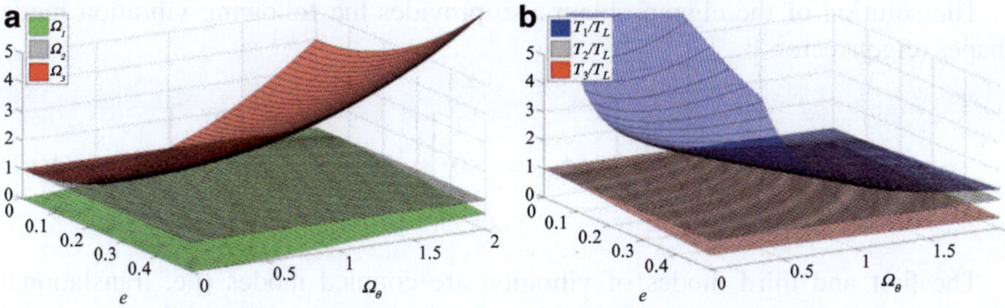

Fig. 10.2 (**a**) The graphical representation of the normalised natural frequencies versus e and Ω_θ. (**b**) The graphical representation of the normalised natural periods versus e and Ω_θ

flexibility of the system (ω_L and ω_θ are the uncoupled translational natural frequency of vibration and the uncoupled torsional natural frequency of vibration, defined in a reference system with origin located at CK, respectively), and $[C]$ is the damping matrix (classical damping is assumed).

The parameter Ω_θ represents a physical property of the eccentric system, leading to the following two classes: (1) torsionally stiff systems, $\Omega_\theta \geq 1.0$, and (2) torsionally flexible systems, $\Omega_\theta < 1.0$.

10.3.2 The Eigenproblem

The solution of the eigenvalues problem governing the undamped free vibrations of the system gives the following closed-form expressions of natural frequencies ω_1, ω_2, ω_3, normalised with respect to the uncoupled longitudinal frequency ω_L and squared (Trombetti and Conte 2005):

$$\Omega_1 = \left(\frac{\omega_1}{\omega_L}\right)^2 = \frac{1}{2}\left(1 + \Omega_\theta^2 + 12e^2 - \sqrt{\left(\Omega_\theta^2 + 12e^2 - 1\right)^2 + 48e^2}\right)$$

$$\Omega_2 = \left(\frac{\omega_2}{\omega_L}\right)^2 = 1$$

$$\Omega_3 = \left(\frac{\omega_3}{\omega_L}\right)^2 = \frac{1}{2}\left(1 + \Omega_\theta^2 + 12e^2 + \sqrt{\left(\Omega_\theta^2 + 12e^2 - 1\right)^2 + 48e^2}\right) \tag{10.6a,b,c}$$

Figure 10.2a plots the normalised natural frequencies versus e and Ω_θ showing that (1) $\omega_2 = \omega_L$, (2) ω_1 is generally close to ω_L and (3) ω_3 can be quite larger than ω_L.

The solution of the eigenproblem also provides the following vibration mode shapes (eigenvectors):

$$\{\phi_1\} = \begin{bmatrix} 0 \\ 1 \\ \frac{\Omega_1 - 1}{e\sqrt{12}} \end{bmatrix}; \quad \{\phi_2\} = \begin{bmatrix} 1 \\ 0 \\ 0 \end{bmatrix}; \quad \{\phi_3\} = \begin{bmatrix} 0 \\ 1 \\ \frac{\Omega_3 - 1}{e\sqrt{12}} \end{bmatrix} \tag{10.7}$$

The first and third modes of vibration are coupled modes (i.e. translational component in y-direction coupled with a torsional component), whilst the second mode is purely translational in x-direction, due to the assumption of null eccentricity in y-direction.

From Eq. 10.6a,b,c, the expressions of the natural periods of vibration, normalised with respect to the uncoupled lateral period T_L, can be obtained:

$$\frac{T_1}{T_L} = \frac{1}{\sqrt{\frac{1}{2}\left(1 + {\Omega_\theta}^2 + 12e^2 - \sqrt{\left({\Omega_\theta}^2 + 12e^2 - 1\right)^2 + 48e^2}\right)}}$$

$$\frac{T_2}{T_L} = 1$$

$$\frac{T_3}{T_L} = \frac{1}{\sqrt{\frac{1}{2}\left(1 + {\Omega_\theta}^2 + 12e^2 + \sqrt{\left({\Omega_\theta}^2 + 12e^2 - 1\right)^2 + 48e^2}\right)}} \tag{10.8a,b,c}$$

Figure 10.2b plots the graph of the natural periods of vibration, normalised with respect to T_L versus e and Ω_θ. Inspection of the graph shows that (1) $T_2 = T_L$ for all values of e and Ω_θ, (2) torsionally stiff systems are characterised by a fundamental period T_1 close to T_L and by a third period T_3 quite smaller than T_L and (3) torsionally flexible systems are characterised by a fundamental period T_1 that can achieve value also larger than 5 times T_L and by a third period T_3 close to T_L. It will be shown that the consistent increase of the fundamental period, here defined as "period shifting", strongly affects the displacement response of torsionally flexible systems.

10.3.3 The Modal Contribution Factors

In order to understand how each mode of vibration contributes to the dynamic response of the system, the closed-form expressions of the modal contribution

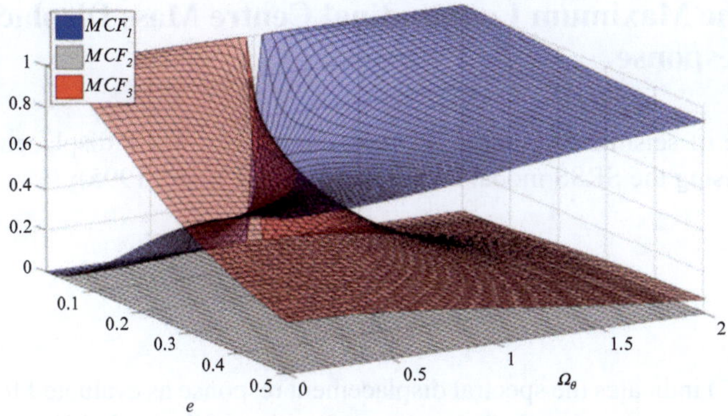

Fig. 10.3 Modal contribution factors versus e and Ω_θ

factors MCF_i, $i = 1,2,3$, activated by a dynamic input characterised by influence vector {0,1,0} (i.e. input only along the y-direction), have been derived:

$$\text{MCF}_1 = \frac{1}{1 + \left(\frac{\Omega_1 - 1}{\sqrt{12}}\right)^2}$$

$$\text{MCF}_2 = 0$$

$$\text{MCF}_3 = \frac{1}{1 + \left(\frac{\Omega_3 - 1}{\sqrt{12}}\right)^2} \tag{10.9a,b,c}$$

Figures 10.3 plots the modal contribution factors versus e and Ω_θ. Inspection of the graph leads to the following observations:

- $\text{MCF}_2 = 0$ for all values of e and Ω_θ results from the assumptions of null eccentricity in the y-direction and influence vector along the y-direction.
- Torsionally stiff systems are principally governed by the first mode of vibration T_1 that, as showed in the previous section, is close to the second period of vibration, T_2, which in turn is equal to the uncoupled lateral period, T_L.
- Torsionally flexible systems with small eccentricity ($e < 0.1$) are mainly governed by the third mode of vibration that is approximately equal to T_L; torsionally flexible systems with high eccentricity ($e > 0.3$) are substantially governed by the first mode of vibration that may be considerably higher than T_L; for torsionally flexible systems characterised by eccentricity e comprised between 0.1 and 0.3 both the first, T_1, and third, T_3, natural periods of vibration contribute to the dynamic response of the system.

10.4 The Maximum Longitudinal Centre Mass Displacement Response

In the case of seismic excitation, the maximum centre mass displacement can be predicted using the SRSS modal combination rule (Chopra 1995):

$$u_{y,\mathrm{CM,max}} \cong \sqrt{\sum_{i=1}^{3} (S_\mathrm{d}(T_i) \cdot \mathrm{MCF}_i)^2} \tag{10.10}$$

where $S_\mathrm{d}(T_i)$ indicates the spectral displacement response as evaluated for period T_i. Under the assumption that $S_\mathrm{d}(T)$ is a linear function of the period T,

$$S_\mathrm{d}(T) = \varphi \cdot T \tag{10.11}$$

where φ represents the angular coefficient in the response spectrum diagram, Eq. 10.10 becomes:

$$u_{y,\mathrm{CM,max}} \cong \varphi T_L \sqrt{\frac{\mathrm{MCF}_1}{\Omega_1}^2 + \mathrm{MCF}_2^2 + \frac{\mathrm{MCF}_3}{\Omega_3}^2} \tag{10.12}$$

Equation 10.12 allows to obtain the following closed-form expression of the centre mass displacement amplification, δ, as a function of e and Ω_θ:

$$\delta = \frac{u_{y,\mathrm{CM,max}}}{u_{y,\mathrm{CM,max,N-E}}} \cong 12e^2 \sqrt{\frac{1}{\Omega_1\left[12e^2 + (\Omega_1 - 1)^2\right]^2} + \frac{1}{\Omega_3\left[12e^2 + (\Omega_3 - 1)^2\right]^2}} \tag{10.13}$$

It has to be noted that, for the sake of conciseness, Eq. 10.13 is not directly expressed in terms of Ω_θ, but in terms of the normalised frequencies Ω_1 and Ω_3. δ is plotted in Fig. 10.4 as a function of e and Ω_θ. Inspection of Fig. 10.5 clearly reveals that:

- For a wide region of e and Ω_θ, δ is close to 1.
- For high values of eccentricity e coupled with low values of Ω_θ, the displacement amplification δ can achieve values larger than 5.

A more in-depth observation of the graph allows to introduce the following classes of eccentric systems:

1. Class 1: torsionally stiff systems, $\Omega_\theta \geq 1.0$. The displacement response of these systems is governed by the first natural period of vibration T_1, close to T_L. In this case, the centre mass displacement can be approximated by the displacement of the equivalent not-eccentric system. An example of such systems is given by frame structures with perimeter shear walls.

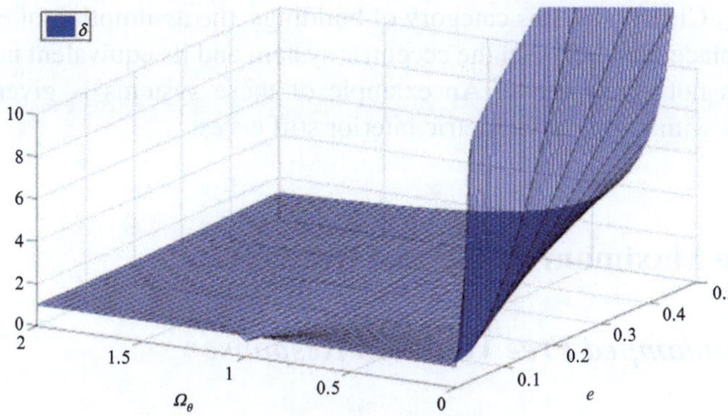

Fig. 10.4 The graphical representation of δ versus e and Ω_θ

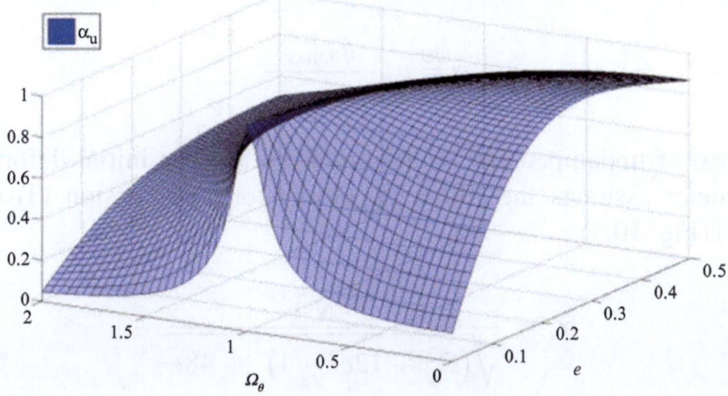

Fig. 10.5 The graphical representation of α_u versus e and Ω_θ

2. Class 2: systems with $\Omega_\theta \cong 1.0$. The displacement response of these systems results from both the contributions of the first and third mode of vibration. In detail, (1) for low values of eccentricity ($e < 0.2$), the centre mass displacement is slightly lower than the displacement of the equivalent not-eccentric system, and (2) for high values of eccentricity ($e > 0.2$), the centre mass displacement is slightly higher than the corresponding displacement of the equivalent not-eccentric system. An example of these systems is given by frame structures.
3. Class 3: low-eccentric torsionally flexible systems, $\Omega_\theta < 1.0$ and $e < 0.1$. The displacement response of these systems is mainly governed by the third mode of vibration that is approximately equal to the uncoupled lateral period T_L. For this class the centre mass displacement can be approximated by the displacement of the equivalent not-eccentric system. An example of these systems is given by structures with slightly asymmetric interior stiff cores.
4. Class 4: high-eccentric torsionally flexible systems, $\Omega_\theta < 1.0$ and $e > 0.3$. The displacement response of these systems is mainly governed by the first mode of vibration T_I that may be considerably higher than the uncoupled lateral

period T_L. Clearly, for this category of buildings, the assumption of equal centre mass displacement between the eccentric system and its equivalent not-eccentric system is not conservative. An example of these systems is given by frame structures with highly asymmetric interior stiff cores.

10.5 The Maximum Rotational Response

10.5.1 Undamped Free Vibration Response

In a previous research work (Trombetti and Conte 2005), the authors identified a rotational parameter called "alpha", governing the maximum rotational response of eccentric systems:

$$\alpha \overset{\text{def}}{=} \rho_m \frac{u_{\theta,\max}}{u_{y,\mathrm{CM},\max}} \tag{10.14}$$

In the case of undamped free vibrations from a given initial deformation, the alpha parameter assumes the following closed-form expression (Trombetti and Conte 2005) (Fig. 10.5):

$$\alpha_u = \frac{4e\sqrt{3}}{\sqrt{\left(\Omega_\theta^2 + 12e^2 - 1\right)^2 + 48e^2}} \tag{10.15}$$

where the subscript u indicates "undamped conditions".

Figure 10.5 shows that α_u is bounded between zero and one. The above-introduced rotational parameter allows to express the maximum rotational response as follows:

$$u_{\theta,\max} = \frac{\alpha_u}{\rho_m} \cdot u_{y,\mathrm{CM},\max} \tag{10.16}$$

upper-bounded by the following value:

$$u_{\theta,\max} \le \frac{u_{y,\mathrm{CM},\max}}{\rho_m} \tag{10.17}$$

10.5.2 Damped Seismic Response

In the case of damped systems subjected to seismic excitation, the alpha parameter is indicated as:

$$\alpha_{\text{d,eqke}} \overset{\text{def}}{=} \rho_m \left. \frac{u_{\theta,\max}}{u_{y,\text{CM},\max}} \right|_{\text{d,eqke}} \tag{10.18}$$

where the subscript d,eqke indicates "damped conditions and earthquake input".

By posing:

$$A \overset{\text{def}}{=} \frac{\alpha_{\text{d,eqke}}}{\alpha_u} \tag{10.19}$$

the maximum rotational response experienced by a damped eccentric system under seismic excitation can be expressed by the following simple relationship:

$$u_{\theta,\max} = \frac{A \cdot \alpha_u \cdot u_{y,\text{CM},\max}}{\rho_m} \tag{10.20}$$

Parameter A should be obtained and calibrated by means of extensive numerical simulations, which are currently under development.

10.6 The Combination of the Maximum Displacement and Rotational Response

10.6.1 Undamped Free Vibration Response

The solution of the equations of motion of the studied eccentric system, in the case of undamped free vibrations from a given initial displacement a along the y-direction, is given by (Trombetti and Conte 2005):

$$\begin{aligned} u_y(t) &= a \cdot (R_1 \cos(\omega_1 t) + R_3 \cos(\omega_3 t)) \\ u_x(t) &= 0 \\ u_\theta(t) &= \frac{a}{\rho_m} \cdot \frac{\alpha_u}{2} (\cos(\omega_1 t) - \cos(\omega_3 t)) \end{aligned} \tag{10.21a,b,c}$$

where R_1 and R_3 are defined as follows (Fig. 10.6):

$$\begin{aligned} R_1 &= \frac{1 - \Omega_3}{\Omega_1 - \Omega_3} \\ R_3 &= \frac{\Omega_1 - 1}{\Omega_1 - \Omega_3} \end{aligned} \tag{10.22a,b}$$

It can be demonstrated that R_1 and R_3 are coincident with MCF_1 and MCF_3.

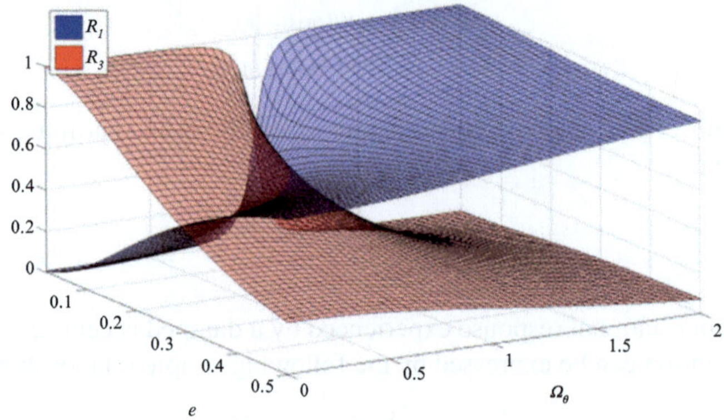

Fig. 10.6 The graphical representation of R_1 and R_3 versus e and Ω_θ

Careful inspection of Eq. 10.21a,b,c leads to the following observations regarding the simultaneity of translational and rotational responses:

- The maximum longitudinal displacement is developed for $\omega_1(t) = n\pi$ and $\omega_3(t) = m\pi$ (with n and m both odd or both even) and is equal to a. The corresponding rotation is zero.
- The maximum rotation is developed for $\omega_1(t) = n\pi$ and $\omega_3(t) = (m+1)\pi$ (with n and m both even or odd) and is equal to $\frac{a}{\rho_m}\alpha_u$. The corresponding longitudinal displacement, $u_{y,\mathrm{CM}@u_{\theta,\max}}$, is equal to $a(R_1–R_3)$.

Based on these observations, two limit assumptions (HP1 and HP2) are introduced:

1. The maximum corner displacement is calculated supposing a full correlation between the maximum rotational response and maximum centre mass displacement response (HP1):

$$u_{y,B,\max,\mathrm{HP1}} = u_{y,\mathrm{CM},\max} + u_{\theta,\max}\frac{L}{2} \tag{10.23}$$

which can be easily rewritten as:

$$u_{y,B,\max,\mathrm{HP1}} = \delta \cdot u_{y,\mathrm{CM},\max,\mathrm{N-E}} \cdot (1 + \alpha_u \cdot \phi) \tag{10.24}$$

2. The maximum corner displacement is calculated combining the maximum rotational response with the centre mass displacement achieved at the instant of maximum rotation (HP2):

$$u_{y,B,\max,HP2} = u_{y,\mathrm{CM}@u_{\theta,\max}} + u_{\theta,\max}\frac{L}{2} \tag{10.25}$$

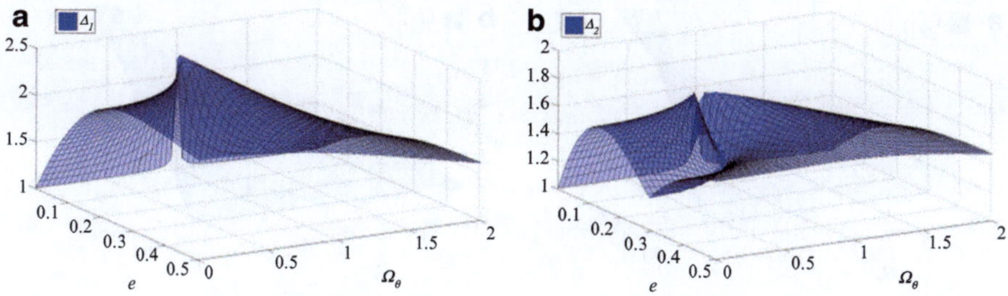

Fig. 10.7 The graphical representations of (**a**) Δ_1 versus *e* and Ω_θ and (**b**) Δ_2 versus *e* and Ω_θ

which can be easily rewritten as:

$$u_{y,B,\max,\mathrm{HP2}} = \delta \cdot u_{y,\mathrm{CM},\max,\mathrm{N-E}} \cdot ((R_1 - R_3) + \alpha_u \cdot \phi) \tag{10.26}$$

It is clear that Eqs. 10.24 and 10.26 represent an upper bound and a lower bound for the maximum corner displacement, respectively:

$$u_{y,B,\max,\mathrm{HP2}} \leq u_{y,B,\max} \leq u_{y,B,\max,\mathrm{HP1}} \tag{10.27}$$

The following closed-form expressions of the corner displacement amplifications result from the two limit assumptions HP1 and HP2:

$$\Delta_1 = \frac{u_{y,B,\max,\mathrm{HP1}}}{u_{y,\mathrm{CM},\max}} = 1 + \alpha_u \phi$$
$$\Delta_2 = \frac{u_{y,B,\max,\mathrm{HP2}}}{u_{y,\mathrm{CM},\max}} = (R_1 - R_3) + \alpha_u \phi \tag{10.28a,b}$$

$$\Delta_{\mathrm{N-E},1} = \frac{u_{y,B,\max,\mathrm{HP1}}}{u_{y,\mathrm{CM},\max,\mathrm{N-E}}} = \delta \cdot \Delta_1 = \delta\,(1 + \alpha_u \phi)$$
$$\Delta_{\mathrm{N-E},2} = \frac{u_{y,B,\max,\mathrm{HP2}}}{u_{y,\mathrm{CM},\max,\mathrm{N-E}}} = \delta \cdot \Delta_2 = \delta\,[(R_1 - R_3) + \alpha_u \phi] \tag{10.29a,b}$$

Note that Δ_1 and Δ_2 (Fig. 10.7a, b) give the corner displacement amplification with respect to the centre mass displacement (i.e. the amplification due to the torsional contribution) based on HP1 and HP2, respectively, whilst $\Delta_{\text{N-E},1}$ and $\Delta_{\text{N-E},2}$ (Fig. 10.8a, b) give the corner displacement amplification with respect to the centre mass displacement of the equivalent not-eccentric system, based on HP1 and HP2, respectively. It is also interesting to evaluate the ratio Δ_2/Δ_1 (equal to the ratio $\Delta_{\text{N-E},2}/\Delta_{\text{N-E},1}$, Fig. 10.9). Careful examination of the graphs plotted in Figs. 10.7, 10.8, and 10.9 leads to the following fundamental observations:

- Both Δ_1 and Δ_2 are larger than one for all values of *e* and Ω_θ. This result justifies the introduction of the assumption HP2 (lower bound). The maximum corner

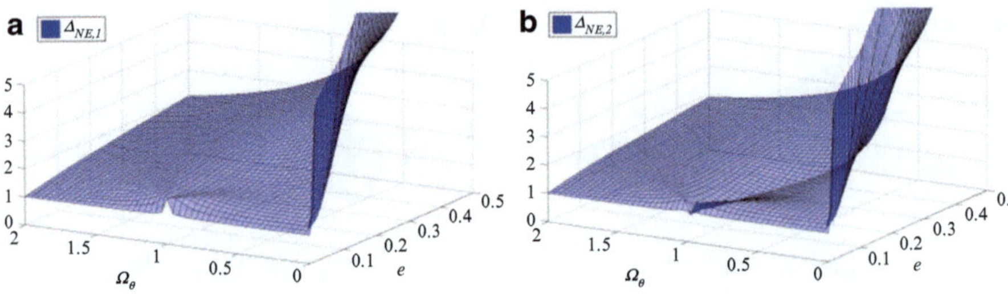

Fig. 10.8 The graphical representations of (**a**) $\Delta_{N-E,1}$ versus e and Ω_θ and (**b**) $\Delta_{N-E,2}$ versus e and Ω_θ

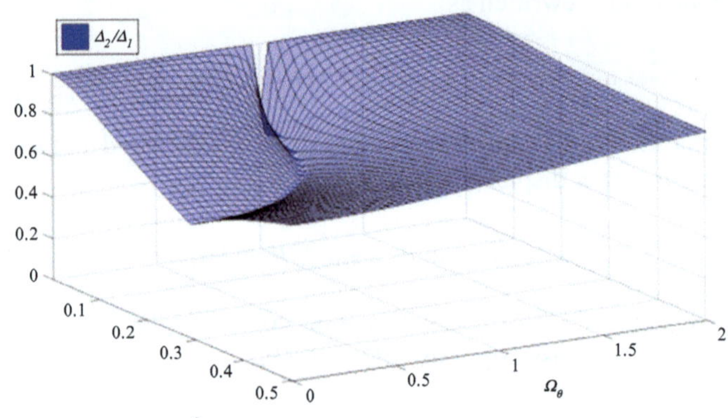

Fig. 10.9 The graphical representations of Δ_1/Δ_2 versus e and Ω_θ

displacement amplifications Δ_1 and Δ_2 are limited to values around 2.3 and 1.6, with reference to HP1 and HP2, respectively.

- Both $\Delta_{\text{N-E},1}$ and $\Delta_{\text{N-E},2}$ are significantly affected by δ for all classes of eccentric structures. In detail, (1) high torsionally stiff systems (i.e. $\Omega_\theta > 1.5$) exhibit maximum corner displacement amplifications approximately equal to 2.5 and 2.0 with reference to HP1 and HP2, respectively; (2) low torsionally stiff systems ($\Omega_\theta \cong 1.0$) exhibit maximum corner displacement amplifications approximately equal to 3.5 and 3.0 with reference to HP1 and HP2, respectively; and (3) torsionally flexible systems exhibit maximum corner displacement amplifications larger than 5.
- The ratio Δ_1/Δ_2 is close to 1 and low dependent on e, for torsionally stiff systems, whilst it is comprised between 0.55 and 1 and strongly dependent on e, for torsionally flexible systems.

It has to be noted that all graphs reported in Figs. 10.7, 10.8, and 10.9 are plotted for a shape factor ϕ equal to 1.22 (corresponding to a square plan, i.e. $L = B$ with reference to Fig. 10.1).

10.6.2 Damped Seismic Response

In the case of seismic excitation, a parameter of simultaneity B is introduced to account for the time correlation between the rotational and displacement seismic responses (see Eq. 10.4).

Parameter B should be obtained and calibrated by means of extensive numerical simulations. However, it should be noted that parameter B is certainly less than 1, and, therefore, from a conservative design point of view, it can be taken equal to 1.

10.7 "Alpha Method" for Prediction of Maximum Displacement at the Flexible Side of Eccentric Systems

In a previous research work (Trombetti and Conte 2005), the authors proposed a simplified method, called "Alpha method", for the prediction of the maximum rotational response of eccentric systems. The original formulation of the method was developed studying the dynamic behaviour of torsionally stiff system and assuming that the maximum centre mass displacement of the eccentric system can be reasonably approximated by the corresponding displacement of the equivalent not-eccentric system.

The results described in this chapter lead to a more comprehensive understanding of the dynamic behaviour of eccentric system. The analytical tools detailed in previous sections allow to extend the original formulation of the "Alpha method" to a generic eccentric system (i.e. torsionally flexible or torsionally stiff system), removing the assumption of equal centre mass displacement between the eccentric system and its equivalent not-eccentric system through the introduction of the δ parameter (Sect. 4, Eq. 10.13).

In the light of all the results reported in previous sections, the following formula for the evaluation of the maximum corner displacement of an eccentric system under seismic excitation is proposed:

$$u_{y,B,\max} = u_{y,\mathrm{CM},\max,\mathrm{N-E}} \cdot \delta \cdot (1 + A \cdot B \cdot \alpha_u \cdot \phi) \tag{10.30}$$

Examination of Eq. 10.30 (i.e. Eq. 10.4) allows the following observations:

- The maximum corner displacement can be expressed as a function of the maximum displacement response of the equivalent not-eccentric system, obtained as the maximum deformation of a SDOF oscillator of undamped natural period T_L, damping ratio ξ and mass equal to the total mass of the structure (e.g., use of displacement response spectrum (Chopra 1995)).
- The formulation of the "Alpha method" requires the introduction of two parameters, named A and B, which characterise the damped response under seismic input, in general depending on T_L, e, ξ, ϕ and Ω_θ.

10.8 Conclusions

This chapter provides a comprehensive insight into the dynamic behaviour of one-storey eccentric systems, aimed at increasing the knowledge about this class of structures as well as providing simple tools for their seismic design. For the specific case of undamped eccentric systems in free vibrations, closed-form expressions for an upper bound and a lower bound of the maximum longitudinal corner displacement have been derived. Based on these results, a simplified approach for the seismic design of eccentric systems, originally proposed by the author for the evaluation of the torsional response of torsionally stiff eccentric systems, has been revised accounting for all classes of eccentric systems.

The main conclusions of the study can be summarised as follows:

- The key parameters governing the dynamic behaviour of eccentric systems are the uncoupled lateral period, T_L; the eccentricity, e; the torsional flexibility parameter, Ω_θ; the shape factor, ϕ and the damping ratio, ξ.
- A physically based approach for the evaluation of the maximum corner displacement has been proposed, leading to a simple design formula. The formula is composed of three contributions: (1) the translational contribution, depending on δ parameter; (2) the torsional contribution, depending on the α_u parameter and on the shape factor ϕ; and (3) the combination of the two translational and torsional responses.
- The translational contribution, given by parameter δ, can be strongly affected by the "period shifting" effect, i.e. the shift in the fundamental period of vibration of the eccentric system with respect to that of the equivalent not-eccentric system. The period shifting mainly affects the class of torsionally flexible systems. For this class of eccentric systems, the estimation of the centre mass displacement by response spectrum analysis using the uncoupled period of vibration of the equivalent not-eccentric system may lead to very unconservative results (the displacement amplification can be larger than 5).
- The torsional contribution is given by two components: (1) the pure rotational response measured by parameter A and parameter α_u that is bounded to one for all classes of eccentric systems and (2) the "lever arm" effect depending on the plan shape of the structures, measured by ϕ.
- The time correlation between the displacement and the rotational response, given by parameter B, is dependent on the eccentricity and torsional flexibility of the systems. In detail, (1) torsionally stiff systems exhibit a high correlation between the two maximum responses, and (2) torsionally flexible systems show a correlation strongly dependent on the eccentricity.

References

Chopra AK (1995) Dynamics of structures. Theory and applications to earthquake engineering. Prentice-Hall, Upper Saddle River

Hejal R, Chopra AK (1987) Earthquake response of torsionally coupled buildings. Report UCB/EERC-87/20. Earthquake Engineering Research Center, University of California, Berkeley, CA

Kan CL, Chopra AK (1977a) Effects of torsional coupling on earthquake forces in buildings. J Struct Div ASCE 103(4):805–819

Kan CL, Chopra AK (1977b) Elastic earthquake analysis of a class of torsionally coupled buildings. J Struct Div ASCE 103(4):821–839

Lu S, Hall WJ (1992) Torsion in two buildings—1987 Whittier narrows earthquake. Earthq Eng Struct Dyn 21(2):387–407

Naeim F, Kelly JM (1999) Design of seismic isolated structures—from theory to practice. Wiley, New York

Nagarajaiah S, Reinhorn AM, Constantinou MC (1993) Torsion in base isolated structures with elastomeric isolation systems. J Struct Eng ASCE 119:2932–2951

Perus I, Fajfar P (2005) On the inelastic torsional response of single-storey structures under bi-axial excitation. Earthq Eng Struct Dyn 34:931–941

Trombetti T (1994) Un approccio semplificato all'analisi dei problemi torsionali negli edifici isolati sismicamente alla base. Giornale del Genio Civile 132(243–267):1994

Trombetti T, Conte JP (2005) New insight into and simplified approach to seismic analysis of torsionally coupled one-storey, elastic systems. J Sound Vib 286:265–312

Trombetti T, Silvestri S, Gasparini G, Pintucchi B, De Stefano M (2008a) Numerical verification of the effectiveness of the "Alpha" method for the estimation of the maximum rotational elastic response of eccentric systems. J Earthq Eng 12(2):249–280

Trombetti T, Silvestri S, Gasparini G (2008b) An upper bound and an estimation for the maximum non-linear rotational response of one storey asymmetric building. In: Proceeding of the fifth European workshop on the seismic behaviour of irregular and complex structures, Catania, Italy

Trombetti T, Silvestri S, Gasparini G (2008c) Closed-form solutions for corrective eccentricity, sensitivity to accidental eccentricità and increase in peak local displacement. In: Proceeding of the fifth European workshop on the seismic behaviour of irregular and complex structures, Catania, Italy

Tso WK (1990) Static eccentricity concept for torsional moment estimations. J Struct Eng ASCE 116(5):1199–1212

Chapter 11
Plan Irregular Structures: Simplified Approach

Dietlinde Köber and Dan Zamfirescu

Abstract A proposed simplified method (called) SESA is used to estimate the displacement amplification due to general torsion. The SESA method uses modal response spectrum analysis and takes into account the inelastic behavior by applying the capacity spectrum method. For single-story structures, the single degree of freedom inelastic system is equated to an elastic one, equivalent in translation. The resulting linear equivalent system is defined by the secant to maximum displacement stiffness, and the viscous damping properties are set through equivalence with the hysteretic damping properties of the initial system. Regular modal analysis is applied to the linear equivalent system by using over damped response spectra. This is how the displacement amplification due to torsion (compared to translational only behavior) is determined. A large parametric study was conducted on single-story structures comparing results from the SESA method and from dynamic nonlinear analysis. For the multistory inelastic structure analyzed in this chapter, a force–displacement relationship was determined by static nonlinear analysis. The corresponding single degree of freedom inelastic system was defined by an idealized force–displacement curve. The response of a "real" multistory plan irregular structure was computed by dynamic nonlinear analysis (target of results), by dynamic nonlinear analysis on the single-story model, and by the simplified SESA method. By applying the SESA method, results give a better estimation of the structural response of irregular structures (influenced by general torsion) than engineers can obtain by using code provisions for the most cases (Köber and Zamfirescu, Sci J Math Model Civ Eng 5(2):32–51, 2009).

D. Köber (✉) • D. Zamfirescu
Reinforced Concrete Department, Technical University of Civil Engineering Bucharest, B-dul Lacul Tei, no.124, RO 020396 Bucharest, Romania
e-mail: kober_dietlinde@yahoo.com; dzam@utcb.ro

O. Lavan and M. De Stefano (eds.), *Seismic Behaviour and Design of Irregular and Complex Civil Structures*, Geotechnical, Geological and Earthquake Engineering 24, DOI 10.1007/978-94-007-5377-8_11,

11.1 Introduction

Coupling between translation and torsion of plan-asymmetric structures causes ununiformed displacements in structural elements.

Nowadays code provisions for plan irregular structures are based on elastic calculation although seismic response is nonlinear and inelastic torsion differs significantly from the elastic one.

The authors applied the SESA method to torsional flexible (twist is not restricted with structural elements placed perpendicular to seismic direction) and torsional stiff (twist is restricted) structures. The main parameter was the eccentricity between the mass center and the stiffness center (that coincides with the resistance center). The results were obtained for elastic behavior and two intensity levels of seismic input for the inelastic behavior. In order to validate the method, the results were compared to the ones obtained by three-dimensional dynamic analysis using seismic records as well as a set of spectrum-compatible accelerograms. A large parametric study was conducted in order to validate the SESA method for different plan assemblages and structural characteristics. The results of the parametric study were compared in terms of displacement values at different points of the structure. Statistic evaluation of results was made, gathering results with respect to plan dimensions of the analyzed structures, seismic input intensity, and corner period.

Nevertheless, most real structures are multistory buildings and have a more complex response than single-story models used often for research.

Therefore, the response of a "real" multistory plan irregular central core structure was computed by dynamic nonlinear analysis (target of results), by dynamic nonlinear analysis on the single-story model, and by the simplified SESA method. The results were obtained for ultimate limit state. Spectrum-compatible accelerograms and seismic records were used as seismic input. The results of the study were compared in terms of displacement values at different points of the structure.

Code regulations concerning torsional sensitivity were checked for the analyzed structures.

11.2 Simplified Method for the Estimation of the Effects of General Torsion (SESA)

SESA evaluates the elastic structural response under seismic action by modal analysis based on response spectra of an irregular single-story system. It uses the capacity spectrum method and transforms the real structure having nonlinear behavior into an elastic structural system, equivalent in translation. This means equivalence of hysteretic and viscous damping. That is how the simplified method establishes the displacement gain due to general torsion and maintains the simplicity of the spectral analysis (elastic calculation) (Köber and Zamfirescu 2009).

Response values are determined by modal analysis and assembled using the CQC rule (Chopra 2007)

$$\begin{pmatrix} u_{yi} \\ r u_{\theta i} \end{pmatrix} = \begin{pmatrix} \phi_{yi} \\ \phi_{\theta i} \end{pmatrix} |\eta_i(t)|_{\max} = \begin{pmatrix} \phi_{yi} \\ \phi_{\theta i} \end{pmatrix} \frac{P_i^*}{M_i^* \omega_i^2} \mathrm{SA}_i \tag{11.1}$$

where u_{yi} and $u_{\theta i}$ are modal displacement and rotation, r is the gyration ratio, ϕ_{yi} and $\phi_{\theta i}$ are modal coordinates, M_i^* is the modal mass, P_i^* is the modal participation factor, and ω_i is the natural frequency. SA_i is the pseudo-acceleration related to equivalent period T_i and damping ratio ξ.

The equivalent structure is based on the approach of the nonlinear displacement of the real structure by the elastic displacement of an equivalent structure. The stiffness of the equivalent structure is defined by the secant stiffness to the point of maximum displacement.

It is worth to notice that the simplified method is entirely consistent with the assumptions used for capacity spectrum method. Consequently, the SESA method results for torsional amplification are decisively influenced by the inaccuracy given by substitution of the translational inelastic behavior by a translation equivalent elastic structure. The equivalence process has shortcomings particularly for periods lower than the characteristic period of ground motion (T_c).

By applying the SESA method, results show a relatively good match to the structural response determined by dynamic nonlinear analysis and a better estimation of the structural response of irregular structures (influenced by general torsion) than engineers can obtain by using code provisions for the most cases.

11.3 Parametric Study on Single-Story Plan Irregular Structures

11.3.1 Analyzed Structures

In this study, there were analyzed two types of structures that differ fundamentally regarding their torsional behavior: torsional flexible (twist unrestrained) structures and torsional stiff (twist restrained) structures (see Fig. 11.1).

Both analyzed structures are idealized single-story structures with a rigid diaphragm floor and columns and walls as vertical structural elements. The vertical structural elements are disposed symmetric about the x-axis and y-axis. The structural mass is lumped at the center of mass (CM).

The structures of Fig. 11.1 are symmetric structures, being characterized through a coincidence between the center of stiffness (CR), the center of mass (CM), and the center of resistance (CF). The corresponding eccentric systems are obtained by translating gradually CR and CF along the y-axis, from its initial position up to $\pm 20\%$ of the plan dimension of the structure normal to the direction of the seismic input.

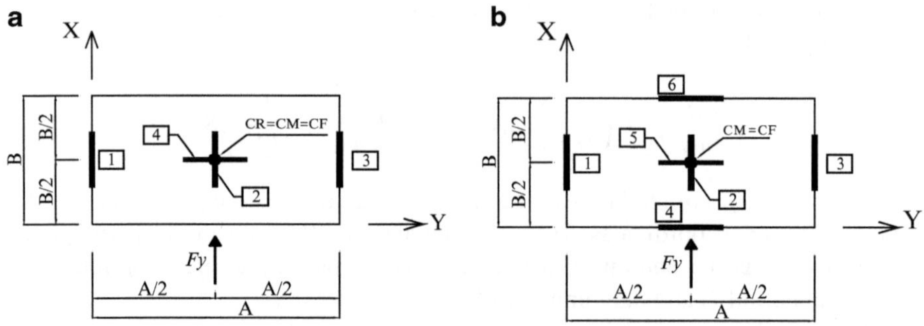

Fig. 11.1 (**a**) Symmetric torsional flexible (unrestrained) structure (TL). (**b**) Symmetric torsional stiff (restrained) structure (TI)

The floor is considered to be a rigid diaphragm. The total weight (*G*) of the floor is 4,840 kN (considering a uniform load $p = 10$ kN/m^2). The total strength of the structural elements is 0.15 G, say $F_y = 726$ KN in each direction.

The structural walls are modeled as elastic-perfectly plastic springs acting on *x* and *y* direction.

The stiffness of the structural elements was chosen so that the initial translational period of the structure is 0.5s, 0.7s, or 1.6s. The strength of walls *P*2 and *P*4 remains constant.

The torsional restrained structure differs from the torsional flexible structure (see Fig. 11.1), by considering at both sides of the structure walls in *y* direction.

11.3.2 Seismic Input

The seismic input considered in this study is given both by original (recorded) accelerograms as well as by spectrum-compatible ones, acting along the *x*-axis. Therefore, design spectra from (P100-1/ 2006) and (EN 1998-1: 2004) (for soil type B) were used.

The results were obtained for elastic behavior (serviceability limit state, SLS) and two intensity levels of seismic input for the inelastic behavior. Therefore, each accelerogram was scaled for two levels of strength: 0.2 g (ultimate limit state, SLU) and 0.4 g (survivability limit state, SLSV). For each limit state, the authors considered one recorded accelerogram and five spectrum-compatible ones (see Table 2.1) for four characteristic ground periods: 0.5s, 0.7s, 1.0s, and 1.6s.

11.3.3 Parameters Under Consideration

From all known parameters that affect the elastic as well as the nonlinear response of plan-asymmetric structures under seismic action, (Goel and Chopra 1990), the authors have chosen the following:

Table 11.1 Cases analyzed for the parametric study

Structure	In-plane dimensions [m]	Natural lateral period [s]	Limit state	Accelerograms
TL	11 × 44	0.3	SLS, SLU, SLSV	1+5
		0.7	SLS, SLU, SLSV	1+5
		1.6	SLS, SLU, SLSV	1+5
	22 × 22	0.3	SLS, SLU, SLSV	1+5
		0.7	SLS, SLU, SLSV	1+5
		1.6	SLS, SLU, SLSV	1+5
TI	11 × 44	0.3	SLS, SLU, SLSV	1+5
		0.7	SLS, SLU, SLSV	1+5
		1.6	SLS, SLU, SLSV	1+5
	22 × 22	0.3	SLS, SLU, SLSV	1+5
		0.7	SLS, SLU, SLSV	1+5
		1.6	SLS, SLU, SLSV	1+5

ω_y, the natural lateral frequency of the corresponding symmetric system. Note that to each asymmetric system, there corresponds a symmetric system for which the center of mass, the center of stiffness, and the center of resistance coincide.

ω_θ/ω_y, ratio between the torsional and the lateral natural frequency of the corresponding symmetric system.

e/r, ratio between the stiffness eccentricity (distance between the center of mass and the center of stiffness) and the radius of gyration of the deck.

ξ, critical damping ratio.

γ_x, ratio between the torsional stiffness of structural elements situated perpendicular to the direction of seismic input and the total structural torsional stiffness (K_θ).

e_p, strength eccentricity (distance between the center of mass and the center of resistance).

u_y, yield displacement of the corresponding symmetric system.

Considering the variation of these parameters, the authors generated the cases for the parametric study. The analyzed cases for symmetric structures (364 cases) are given in Table 11.1. Asymmetric structures are derived from the symmetric ones by translating gradually CR and CF from its initial position. Therefore, the total number of considered situations is 14,596.

11.3.4 Results of the Parametric Study

The target of this comparative study was to identify how well the SESA method based on modal analysis and over damped response spectra can estimate the seismic response obtained by dynamic nonlinear calculation. For validation, the results of the SESA method were compared to the ones obtained by three-dimensional dynamic analysis in terms of displacement values at different points of the structure: total displacement of the center of mass (u_x^{CM}), structural rotation (θ), and displacements of walls $P1$ and $P3$ (u_x^{P1}, u_x^{P3}).

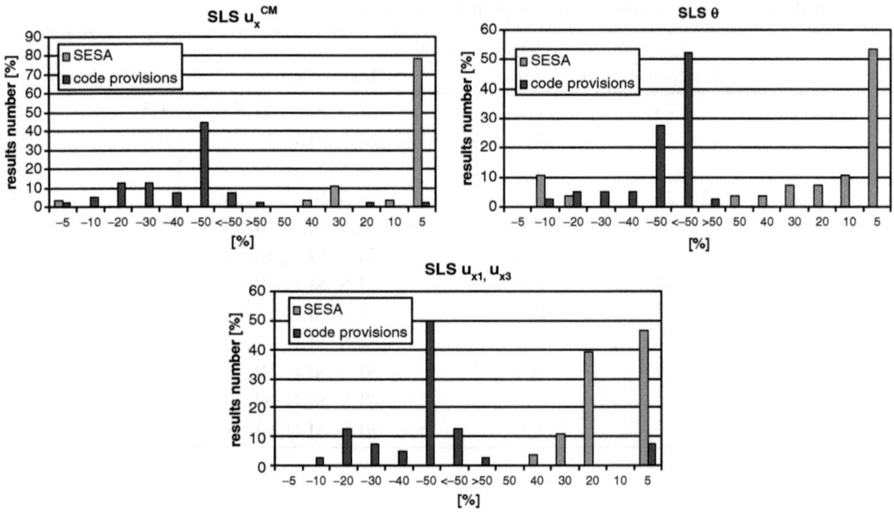

Fig. 11.2 Serviceability limit state

In order to sustain the applicability of the SESA method for design, results were compared also to regular code provisions. The seismic design force used was $F_b = F_y/1.5$, where F_y represents the yield force of the structure. The translation and rotation of CM were determined as follows:

$$u_x^{\mathrm{CM}} = cq\frac{F_b}{K_t} + \theta e; \theta = cq\frac{F_b e}{K_\theta} \tag{11.2}$$

where c represents displacement amplifying coefficient for structural periods lower than the characteristic period of the ground, q is the behavior factor, and K_t is the structural lateral stiffness. For other terms, see Chapter 2.3. At SLS, the factor c turns into 0.5 in order to consider the reduced strength of the seismic input. Although the code method can be used only for the main limit state, in this study it was applied also for SLSV (constant F_y/G ratio and unreal values for q).

The displacements of walls $P1$ and $P3$ are to be determined by

$$u_x^{P1} = u_x^{\mathrm{CM}} \pm \theta \cdot d_1; u_x^{P3} = u_x^{\mathrm{CM}} \pm \theta \cdot d_3 \tag{11.3}$$

where d_1 and d_3 are distances between walls P_1, P_3, and CM (Fig. 11.2).

Statistic evaluation of the results was made in order to identify whether the SESA method overestimates (graphics right side, positive percentage) or underestimates (graphics left side, negative percentage) dynamic nonlinear analysis results. For comparison, results obtained by applying code provisions are represented too. Note that only amplified wall displacements due to general torsion were taken into account. Also, statistics are based on results obtained for $e \leq \pm 10\%$.

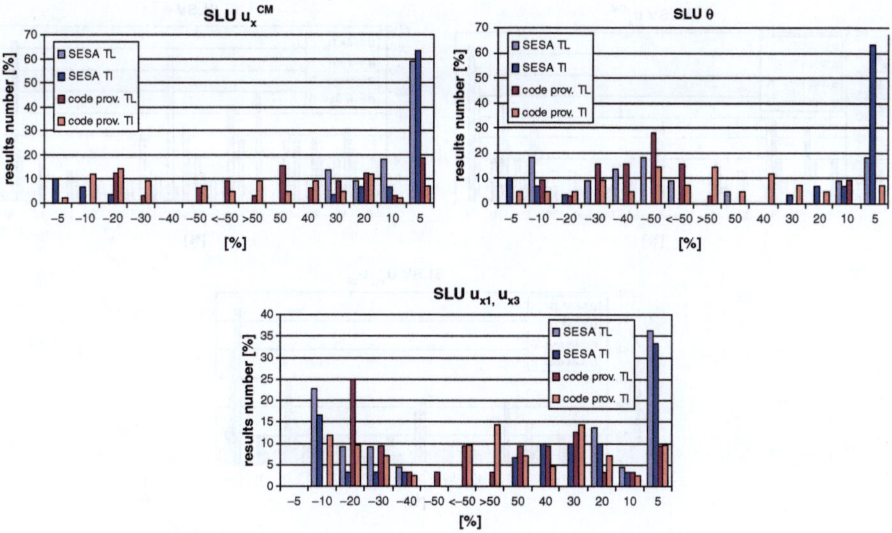

Fig. 11.3 Ultimate limit state

For SLS, code provisions mostly underestimate displacements and rotations given by dynamic nonlinear analysis, whereas the SESA method mostly overestimates them. As expected, results are better for u_x^{CM} and θ as for u_{x1} and u_{x3}. The analyzed situations show that static elastic calculation is not a proper method for plan irregular structures, not even in the elastic range of behavior.

Figure 11.3 shows results for SLU, separately for TL and TI. Walls situated perpendicular to the direction of seismic input participate with half of their lateral stiffness to the rotational stiffness of the structure.

For SLU, results obtained by applying code provisions are well distributed over the entire range of percentages. That means they are equally underestimated and overestimated. Over 60% of CM displacements and rotations are overestimated with less than 5% by the SESA method. As expected, this percentage drops for wall displacements. Only for rotations the SESA method indicates differences up to 30% between TL and TI structures. Also, code provisions indicate better results for TI structures than for TL ones.

Figure 11.4 shows results for SLSV, separately for TL and TI. The variation of results is greater at SLSV compared to SLU, both for SESA method and code provisions. Still, estimation of dynamic nonlinear results is better when applying the SESA method. Up to 40% of the results are overestimated by less than 5%. As expected, rotations scatter more for TL structure than for TI.

Figures 11.5–11.8 show results gathered by characteristic periods (T_c) of the seismic input.

SESA method applied for $T_c = 0.5$s uses the horizontal elastic response spectrum of (EN 1998-1: 2004), leads to greater errors in the equivalence process, and

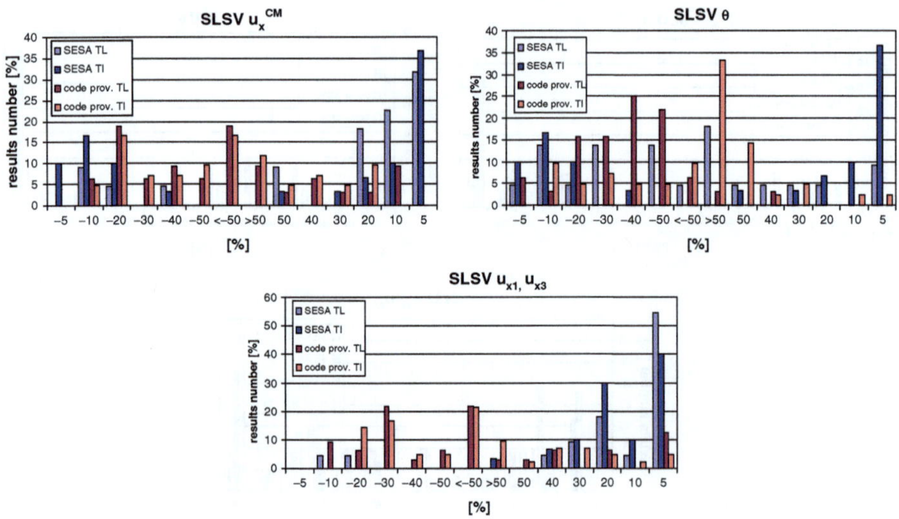

Fig. 11.4 Survivability limit state

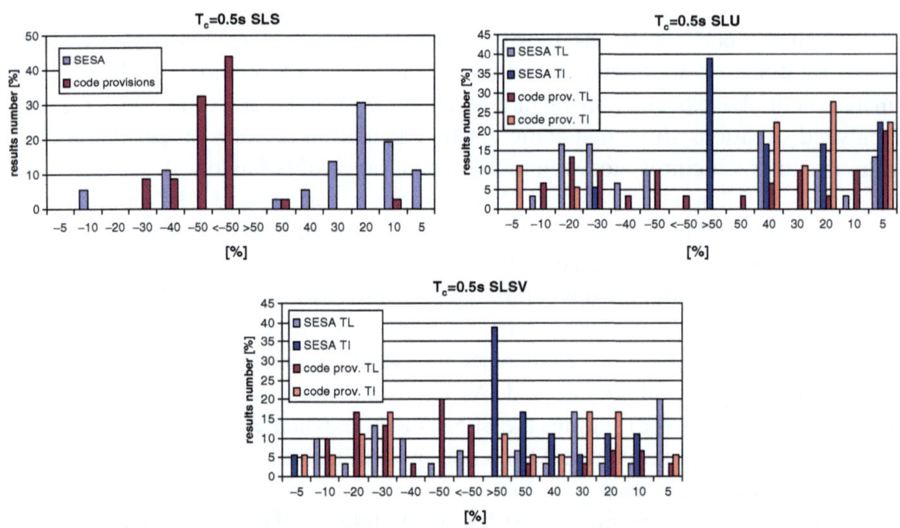

Fig. 11.5 Results for $T_c = 0.5$s

therefore to greater differences compared to the dynamic nonlinear calculation. For $T_c = 0.7$s, $T_c = 1.0$s, and $T_c = 1.6$s, an iterative procedure for the determination of the equivalent system was used and results are better (Figs. 11.6 and 11.7).

SESA method results scatter more, the greater the characteristic period of the ground motion, showing that the simplified method is entirely consistent with the

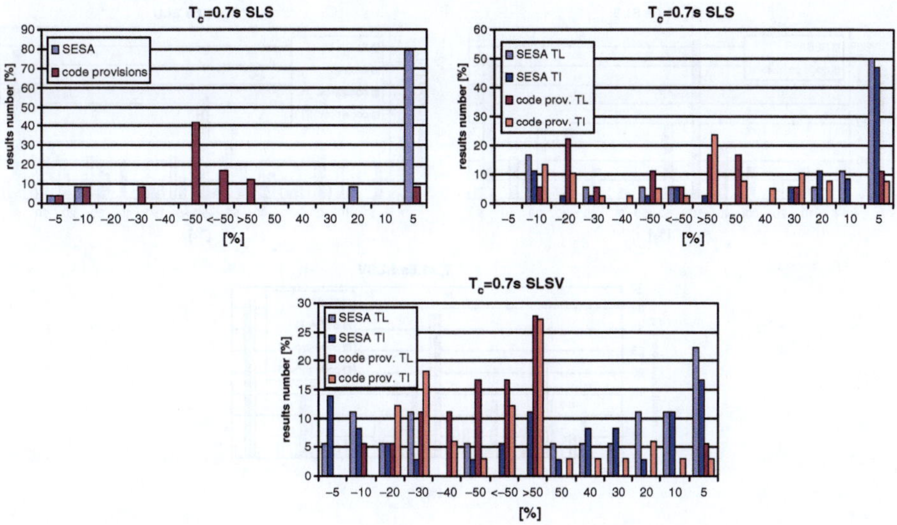

Fig. 11.6 Results for $T_c = 0.7$s

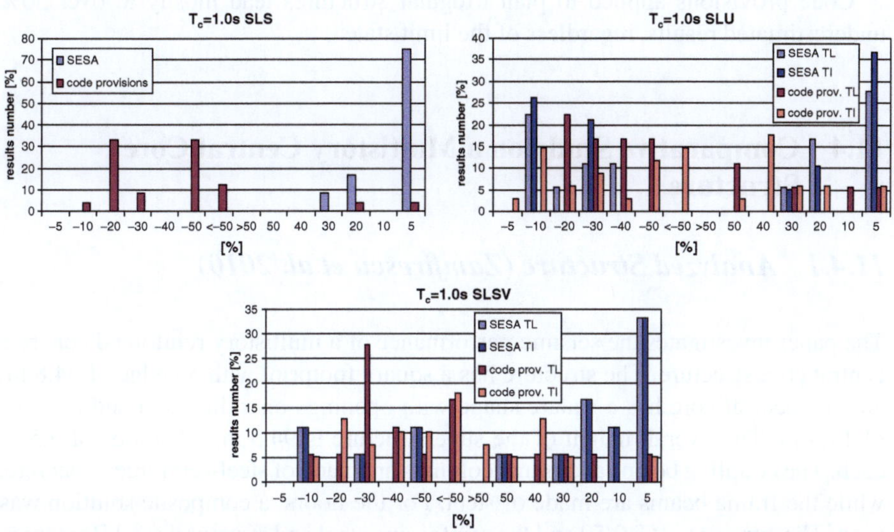

Fig. 11.7 Results for $T_c = 1.0$s

assumptions used for capacity spectrum method. Consequently, the SESA method results for torsional amplification are decisively influenced by the inaccuracy given by substitution of the translational inelastic behavior by a translation equivalent elastic structure. The equivalence process has shortcomings particularly for periods lower than the characteristic period of ground motion (Fig. 11.8).

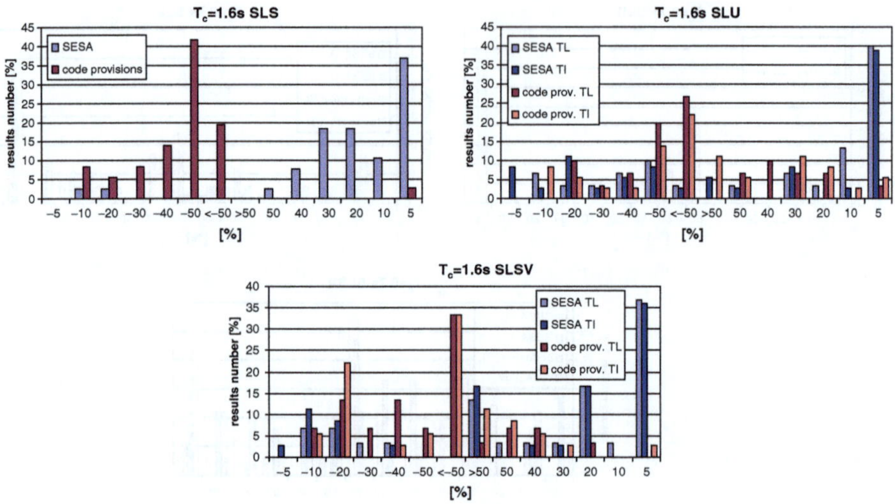

Fig. 11.8 Results for $T_c = 1.6$s

Code provisions applied to plan irregular structures lead mostly to over 50% underestimated results, regardless of the limit state.

11.4 Comparative Study on a Multistory Central Core Structure

11.4.1 Analyzed Structure (Zamfirescu et al. 2010)

The paper investigates the seismic performance of a multistory reinforced concrete central core structure. The structure has a square footprint with an edge of 34.8 m, and the central core has a square shape with openings on *x* direction and an edge of 14.5 m. The overall height of the superstructure is 94.5 m, 27 stories of 3.5 m each. The coupling beams and frame columns are made of steel-reinforced concrete, while the frame beams are made of steel. For the floors, a composite solution was used. The concrete is C40/50 and the reinforcing steel and structural steel Romanian equivalents of S355. The structure is placed in Bucharest area, featured by ground motions having a corner period around 1.6s and a design ground acceleration corresponding to a mean return period (MRI) of 100 years, equal to 0.24 g.

The structure was designed following the Romanian Guideline for Structural Wall Structures assuming an importance factor of 1.2 and a behavior factor 3, corresponding to a torsional sensitive structure for ductility class H. The provisions of the Romanian Guideline is similar to EC8 design provisions for structural walls with the difference that the minimum lateral strength is always associated with the

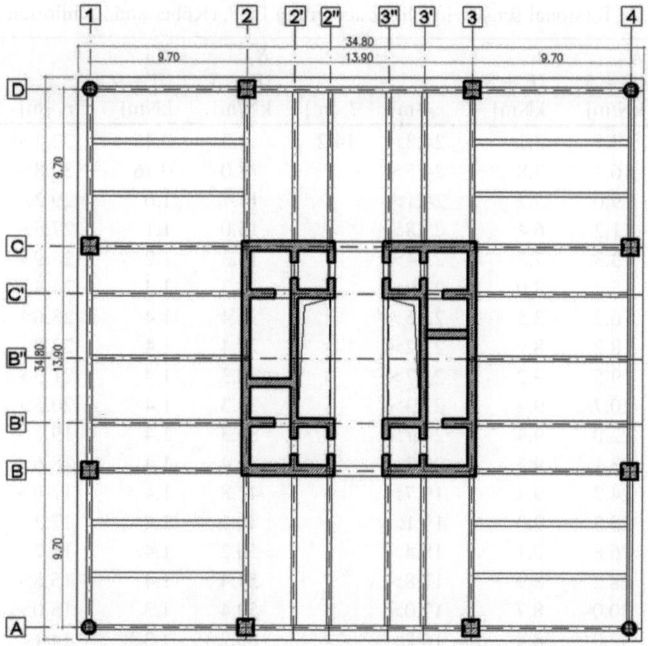

Fig. 11.9 Plan view of a current story

full plastic mechanism, and the credited shear force of walls corresponding to web crush is higher than the one prescribed by the EC8.

The plan assembly of the analyzed structure is shown in Fig. 11.9.

11.4.2 Torsional Sensitivity

EC8 provisions, (EN 1998-1: 2004), were applied to the analyzed structure in order to conclude if it is torsional sensitive or not. Therefore, for each story, the torsional ratio (equal to the square root of the ratio between the torsional stiffness, K_θ, and the translational stiffness, K_t), r, was compared to the radius of gyration, l_s. For a given story, the torsional stiffness was calculated as the ratio between the torsional moment and the story rotation. Similarly, the translational stiffness was determined as the ratio between the story shear and the story drift. Curious from this point of view is the fact that only the first six stories are torsional sensitive (see Table 11.2).

In Tables 11.2 and 11.3, x and y mean the direction of the seismic input.

Romanian Seismic Code provisions, (P100-1/ 2006), were also applied. Therefore, torsional sensitivity appears when the maximum displacement at one edge of the structure, Δ_{edge}, exceeds by more than 1,35 the mean displacement, Δ_{mean}. From this point of view, the analyzed structure is not torsional sensitive (see Table 11.3).

Table 11.2 Torsional sensitivity check according EC8, (Köber and Zamfirescu 2009)

Story	K_{tx} [E+5 kN/m]	$K_{\theta x}$ [E+8 kNm]	r_x [m]	l_s [m]	K_{ty} [E+5 kN/m]	$K_{\theta y}$ [E+9 kNm]	r_y [m]	l_s [m]
25	3.5	2.1	24.8>	14.2	4.3	0.45	32.4>	14.2
24	6.4	3.8	24.5>		8.0	0.76	30.8>	
23	9.0	5.2	24.1>		11.7	1.0	29.2>	
22	11.2	6.4	23.8>		15.0	1.1	27.5>	
21	13.3	7.3	23.4>		18.3	1.2	25.9>	
20	15.2	8.0	23.0>		21.3	1.3	24.4>	
19	16.7	8.5	22.6>		24.4	1.4	23.6>	
18	18.2	8.9	22.2>		27.4	1.4	22.5>	
17	19.5	9.2	21.7>		30.3	1.4	21.5>	
16	20.7	9.4	21.3>		33.3	1.4	20.5>	
15	22.0	9.4	20.7>		36.3	1.4	19.7>	
14	23.1	9.5	20.2>		39.4	1.4	18.6>	
13	24.2	9.4	19.7>		42.8	1.4	17.9>	
12	25.5	9.3	19.1>		46.1	1.4	17.2>	
11	26.8	9.1	18.4>		50.2	1.4	16.7>	
10	28.2	8.9	17.8>		54.4	1.4	15.8>	
9	30.0	8.7	17.0>		59.4	1.3	15.0>	
8	32.0	8.4	16.2>		65.2	1.3	14.1<	
7	34.3	8.0	15.3>		71.9	1.3	13.4<	
6	37.5	7.7	14.3>		81.0	1.3	12.7<	
5	41.4	7.2	13.2<		92.5	1.2	11.6<	
4	47.3	6.7	11.9<		106.9	1.2	10.7<	
3	56.0	6.1	10.5<		130.4	1.1	9.2<	
2	71.0	5.4	8.7<		167.5	1.0	7.9<	
1	103.5	4.1	6.3<		231.6	0.97	6.5<	
P	231.5	2.8	3.5<		464.6	1.4	5.5<	

11.4.3 Simplified Approach

11.4.3.1 Transformation of the Multistory Structure into a Single-Story Structure

The simplified method SESA can be applied only to single-story structures. That is why for the simplified approach, the real multistory structure has to be modeled as a single-story one.

For this purpose, static nonlinear analysis was performed on both main structural directions. The obtained pushover curves have been transformed into bilinear force–displacement relationships, considering the tangent initial stiffness and the displacement target indicated by response spectra for a given accelerogram. Force–displacement relationships valid for single-story systems were computed by using factor Γ for displacements (see Eq. 11.4) and factor ε for forces (see Eq. 11.5), in order to use the real structural mass for calculation:

Table 11.3 Torsional sensitivity check according to Romanian Seismic Code (Goel and Chopra 1990)

Story	Δ^x_{edge}[m]	Δ^x_{mean}[m]	$\Delta^x_{edge}/\Delta^x_{mean}$[−]	Limit	Δ^y_{edge}[m]	Δ^y_{mean}[m]	$\Delta^y_{edge}/\Delta^y_{mean}$[−]	Limit
25	0.0141	0.0138	1.02 <	1.35	0.0155	0.0155	1.00 <	1.35
24	0.0144	0.0142	1.02 <		0.0157	0.0156	1.00 <	
23	0.0148	0.0146	1.02 <		0.0156	0.0156	1.00 <	
22	0.0154	0.0151	1.02 <		0.0158	0.0157	1.00 <	
21	0.0158	0.0155	1.02 <		0.0158	0.0157	1.00 <	
20	0.0163	0.0160	1.02 <		0.0158	0.0158	1.00 <	
19	0.0169	0.0166	1.02 <		0.0158	0.0157	1.00 <	
18	0.0173	0.0170	1.02 <		0.0157	0.0156	1.00 <	
17	0.0177	0.0174	1.02 <		0.0156	0.0155	1.00 <	
16	0.0181	0.0178	1.02 <		0.0154	0.0153	1.00 <	
15	0.0184	0.0180	1.02 <		0.0152	0.0151	1.00 <	
14	0.0186	0.0182	1.02 <		0.0149	0.0148	1.00 <	
13	0.0187	0.0184	1.02 <		0.0145	0.0144	1.00 <	
12	0.0187	0.0183	1.02 <		0.0141	0.0141	1.00 <	
11	0.0185	0.0182	1.02 <		0.0135	0.0135	1.00 <	
10	0.0183	0.0180	1.02 <		0.0130	0.0129	1.00 <	
9	0.0178	0.0175	1.02 <		0.0123	0.0122	1.00 <	
8	0.0172	0.0169	1.02 <		0.0115	0.0115	1.00 <	
7	0.0164	0.0161	1.02 <		0.0107	0.0107	1.00 <	
6	0.0154	0.0151	1.02 <		0.0097	0.0097	1.00 <	
5	0.0141	0.0139	1.01 <		0.0087	0.0086	1.01 <	
4	0.0126	0.0123	1.02 <		0.0076	0.0076	1.00 <	
3	0.0107	0.0105	1.01 <		0.0063	0.0063	1.00 <	
2	0.0085	0.0084	1.01 <		0.0050	0.0049	1.01 <	
1	0.0059	0.0058	1.02 <		0.0036	0.0036	1.00 <	
P	0.0026	0.0026	1.00 <		0.0018	0.0018	1.00 <	

$$\Gamma = \frac{\sum m_i \phi_i}{\sum m_i \phi_i^2} \tag{11.4}$$

$$\varepsilon = \frac{(\sum m_i \phi_i)^2}{\sum m_i \sum m_i \phi_i^2} \tag{11.5}$$

In Eqs. (11.4) and (11.5), m_i represents the ith story mass and ϕ_i the modal ordinate.

An iterative procedure was applied until the ultimate displacement equaled the considered displacement target.

For example, the bilinearization process for one of the considered accelerograms in x direction is shown in Fig. 11.10.

The transformation multistory structure to single-story structure is made considering the same translational periods for both models. In order to reach also the same torsional period, structural walls were moved away from the center of mass,

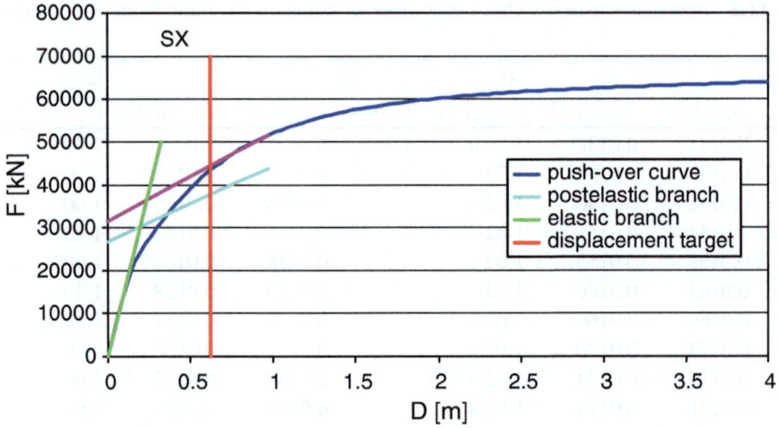

Fig. 11.10 Bilinearization process of the pushover curve

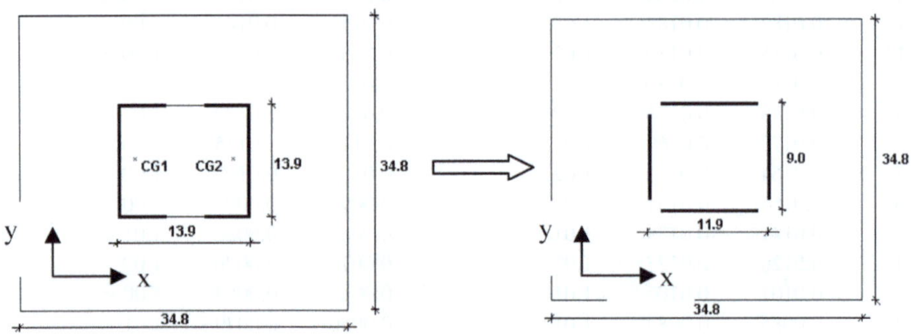

Fig. 11.11 Simplified model for single-story structure

maintaining the same ratio between their initial positions on *x* and *y* direction. Therefore, the single-story model has independent walls along both main directions instead of a central core (see Fig. 11.11).

11.4.4 Seismic Input

The analyzed structure has been designed for Bucharest. For this reason, the N-S recording of Vrancea 1977 earthquake from INCERC station was applied to the structure. It is the most severe earthquake motion recorded in Romania. Supplementary a set of three spectrum-compatible accelerograms was considered, fitting the response spectrum for corner period equal to 1,6s of the Romanian seismic code (P100-1/ 2006). The referred response spectrum is represented in Fig. 11.12.

All accelerograms were scaled at first mode elastic spectral acceleration for MRI 100 years. Calculation for ultimate limit state was made.

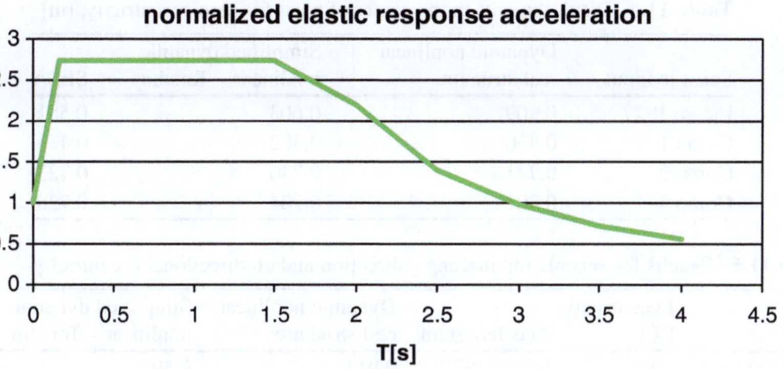

Fig. 11.12 Normalized elastic response spectrum for corner period $T_c = 1{,}6$s, (P100 -1/ 2006)

11.4.5 Comparative Study

Results of dynamic nonlinear analysis performed on the real structure described in paragraph 2 were compared with two simplified approaches. Calculation on the real structure was made in PERFORM 3D v 3.04 by considering a Takeda hysteretic behavior. The simplified approaches consider a bilinear hysteretic behavior with post elastic hardening.

The first simplified approach refers to dynamic nonlinear analysis performed on the equivalent single-story structure. For this approach the Torsdin program, developed at the Technical University of Bucharest, was used.

The second approach refers to the simplified spectral analysis method called SESA.

Results were computed for zero eccentricity (only translation) and eccentricities (stiffness and strength) equal to 10 and 20% of the plan dimension of the structure. This means up to four times the accidental eccentricity proposed by most seismic codes.

Eccentricities were considered first along the *x*-axis of the structure (direction with coupling beams) for seismic input 100% in the *y* direction of the structure.

Then eccentricities along the *x*-axis were coupled with bidirectional seismic input (100% in *y* direction combined with 30% in *x* direction).

Dynamic analysis on the real structure showed that for input along the *y* direction, the torsional rotations are greater than for seismic input along the *x* direction. Therefore, the comparative study concentrates on seismic input along the *y* direction.

The authors compared displacements in the center of mass, torsional rotations, and displacements of the walls parallel to the direction of seismic input (*y* direction).

Displacements in the center of mass for zero eccentricity and seismic input in *y* direction are given in Table 11.4.

Results for translation show a good match between the simplified dynamic nonlinear calculation and the SESA method. Differences between calculation on the

Table 11.4 Displacements in the center of mass for zero eccentricity, [m]

Seismic input	Dynamic nonlinear real structure	Simplified dynamic nonlinear – Torsdin	SESA
Incerc 1977	0.909	0.603	0.581
Comp 1	0.830	0.382	0.413
Comp 2	0.741	0.746	0.723
Comp 3	0.801	0.704	0.728

Table 11.5 Results for seismic input along *y* direction and unidirectional eccentricity

	Eccentricity [%]	Accelerogram	Dynamic nonlinear real structure	Simplified dynamic nonlinear – Torsdin	SESA
CM displ. [m]	10%	Incerc 1977	0.911	0.593	0.603
		Comp	0.869	0.490	0.544
Rotation [rad]		Incerc 1977	0.0028	0.0225	0.020
		Comp	0.0029	0.020	0.016
Disp. stiff side [m]		Incerc 1977	0.861	0.289	0.579
		Comp	0.815	0.253	0.499
Disp. flexible side [m]		Incerc 1977	0.945	0.834	0.833
		Comp	0.907	0.709	0.688
CM displ. [m]	20%	Incerc 1977	0.905	0.57	0.615
		Comp	0.873	0.459	0.552
Rotation [rad]		Incerc 1977	0.0055	0.03	0.034
		Comp	0.0058	0.0207	0.027
Disp. stiff side [m]		Incerc 1977	0.786	0.165	0.618
		Comp	0.742	0.269	0.490
Disp. flexible side [m]		Incerc 1977	0.957	0.962	1.158
		Comp	0.929	0.419	1.039

real structure and the simplified approaches are up to 55%, depending on the applied accelerogram. They may be explained by the influence of the Takeda hysteretic behavior considered only in the dynamic nonlinear calculation for the real structure.

Results for eccentricities of 10 and 20% in *x* direction combined with seismic input in *y* direction are given in Table 11.5. For the spectrum-compatible accelerograms, mean values are computed.

Because the structural walls are situated different in the single-story structure compared to the multistory structure (in order to fit the torsional eigenperiod), floor edge displacements were compared.

Differences between results for the real structure and results for the simplified model are quiet large. By neglecting the error that occurs for zero eccentricity (due to the hysteretic model), differences for edge displacements are up to 50% for Torsdin and 5% for SESA. These differences appear for displacements of the stiff side. Displacements of the flexible side are estimated well, with differences under 5%. Better estimation of maximum displacements (on the flexible side) was observed also for single-story structures, (Köber and Zamfirescu 2009) and is due to the modal combination.

Table 11.6 Results for seismic input 100% along *y* direction, 30% along *x* direction and unidirectional eccentricity

	Eccentricity [%]	Accelerogram	Dynamic nonlinear real structure	Simplified dynamic nonlinear – Torsdin
CM displ. [m]	10%	Incerc 1977	0.901	0.593
		Comp	0.851	0.551
Rotation [rad]		Incerc 1977	0.003	0.0216
		Comp	0.0033	0.0221
Disp. stiff side [m]		Incerc 1977	0.842	0.287
		Comp	0.794	0.165
Disp. flexible side [m]		Incerc 1977	0.941	0.849
		Comp	0.891	0.743
CM displ. [m]	20%	Incerc 1977	0.897	0.570
		Comp	0.853	0.510
Rotation [rad]		Incerc 1977	0.0059	0.033
		Comp	0.0062	0.026
Disp. stiff side [m]		Incerc 1977	0.763	0.156
		Comp	0.719	0.170
Disp. flexible side [m]		Incerc 1977	0.955	0.599
		Comp	0.913	0.867

Displacements of the center of mass are greater for the dynamic nonlinear calculation on the real structure and rotations are greater for the simplified approaches. Therefore, the estimation of the edge displacements is better (comparing real structure and simplified approaches) than that of results in CM.

The SESA method shows a good match of results in CM (displacement and rotation) compared to the simplified dynamic nonlinear calculation. Edge displacements are less accurate because of the modal combination (in this chapter CQC was used).

As expected, results scatter more for greater eccentricity values.

Table 11.6 shows results for eccentricity in *x* direction combined with seismic input 100% in *y* direction and 30% in *x* direction.

Because the SESA method cannot consider bidirectional seismic input, only results for Torsdin and computation on the real structure are compared.

Comparing values from Tables 11.4 and 11.6 shows that considering bidirectional seismic loading affects in a minor instance the analyzed results. Therefore, comments for unidirectional loading are valid also for bidirectional loading.

Structural rotations scatter little for bidirectional loading compared to unidirectional loading. Displacements of the center of mass differ up to 5%. Displacements of the floor edges differ up to 15%.

11.5 Concluding Remarks

For the analyzed single-story structures, up to 45% of the SESA method results are close (differences under 5%) to the dynamic nonlinear analysis results.

Code provisions generally underestimate dynamic nonlinear analysis results, often by more than 50%. Static elastic calculation is not a proper method for plan irregular structures, not even in the elastic range of behavior.

Also, the code method can be used only for the main design limit state, and as it is shown, the amount of displacement amplification of irregular structures depends essentially on the earthquake intensity.

For the real central core multistory structure, simplified modeling by a single-story structure shows major differences (up to 55%) in results even if no eccentricity is assumed. These differences may be explained by the different hysteretic model used for simplified calculation compared to calculation on the real structure.

Displacement values for simplified dynamic nonlinear calculation and for the SESA method scatter up to 70% from the results on the real structure. If the difference up to 55% due to translation behavior is skipped, then the inaccuracy of the simplified approaches is limited to 15%.

Nevertheless, eccentricities considered for the comparative study are up to four times greater than accidental eccentricities given by most seismic codes.

Influence of bidirectional loading is negligible for the analyzed results.

Torsional sensitivity is difficult to define because of the great number of parameters that influence the torsional behavior, especially the nonlinear one. This can be seen also in the contradictory results obtained by applying code provisions.

Applying Romanian Code provisions for the results of the simplified approaches leads to the following conclusion: the analyzed structure is torsional sensitive for unidirectional loading as well as for bidirectional loading.

The authors find the obtained results encouraging and plan to expand the study to more structural types.

References

Chopra AK (2007) Dynamics of structures. Prentice Hall, Upper Saddle River. ISBN 0-13-156174-X

EN 1998-1: 2004. Design of structures for earthquake resistance. General rules, seismic actions and rules for buildings

Goel RK, Chopra AK (1990) Inelastic seismic response of one – story, asymmetric – plan systems. College of Engineering, University of California at Berkeley, Report no. UBC/EERC – 90/14

Köber D, Zamfirescu D (2009) Simplified methods used for evaluation of the displacement gain due to general torsion. Sci J Math Model Civ Eng 5(2):32–51

P100-1/ 2006. Cod de proiectare seismică. Prevederi de proiectare pentru clădiri, 23–29

Zamfirescu D, Damian I, Morariu E (2010) Seismic performance of a tall structure for long predominant periods. In: 14th European conference of earthquake engineering, Ohrid

Chapter 12
Application of Nonlinear Static Method with Corrective Eccentricities to Multistorey Buildings: A Case Study

Melina Bosco, Giovanna A. Ferrara, Aurelio Ghersi, Edoardo M. Marino, and Pier Paolo Rossi

Abstract Nonlinear static methods are not always effective in the assessment of three-dimensional building structures because their estimate of the torsional response is not always reliable. To overcome this shortcoming, some of the authors have proposed a double application of the nonlinear static analysis by means of lateral forces applied to points of the deck characterised by different eccentricities with respect to the centre of mass. These eccentricities, named "corrective eccentricities", were calibrated in such a way that the displacements of the edges of the deck evaluated by the two nonlinear static analyses equal those evaluated by nonlinear dynamic analysis. Analytical expressions of the corrective eccentricities were derived and presented in a previous study based on a parametrical investigation conducted on a wide set of one-storey systems subjected to bidirectional ground motions. In this chapter, the peculiarities of the application of the proposed nonlinear static method to multistorey buildings are discussed first. Second, as an example, the method is applied to predict the seismic response of a five-storey asymmetric building. The structure of the analysed building is constituted by reinforced concrete frames arranged along two orthogonal directions. The cross sections of the structural element are equal at all storeys. To test the effectiveness of the proposed method, the in-plan distributions of the deck displacement and drift are evaluated by nonlinear dynamic analysis and compared to those resulting from the nonlinear static method with corrective eccentricities. The obtained results show that the use of the corrective eccentricities leads to a suitable estimate of the maximum dynamic displacements and drifts.

M. Bosco • G.A. Ferrara • A. Ghersi • E.M. Marino (✉) • P.P. Rossi
Department of Civil and Environmental Engineering, University of Catania,
V.le A. Doria, 6, 95125 Catania, Italy
e-mail: mbosco@dica.unict.it; gferrara@dica.unict.it; aghersi@dica.unict.it;
emarino@dica.unict.it; prossi@dica.unict.it

O. Lavan and M. De Stefano (eds.), *Seismic Behaviour and Design of Irregular and Complex Civil Structures*, Geotechnical, Geological and Earthquake Engineering 24, DOI 10.1007/978-94-007-5377-8_12,

12.1 Introduction

It is widely recognised that nonlinear static methods usually do not provide a reliable prediction of the torsional response of asymmetric structures (Chopra and Goel 2004; Fajfar et al. 2005; Marušić and Fajfar 2005). Since only experts in seismic engineering are able to analyse three-dimensional structures by nonlinear dynamic analysis, the scientific community is greatly interested in the possibility of improving the estimate of the seismic response of asymmetric structures by nonlinear static methods (Marušić and Fajfar 2005; Lucchini et al. 2009) and (De Stefano and Pintucchi 2010). On this topic, some of the authors have recently proposed a nonlinear static method for asymmetric structures (Bosco et al. 2009a). This method aims at obtaining an estimate of the maximum seismic response of asymmetric one-storey systems by means of two pushover analyses. The two nonlinear static analyses are necessary because the in-plan distribution of the maximum dynamic displacements is not linear and, thus, it cannot be properly approximated by a single nonlinear static analysis. In particular, according to this method, the nonlinear static analysis is performed by applying the lateral forces to two points of the deck different from the centre of mass CM. The coordinates of the points CF_i where the forces F are applied were calibrated in Bosco et al. (2012) by a parametrical study on one-storey systems subjected to bidirectional ground motions. The positions of the points CF_i were adjusted in such a way that the displacements evaluated at the two edges of the deck by the two nonlinear static analyses were equal to the displacements provided by the nonlinear dynamic analysis (Fig. 12.1). The difference between the abscissa of the point CF_i and that of CM is named "corrective eccentricity" e (with $i = 1$ or 2 for the left and the right side of the deck,

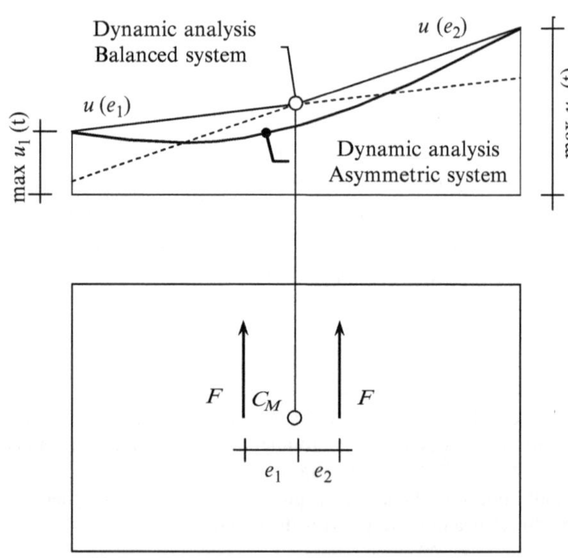

Fig. 12.1 Maximum dynamic displacements versus displacements determined by the nonlinear static method with corrective eccentricities

respectively). On the basis of the results of this parametrical study, the value of each corrective eccentricity was expressed by analytical relations as a function of the four structural parameters which influence the seismic elastic and inelastic response of asymmetric systems. The parameters which affect the torsional response in the elastic range of behaviour (Hejal and Chopra 1987) are the rigidity eccentricity e_r (distance between CM and the centre of rigidity CR) and the ratio Ω_θ of the torsional to lateral frequencies of the corresponding torsionally balanced system (obtained by shifting CM into CR). The parameters which affect the torsional response in the inelastic range of behaviour (Goel and Chopra 1990) are the strength eccentricity e_s (distance between CM and the centre of strength CS) and the ratio R_μ of the elastic shear force to the actual strength of the corresponding torsionally balanced system.

The corrective eccentricities e_i were written as a linear function of the rigidity and strength eccentricities

$$e_i = a_i e_s + b_i e_r \quad i = 1 \text{ or } 2 \tag{12.1}$$

where the coefficients a_i and b_i depend on Ω_θ and R_μ according to the following expressions:

$$a_1 = \begin{cases} c_2^1\, R_\mu^2 + c_1^1\, R_\mu + c_0^1 & R_\mu \le 2 \\ c_2^2\, R_\mu^2 + c_1^2\, R_\mu + c_0^2 & R_\mu > 2 \end{cases}$$

$$c_2^1 = -0.25 c_1^1 \qquad\qquad c_2^2 = -0.25 c_1^2$$

$$c_1^1 = \begin{cases} -0.752\quad \Omega_\theta + 1.373 & \Omega_\theta < 0.85 \\ 1.556\quad \Omega_\theta - 0.589 & 0.85 \le \Omega_\theta \le 1.15 \\ -2.234\quad \Omega_\theta + 3.770 & \Omega_\theta > 1.15 \end{cases} \qquad c_1^2 = 0.273\ \Omega_\theta - 0.182$$

$$c_0^1 = \begin{cases} 1.199\quad \Omega_\theta - 0.834 & \Omega_\theta < 0.85 \\ -2.020\quad \Omega_\theta + 1.902 & 0.85 \le \Omega_\theta \le 1.15 \\ 0.508\quad \Omega_\theta - 1.004 & \Omega_\theta > 1.15 \end{cases} \qquad c_0^2 = c_0^1 + c_1^1 - c_1^2 \tag{12.2}$$

$$a_2 = \begin{cases} c_2^1\, R_\mu^2 + c_1^1\, R_\mu + c_0^1 & R_\mu \le R_{\mu V} \\ c_2^2\, R_\mu^2 + c_1^2\, R_\mu + c_0^2 & R_{\mu V} < R_\mu \le 5 \\ m\ R_\mu + t & R_\mu > 5 \end{cases}$$

$$R_{\mu V} = -0.6\ \Omega_\theta + 2.86$$

$$c_2^1 = -0.50 \frac{c_1^1}{R_{\mu V}} \qquad\qquad c_2^2 = -0.5 \frac{c_1^2}{R_{\mu V}}$$

$$c_1^1 = \begin{cases} 0.946\ \Omega_\theta + 0.314 & \Omega_\theta \le 0.95 \\ 1.213 & \Omega_\theta > 0.95 \end{cases} \qquad c_1^2 = 0.606\ \Omega_\theta - 0.396$$

$$c_0^1 = \begin{cases} -0.171 & \Omega_\theta < 0.75 \\ -0.769\ \Omega_\theta + 0.406 & 0.75 \le \Omega_\theta \le 0.95 \\ -0.255\ \Omega_\theta - 0.083 & \Omega_\theta > 0.95 \end{cases} \quad c_0^2 = c_0^1 + \frac{R_{\mu V}}{2}\left(c_1^1 - c_1^2\right)$$

$$m = \begin{cases} 0.074 & \Omega_\theta \le 1.05 \\ -0.720\ \Omega_\theta + 0.831 & \Omega_\theta > 1.05 \end{cases} \qquad t = 25\ c_2^2 + 5\ c_1^2 + c_0^2 - 5\ m \tag{12.3}$$

$$b_1 = \begin{cases} \alpha\ R_\mu^\beta & R_\mu \le 2 \\ m\ R_\mu + t & R_\mu > 2 \end{cases}$$

$$\alpha = \begin{cases} 0.756 & \Omega_\theta < 0.90 \\ -2.521\ \Omega_\theta + 3.025 & 0.90 \le \Omega_\theta \le 1.20 \\ 0 & \Omega_\theta > 1.20 \end{cases} \qquad \beta = -0.881\ \Omega_\theta - 0.015$$

$$m = \begin{cases} -0.085\ \Omega_\theta + 0.010 & \Omega_\theta < 1.00 \\ 0.373\ \Omega_\theta - 0.447 & 1.00 \le \Omega_\theta \le 1.20 \\ 0 & \Omega_\theta > 1.20 \end{cases} \qquad t = 2^\beta \alpha - 2\ m \tag{12.4}$$

$$b_2 = \begin{cases} c_2^1\ R_\mu^2 + c_1^1\ R_\mu + c_0^1 & R_\mu \le 3 \\ c_2^2\ R_\mu^2 + c_1^2\ R_\mu + c_0^2 & R_\mu > 3 \end{cases}$$

$$c_2^1 = -\frac{c_1^1}{6} \qquad c_2^2 = -\frac{c_1^2}{12}$$

$$c_1^1 = \begin{cases} 9.91\ \ \Omega_\theta^2 - 14.6\ \Omega_\theta + 4.46 \\ -10.4\ \Omega_\theta^2 + 25.7\ \Omega_\theta - 15.5 \end{cases}$$

$$c_0^1 = \begin{cases} -15.6\ \quad 22.2\ \quad \Omega_\theta - 6.74 & \Omega_\theta \le 1.00 \\ 16.3\ \ \Omega_\theta^2 - 40.6\ \Omega_\theta + 24.1 & \Omega_\theta > 1.00 \end{cases}$$

$$c_1^2 = \begin{cases} -2.59\ \Omega_\theta^2 + 3.28\ \Omega_\theta - 1.16 \\ 2.12\ \ \Omega_\theta^2 - 4.90\ \Omega_\theta + 2.31 \end{cases}$$

$$c_0^2 = c_0^1 + 3\left(\frac{c_1^1}{2} - \frac{3}{4}c_1^2\right) \tag{12.5}$$

Unfortunately, results obtained by one-storey systems need to be validated by multistorey systems (Anagnostopoulos et al. 2010; De Stefano et al. 2006; Ghersi et al. 2007). Further, the determination of e_r, Ω_θ, e_s and R_μ is not trivial when the proposed method is applied to multistorey structures. For these reasons, in this chapter the aspects concerning the application of the proposed nonlinear static method to multistorey buildings are first discussed. Then, as an example, the method is applied to a five-storey asymmetric building. The structure of the building is constituted by reinforced concrete frames arranged along two orthogonal directions. The cross sections of the structural elements are equal at all storeys. To test the effectiveness of the proposed method, the deck displacements and the drifts at each storey are evaluated by nonlinear dynamic analysis and compared to those obtained by the proposed nonlinear static method.

12.2 Parameters Affecting the Elastic Response of Asymmetric Multistorey Buildings

12.2.1 Rigidity Eccentricity e_r

The rigidity eccentricity is defined as the distance between the centre of mass CM and the centre of rigidity CR of the structural system. The determination of this latter point does not cause any problem in one-storey systems. Indeed, as remarked by several researchers (Hejal and Chopra 1987; Tso 1990), in such structures the centre of rigidity, rigorously defined as the point of the floor to which a static force must be applied to cause translation of deck without rotation, can be univocally determined. Also, this point is coincident with the shear centre (point through which the resultant of the storey shear forces passes when the floor is subjected to translation) and with the centre of twist (point that remains stationary in plan when the structure is subjected to torque loading).

Contrary to what has been stated with reference to one-storey systems, the determination of the centre of rigidity may cause troubles when dealing with multistorey buildings. Indeed, only for a particular class of buildings, the regularly asymmetric buildings, the generalised centres of rigidity and twist and the generalised shear centres are independent of the lateral load distribution and are lined up along a single vertical axis named "elastic axis" (Hejal and Chopra 1987). Unfortunately, real asymmetric structures rarely fulfil the strict conditions that characterise the above-mentioned buildings and have generalised centres of rigidity which are strongly dependent on the lateral load distribution. An interesting solution to the problem has been proposed by Makarios and Anastassiadis (1998a, b) who suggested using, as a reference for the calculation of the structural eccentricity, the optimum torsion axis. This axis is defined as the vertical line that joins the points of the floors where the equivalent seismic forces must be applied in order to minimise the sum of the squares of the deck rotations:

$$\Theta = \sum_{i=1}^{n} \theta_i^2 = \text{minimum} \tag{12.6}$$

This axis coincides with the elastic axis in regularly asymmetric systems, and its position is little influenced by the distribution of the horizontal forces in non-regularly asymmetric systems (Makarios and Anastassiadis 1998a, b). Further, Anastassiadis and Makarios demonstrated that the optimum torsion axis can be approximately determined as the vertical line passing through the centre of twist of the storey at the level z equal to 0.80 times the height of the building (Makarios and Anastassiadis 1998a). An analytical procedure to evaluate the exact location of the optimum torsion axis has also been proposed by Marino and Rossi (2004). In particular, given an orthogonal reference system Oxyz, with x- and y-axes parallel to the principal axes of the structural system, the coordinates of the optimum torsion axis are

$$\mathbf{x}_{\mathrm{CR}} = -\frac{\boldsymbol{\theta}_{\mathrm{F}y}^{T}\boldsymbol{\theta}_{M}}{\boldsymbol{\theta}_{M}^{2}}$$

$$\mathbf{y}_{\mathrm{CR}} = \frac{\boldsymbol{\theta}_{\mathrm{F}x}^{T}\boldsymbol{\theta}_{M}}{\boldsymbol{\theta}_{M}^{2}} \tag{12.7}$$

where $\boldsymbol{\theta}_{\mathrm{F}x}$ and $\boldsymbol{\theta}_{\mathrm{F}y}$ are vectors containing the deck rotations produced by a distribution of seismic forces $\boldsymbol{F}$ applied at the origin of the reference system and acting along the x- and the y-axes, respectively, and $\boldsymbol{\theta}_{M}$ is a vector containing the deck rotations produced by a distribution of torsional couples $\boldsymbol{M}$ equal to $\boldsymbol{F} \times 1$.

The heightwise distribution of the centres of rigidity can also be calculated by the procedure proposed in Bosco et al. (2008a). The position of the centres of rigidity evaluated according to this procedure is little dependent on the distribution of the horizontal forces. Further, the heightwise distribution of the centres of rigidity gives an idea of the irregularity in the elevation of the building.

Finally, given an assigned distribution of lateral forces, the centre of rigidity can be calculated at each storey as the centroid of the stiffness k_j of each frame defined as the ratio of the storey shear to the storey drift:

$$x_{\mathrm{CR}}^{i} = \frac{\sum_{j=1}^{ny} k_{y,j}^{i} x_j}{\sum_{j=1}^{ny} k_{y,j}^{i}}; \quad y_{\mathrm{CR}}^{i} = \frac{\sum_{j=1}^{nx} k_{x,j}^{i} y_j}{\sum_{j=1}^{nx} k_{x,j}^{i}} \tag{12.8}$$

In Eq. 12.8, n_x and n_y are the number of frames arranged along the x- or y-direction, respectively, while y_j and x_j are the coordinates of the frames along the x- or y-direction.

12.2.2 Ratio Ω_θ of the Torsional to Lateral Frequencies of the Torsionally Balanced System

Like the rigidity eccentricity, the uncoupled torsional to lateral frequency ratio can be univocally determined only for one-storey systems and regularly asymmetric systems. In this special class of buildings, like in one-storey systems, the parameter Ω_θ may be calculated as the ratio of the radius of gyration of the lateral stiffness r_k

of the resisting elements about a vertical axis passing through the rigidity centre with respect to the radius of gyration of the mass r_m about a vertical axis passing through the mass centre. With reference to real structures, Makarios and Anastassiadis (1998b) calculated the radius of gyration of the lateral stiffness r_k by the relation:

$$\mathrm{r_{kx}} = \sqrt{\frac{u_y\,(\mathrm{CR})}{\theta_M\,(0.8\;H)}};\quad \mathrm{r_{ky}} = \sqrt{\frac{u_x\,(\mathrm{CR})}{\theta_M\,(0.8\;H)}} \tag{12.9}$$

In these relations, u_y (CR) and u_x (CR) are the displacements of the point of the deck corresponding to the optimum torsion axis at the level $z = 0.80\,H$ due to lateral forces F applied to the optimum torsion axis and acting along the y- and x-directions, respectively; θ_M is the deck rotation at the same level produced by a distribution of torsional couples proportional to the lateral forces F.

The heightwise distribution of the radius of gyration of the lateral stiffness of the resisting elements can also be calculated as suggested in Bosco et al. (2008a).

Finally, for a given distribution of lateral forces, once the stiffness of each frame k_j and the position of CR defined as the centroid of the stiffness k_j have been determined as shown in Sect. 12.2.1, r_{kx} and r_{ky} can also be calculated at each storey as

$$r_{kx}^i = \sqrt{\left[\sum_{j=1}^{nx} k_{x,j}^i \left(y_j - y_{\mathrm{CR}}^i\right)^2 + \sum_{j=1}^{ny} k_{y,j}^i \left(x_j - x_{\mathrm{CR}}^i\right)^2\right] \Big/ \sum_{j=1}^{ny} k_{y,j}^i}$$

$$r_{ky}^i = \sqrt{\left[\sum_{j=1}^{nx} k_{x,j}^i \left(y_j - y_{\mathrm{CR}}^i\right)^2 + \sum_{j=1}^{ny} k_{y,j}^i \left(x_j - x_{\mathrm{CR}}^i\right)^2\right] \Big/ \sum_{j=1}^{nx} k_{x,j}^i} \tag{12.10}$$

12.3 Parameters Affecting the Inelastic Response of Asymmetric Multistorey Buildings

12.3.1 Strength Eccentricity e_s

The strength eccentricity is defined as the distance between the centre of mass CM and the centre of strength CS of the structural system. For one-storey systems with vertical resisting elements with an elastic-perfectly plastic force-displacement relationship, the centre of strength CS can be calculated as the centroid of the yield strength of the resisting elements.

More complex is the evaluation of the centre of strength in multistorey buildings. Although a direct reference to the strength centre is not explicitly made, in some studies the whole building is schematised by an equivalent one-storey model in which the properties of each resisting element are determined from the base shear force versus top displacement curves obtained by pushover analyses of the corresponding plane frames (Anagnostopoulos et al. 2010). Consequently, the strength centre can be determined as the centroid of the lateral strengths of these single-storey resisting elements. According to the authors, the heightwise distribution of the centres of strength CS can be obtained by a pushover analysis of the building forced to translate along one direction. At each step of the pushover analysis, the centre of strength at the *i*th storey is calculated as

$$x_{\mathrm{CS}}^{i}=\frac{\sum_{j=1}^{ny}V_{y,j}^{i}x_{j}}{\sum_{j=1}^{ny}V_{y,j}^{i}};\quad y_{\mathrm{CS}}^{i}=\frac{\sum_{j=1}^{nx}V_{x,j}^{i}y_{j}}{\sum_{j=1}^{nx}V_{x,j}^{i}} \tag{12.11}$$

where $V_{x,j}^{i}$ and $V_{y,j}^{i}$ are the storey shear force resisted at the *i*th storey by the *j*th frame arranged along the *x*- and *y*-directions, respectively. Note that the strength eccentricity at each storey varies at each step of the pushover analysis because of the progressive yielding of beams and columns. The first storey strength eccentricity corresponding to the top displacement demand u_{req} can be assumed as representative of the strength eccentricity of the whole building.

12.3.2 Ratio R_{μ} of the Elastic Strength Demand to the Actual Strength of the Torsionally Balanced System

The ratio R_{μ} is an index of the expected plastic deformations of the system during earthquakes (large values of R_{μ} correspond to large expected plastic deformation demands). For one-storey systems, R_{μ} is defined as

$$R_{\mu,x}=\frac{m\,S_{\mathrm{ae},\,x}}{S_{x}};\quad R_{\mu,y}=\frac{m\,S_{\mathrm{ae},\,y}}{S_{y}} \tag{12.12}$$

where m is the mass of the deck, S_{ae} is the spectral pseudo-acceleration of the corresponding balanced system and S is the total lateral strength of the resisting elements.

For multistorey structures, the factor R_{μ} can be evaluated for both directions as the ratio of the required elastic base shear to the base shear corresponding to the displacement demanded by the earthquake. The elastic base shear V_{el} can be calculated by a modal response spectrum analysis of the balanced system by means of the elastic response spectrum; the base shear corresponding to the displacement demand can be calculated by the pushover analysis of the building forced to translate along one direction (see Sect. 12.3.1).

12.4 Application of the Proposed Method to a Case Study

In this section the proposed nonlinear static method is applied to evaluate the nonlinear seismic response of a five-storey building. This is an L-shaped building with reinforced concrete framed structure. It is obtained by modifying the irregular building studied in the context of the Italian national research project ReLuis (Bosco et al. 2008b). The reinforced concrete frames are arranged along two orthogonal directions, and their structural members are characterised by equal cross sections at all the storeys. The mass centres are lined up along a vertical axis ($x_{CM} = 7.75$ m, $y_{CM} = 8.75$ m). The layout, the position of the centre of mass CM and the size of the beam and column cross sections are summarised in Fig. 12.2.

The storey height is equal to 3.5 m; the mass and the radius of gyration of the mass r_m with respect to the centre of mass are 250 t and 7.91 m (0.396 L), respectively. The building is founded on soft soil (soil C according to Eurocode 8).

The seismic response of this building is first obtained by nonlinear dynamic analysis. The seismic input is represented by a set of ten pairs of artificial accelerograms which act simultaneously along x- and y-directions. The accelerograms are generated by the SIMQKE computer program and are modelled by a trapezoidal envelope function with initial, central (stationary part) and final parts of 3, 22.5 and 5 s, respectively. The parameters for the calculation of the corrective eccentricities (e_r, e_s, Ω_θ and R_μ) are determined, and the seismic response of the structure is evaluated by the proposed nonlinear static procedure. Note that nonlinear static methods can lead to errors which are not negligible also for plane frames (Bosco et al. 2009b). When these methods are applied to in-plan asymmetric structures, additional errors are committed due to the inaccuracy in the estimate of the torsional response. The use of the corrective eccentricities proposed in this chapter aims at improving the sole prediction of the torsional response. Therefore, to eliminate the errors committed for plane systems, the displacement demand of the mass centre of the top storey of the building u_{req} is determined as the average of the ten maximum displacements evaluated by the nonlinear dynamic analysis of the corresponding balanced system (i.e. the planar system). Then, each pushover analysis is arrested when the displacement of the centre of mass of the asymmetric system is equal to this displacement.

Finally, the obtained seismic response is compared to that resulting from nonlinear dynamic analysis. Both the pushover and nonlinear dynamic analyses are performed by SAP2000 computer program.

12.4.1 Evaluation of e_r, Ω_θ, e_s and R_μ

The analysed building is not strictly regularly asymmetric and, thus, it does not have an elastic axis. To evaluate the rigidity eccentricities, in this chapter, the optimum torsion axis is assumed as a reference axis. Its position ($x_{CR} = 6.61$ m, $y_{CR} = 8.10$ m)

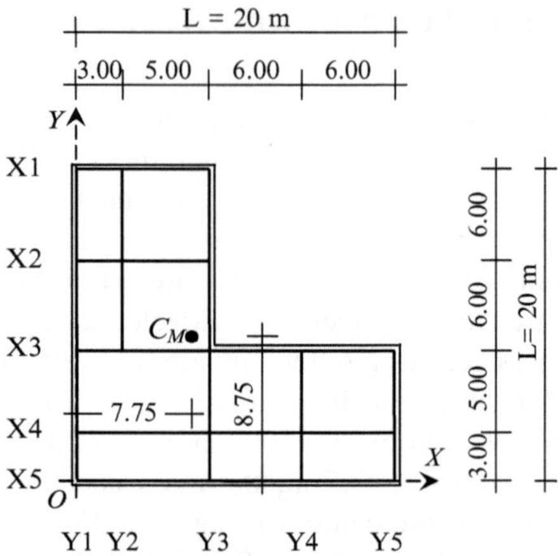

Beams parallel to the x-axis

Position	Y1 – Y2	Y2 – Y3	Y3 – Y4	Y4 – Y5
X1	40×50	40×50		
X2	40×50	40×50		
X4	40×60		40×50	40×50
X5	40×60		40×50	40×50

Beams parallel to the y-axis

Position	Y1	Y2	Y3	Y4	Y5
X1 – X2	50×25	50×25	50×25		
X2 – X3	50×25	50×25	50×25		
X3 – X4	40×50		40×50	50×25	50×25
X4 – X5	40×50		40×50	50×25	50×25

Columns

Position	Y1	Y2	Y3	Y4	Y5
X1	40×40	55×40	55×40		
X2	40×40	55×40	55×40		
X3	55×40	55×40	70×40	55×40	55×40
X4	55×40		70×40	55×40	55×40
X5	55×40		55×40	40×40	40×40

Fig. 12.2 Characteristics of the building: plan layout and member cross sections

Table 12.1 Elastic parameters of the analysed building

Storey	x_{CR} (m)	y_{CR} (m)	r_{kx} (m)	r_{ky} (m)
5	6.28	7.89	12.05	7.64
4	6.42	8.04	11.94	7.71
3	6.46	8.14	11.91	7.74
2	6.60	8.16	11.86	7.77
1	7.00	8.06	11.63	7.83
Average	6.56	8.08	11.87	7.75
Optimum torsion axis	6.61	8.10	11.84	7.77

is calculated, according to the procedure proposed by Marino and Rossi (2004), considering an inverted triangular distribution of forces. Further, to evaluate the heightwise irregularity of the building, the centre of rigidity is calculated at each storey as suggested by Bosco et al. (2008a). The coordinates of the centre of rigidity are listed in Table 12.1 and compared to those of the optimum torsion axis. Note that the rigidity centres, even though they are not strictly lined up along a vertical axis, present a small variation along the height and are close to the optimum torsion axis. The rigidity eccentricities are equal to $e_{rx} = -1.14$ m (about −0.05 L) and $e_{ry} = -0.65$ m (about −0.03 L). Similarly, almost uniform values of the radii of gyration of the lateral stiffness r_k are obtained from the procedure proposed by Bosco et al. (2008a); their average value is basically coincident with that obtained by Eq. 12.9 (see Table 12.1). Finally, the ratios $\Omega_{\theta x}$ and $\Omega_{\theta y}$ of the torsional to lateral frequencies of the corresponding torsionally balanced system are calculated as the ratios of the radii of gyration of the stiffness r_{kx} and r_{ky} to r_m and are equal to 1.50 and 0.98, respectively.

To evaluate the centre of strength, two pushover analyses are performed. The building is forced to translate along either the *x*- or *y*-direction. The pushover analysis is performed by applying a distribution of forces proportional to the first mode of vibration. At each step of the pushover analysis, the centre of strength at the *i*th storey is calculated by Eq. 12.11 and the strength eccentricities are calculated as

$$e^i_{sx} = x^i_{CS} - x_{CM};\, e^i_{sy} = y^i_{CS} - y_{CM} \tag{12.13}$$

The strength eccentricities calculated at each step of the pushover analysis are plotted versus the top displacement u_{top} in Fig. 12.3. Note that for small values of u_{top}, the strength eccentricity varies because of the progressive yielding of beams and columns. When the structure is pushed well into the plastic range of behaviour, the strength eccentricity attains a constant value: the centre of strength does not reach a stationary position only at the fifth storey of the building when the seismic forces are applied along the *x*-direction. This is because only a few elements of this storey are in the plastic range of behaviour when the top displacement is large. The first storey strength eccentricities corresponding to the top displacement demand u_{req} ($u_{req,y} = 147.62$ mm, $u_{req,x} = 112.34$ mm) are $e_{sx} = 0.174$ m (about 0.01 L) and $e_{sy} = -1.140$ m (about −0.06 L) for seismic forces acting along the *y*- and *x*-directions, respectively.

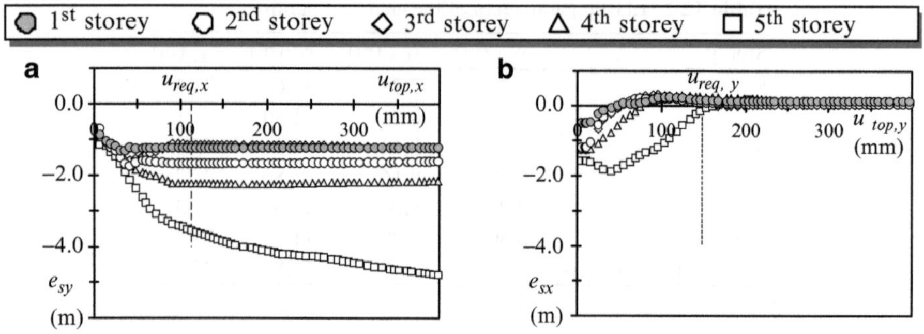

Fig. 12.3 Strength eccentricities at each step of the pushover analysis: seismic forces in (**a**) *x*-direction and (**b**) *y*-direction

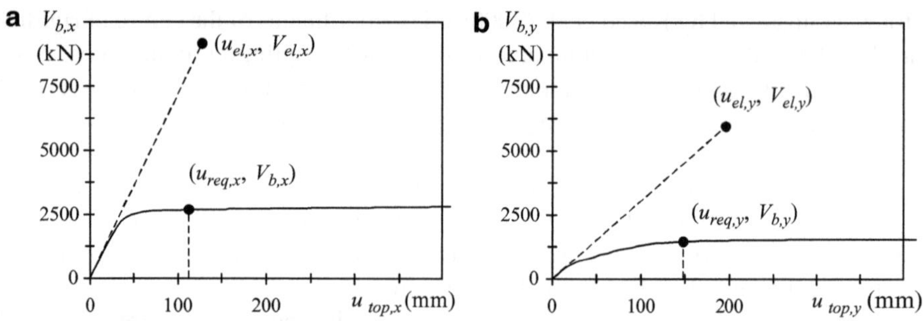

Fig. 12.4 Elastic and plastic required base shears: seismic forces in (**a**) *x*-direction and (**b**) *y*-direction

To evaluate the factor R_μ, the elastic and the plastic base shear forces of the balanced structure are evaluated (Fig. 12.4). The required elastic base shear is calculated in both directions ($V_{el,x} = 9{,}175$ kN, $V_{el,y} = 5{,}953$ kN) by modal response spectrum analysis of the balanced structure. The analysis is performed with reference to the elastic response spectrum stipulated in Eurocode 8 for soil C and characterised by a peak ground acceleration equal to 0.35 g. The plastic base shear corresponding to the displacement demand ($V_{b,x} = 2{,}679$ kN, $V_{b,y} = 1{,}464$ kN) is obtained by the pushover analysis of the building forced to translate along the considered direction. The obtained values of $R_{\mu,x}$ and $R_{\mu,y}$ are 3.42 and 4.07, respectively.

12.4.2 Prediction of the Seismic Response of the Structure

The seismic response of the asymmetric structure is calculated first by nonlinear dynamic analysis. Then, the structure is analysed by the proposed nonlinear static method. Based on the values of the parameters calculated in Sect. 12.4.1, Eqs. 12.1, 12.2, 12.3, 12.4 and 12.5 provide the corrective eccentricities which are

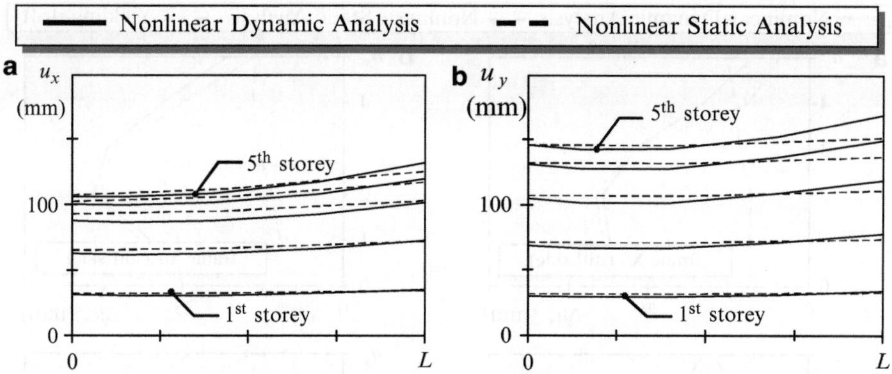

Fig. 12.5 Comparison between the in-plan distribution of absolute displacements provided by nonlinear dynamic analysis and by the proposed nonlinear method: seismic forces in (**a**) *x*-direction and (**b**) *y*-direction

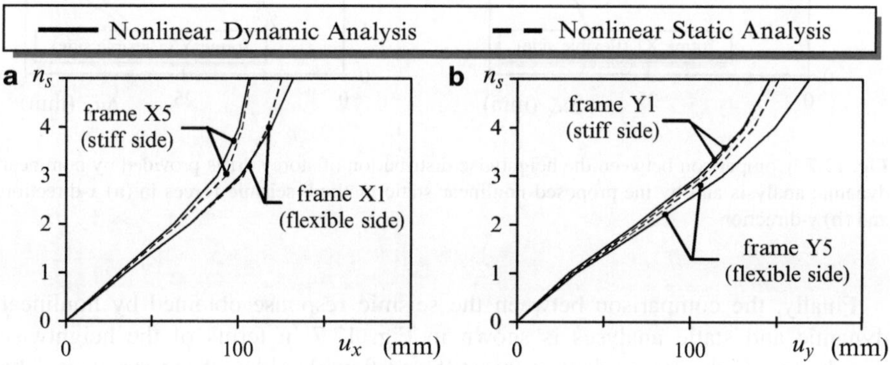

Fig. 12.6 Comparison between the heightwise distribution of absolute displacements provided by nonlinear dynamic analysis and by the proposed nonlinear static method

$e_{y1} = -1.056$ m (about −0.05 L), $e_{y2} = -0.729$ m (about −0.04 L), $e_{x1} = -0.011$ m (almost negligible) and $e_{x2} = 1.030$ m (0.05 L) for seismic forces along *x*- and *y*-directions, respectively.

The comparison between the seismic response obtained by nonlinear dynamic analysis and that predicted by the proposed nonlinear static method is shown in Figs. 12.5 and 12.6 in terms of the in-plan distribution of the deck displacements at each storey and the heightwise distribution of the displacements of the stiff and the flexible edges of the structure, respectively. Both the figures show that the results of nonlinear dynamic analysis are predicted well by the envelope of the deck displacements provided by the two pushover analyses. The maximum error is committed on the flexible side of the top storey for seismic forces applied along the *y*-direction and is about 10% (Figs. 12.5b and 12.6b).

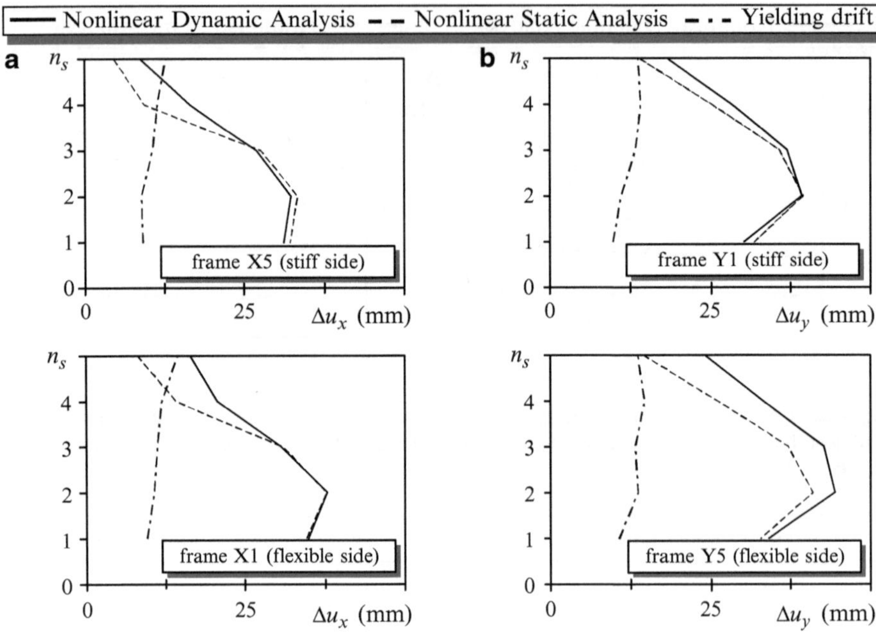

Fig. 12.7 Comparison between the heightwise distribution of storey drifts provided by nonlinear dynamic analysis and by the proposed nonlinear static method: seismic forces in (**a**) x-direction and (**b**) y-direction

Finally, the comparison between the seismic response obtained by nonlinear dynamic and static analyses is shown in Fig. 12.7 in terms of the heightwise distribution of the storey drift at the stiff and flexible sides of the structure. The storey drift along the x-direction is predicted accurately by the proposed method for both sides of the deck at the 1st, 2nd and 3rd storeys, while it is underestimated at the upper two storeys (Fig. 12.7a). The storey drifts along the y-direction of the flexible side are generally slightly underestimated (Fig. 12.7b).

The storey drifts corresponding to the yielding of each column of the outermost frames are calculated. These storey drifts are evaluated by considering an idealised bilinear relationship for the force-displacement curve obtained according to the procedure proposed in Bosco et al. (2008c). The axial force of the columns is assumed equal to that produced by the gravity load of the seismic combination. The heightwise distribution of the minimum value of the storey drift at yielding calculated for the columns of the stiff and the flexible sides of the analysed structure is plotted in Fig. 12.7. Note that storey drifts obtained by the proposed method are underestimated only at the storeys where the structure is almost entirely in the elastic range of behaviour.

12.5 Conclusions

This chapter describes a nonlinear static method for the evaluation of the seismic response of asymmetric systems and discusses about its application to asymmetric multistorey buildings. The method requires a double application of the nonlinear static analysis. In this chapter, the analysis of a five-storey L-shaped structure is presented as an example. The comparison between the in-plan distribution of deck displacements at each storey and storey drifts evaluated by the nonlinear dynamic analysis and those evaluated by the proposed method shows the effectiveness of the proposed method. To generalise this conclusion on the effectiveness of the corrective eccentricities, the analysis of a reasonably wide set of multistorey structures would be required.

References

Anagnostopoulos SA, Alexopoulou C, Stathopoulos KG (2010) An answer to an important controversy and the need for caution when using simple models to predict inelastic earthquake response of buildings with torsion. Earthq Eng Struct Dyn 39:521–540

Bosco M, Marino EM, Rossi PP (2008a) An analytical approach for the evaluation of the in-plan irregularity of real buildings. In: Proceedings of the 5th European workshop on the seismic behavior of Irregular and Complex Structures, Catania, Italy

Bosco M, Camata G, De Stefano M, Ghersi A, Lucchini A, Maddaloni G, Magliulo G, Marino I, Martinelli E, Monti G, Petti L, Saetta A, Spacone E, Trombetti T (2008b) Linee guida per le analisi nonlineari di edifici esistenti in cemento armato. In: Proceedings of the conference "Valutazione e riduzione della vulnerabilità sismica di edifici esistenti in c.a.", Rome, Italy (in Italian)

Bosco M, Ghersi A, Leanza S (2008c) Force-displacement relationships for R/C members in seismic design. In: Proceeding of the 14th world conference on earthquake engineering, Beijing, China

Bosco M, Ferrara G, Ghersi A, Marino EM (2009a) Corrective eccentricities for the evaluation of the seismic response of one-storey asymmetric systems by nonlinear static analysis. Environ Semeiot 2:80–94. doi:10.3383/es.2.2.2

Bosco M, Ghersi A, Marino EM (2009b) On the evaluation of seismic response of structures by nonlinear static methods. Earthq Eng Struct Dyn 38:1465–1482

Bosco M, Ghersi A, Marino EM (2012) Corrective eccentricities for assessment by the nonlinear static method of 3D structures subjected to bidirectional ground motions. Earthq Eng Struct Dyn 41:1751–1773

Chopra AK, Goel RK (2004) A modal pushover analysis procedure to estimate seismic demands for unsymmetric-plan buildings. Earthq Eng Struct Dyn 33:903–927

De Stefano M, Marino EM, Rossi PP (2006) Effect of overstrength on the seismic behaviour of multi-storey regularly asymmetric buildings. Bull Earthq Eng 4:23–42

De Stefano M, Pintucchi B (2010) Predicting torsion-induced lateral displacements for push-over analysis: influence of torsional system characteristics. Earthq Eng Struct Dyn 39:1369–1394

Fajfar P, Marušić D, Perus I (2005) Torsional effects in the pushover-based seismic analysis of buildings. J Earthq Eng 9:831–854

Ghersi A, Marino EM, Rossi PP (2007) Static versus modal analysis: influence on inelastic response of multi-storey asymmetric buildings. Bull Earthq Eng 5:511–532

Goel RK, Chopra AK (1990) Inelastic seismic response of one-story asymmetric-plan systems, Report no. UCB/EERC-90/14, Earthquake Engineering Research Center, Berkeley, CA, USA

Hejal R, Chopra AK (1987) Earthquake response of torsionally-coupled buildings, Report no. UBC/EERC-87/20, Earthquake Engineering Research Center, Berkeley, CA, USA

Lucchini A, Monti G, Kunnath S (2009) Seismic behavior of single-story asymmetric-plan buildings under uniaxial excitation. Earthq Eng Struct Dyn 38:1053–1070

Makarios T, Anastassiadis K (1998a) Real and fictitious elastic axes of multi-storey buildings: theory. Struct Des Tall Spec Build 7:33–55

Makarios T, Anastassiadis K (1998b) Real and fictitious elastic axes of multi-storey buildings: applications. Struct Des Tall Spec Build 7:57–71

Marino EM, Rossi PP (2004) Exact evaluation of the location of the optimum torsion axis. Struct Des Tall Spec Build 13:277–290

Marušić D, Fajfar P (2005) On the inelastic seismic response of asymmetric buildings under bi-axial excitation. Earthq Eng Struct Dyn 34:943–963

Tso WK (1990) Static eccentricity concept for torsional moment estimations. J Struct Eng 116:1199–1212

Chapter 13
A Multimode Pushover Procedure for Asymmetric Buildings Under Biaxial Seismic Excitation

Grigorios E. Manoukas, Asimina M. Athanatopoulou, and Ioannis E. Avramidis

Abstract In this chapter, a new multimode pushover procedure is presented in order to achieve an approximate estimation of structural performance of asymmetric buildings under biaxial seismic excitation. The steps of the proposed methodology are quite similar to those of the well-known modal pushover analysis. However, the properties of the equivalent single-degree-of-freedom (E-SDOF) systems are modified properly, in order to take into account multidirectional seismic effects. The main scope of this methodology is to avoid additional errors introduced by simplified superposition formulas adopted by codes. In addition, a reduction of the computational cost as compared to other multimode procedures is achieved. After a brief outline of the theoretical background, the sequence of steps needed for the implementation of the proposed methodology is presented. Finally, the accuracy of the proposed methodology is evaluated by an extensive parametric study, which shows that, in general, it provides a useful and conservative design tool.

13.1 Introduction

Static pushover analysis (SPA) is a widely accepted procedure for the approximate estimation of the inelastic performance of buildings under strong earthquake ground motions. In the last decades, a series of more or less similar variants of the procedure have been developed, some of which have been already adopted by several seismic codes and prestandards (American Society of Civil Engineers 2008; Applied Technology Council (ATC) 1996; European Committee for Standardization 2004, etc.). Engineering practitioners often prefer to apply SPA in order to avoid the significant computational cost and the various inherent disadvantages of a more

G.E. Manoukas • A.M. Athanatopoulou (✉) • I.E. Avramidis
Aristotle University of Thessaloniki, University Campus, 54124 Thessaloniki, Macedonia, Greece
e-mail: grman7@otenet.gr; minak@civil.auth.gr; avram@civil.auth.gr

O. Lavan and M. De Stefano (eds.), *Seismic Behaviour and Design of Irregular and Complex Civil Structures*, Geotechnical, Geological and Earthquake Engineering 24, DOI 10.1007/978-94-007-5377-8_13,

accurate inelastic dynamic analysis. Thus, SPA, or nonlinear static procedure (NSP) as it is named in seismic codes, became a very popular and useful tool for the earthquake-resistant design of new as well as the seismic rehabilitation of existing buildings.

However, as it has already been stressed by many researchers (e.g., Krawinkler and Seneviratna 1998; Goel and Chopra 2004), this procedure involves many shortcomings and can provide reasonable results only for low- and medium-rise planar systems. This is mainly due to the fact that the determination of the structure's response is based on the assumption that its dynamic behavior depends only on a single elastic vibration mode. In addition, this elastic mode is supposed to remain constant despite the successive formation of plastic hinges during the seismic excitation. Also, the choice of roof displacement as reference displacement for the determination of the pushover curve is arbitrary, and it is doubtful whether the capacity curve is the most meaningful index of the nonlinear response of a structure, especially for irregular and spatial systems. Therefore, various advanced pushover procedures have been proposed to overcome some of these shortcomings, e.g., modal pushover analysis (MPA) (Chopra and Goel 2001).

In the special case of asymmetric buildings under biaxial seismic excitation, additional errors are introduced due to the application of simplified formulas adopted by codes for taking into account multidirectional seismic effects (e.g., American Society of Civil Engineers 2008, Section 3.2.7; European Committee for Standardization 2004, Section 4.3.3.5.1(6)). These formulas are based on the superposition principle, although this approach lacks a theoretical basis in the domain of inelastic response. In the literature, only few investigations concerning this issue can be found. Some researchers (e.g., Fajfar et al. 2005) apply SPA independently in two horizontal directions and use one of the aforementioned formulas to combine the action effects, while others utilize more complicated concepts (e.g., Fujii 2007; Lin and Tsai 2008).

The objective of this chapter is the presentation and preliminary evaluation of a new multimode pushover procedure for the approximate estimation of the seismic response of asymmetric buildings under biaxial excitation. Its main idea is that the seismic response of an asymmetric multi-degree-of-freedom (MDOF) system with N degrees of freedom under biaxial excitation can be related to the responses of N "modal" equivalent single-degree-of-freedom (E-SDOF) systems under uniaxial excitation. The whole procedure is quite similar to the well-known MPA (Chopra and Goel 2001) as extended for asymmetric buildings (Chopra and Goel 2004). However, the properties of the E-SDOF systems are modified properly, in order to take into account multidirectional seismic effects. The main scope of this methodology is to avoid additional errors introduced by applying simplified superposition formulas. Furthermore, a reduction of the computational cost as compared to other multimode procedures is achieved.

First, the theoretical background and the assumptions of the proposed methodology are presented and briefly discussed. Second, the steps to be followed for its implementation are systematically presented. The accuracy of the proposed methodology is evaluated by an extensive parametric study. The whole investigation shows that, in general, the proposed methodology provides a useful and conservative alternative for the simplified inelastic seismic analysis of asymmetric buildings. Finally, this chapter closes with comments on results and conclusions.

13.2 Theoretical Background

The linear elastic response of an L-story MDOF system with N degrees of freedom (in the usual case of rigid diaphragms $N = 3L$) to independent uniaxial horizontal seismic excitations $\ddot{u}_g(t)_X$ and $\ddot{u}_g(t)_Y$ along X- and Y-axes, respectively, is governed by the following equations:

$$\boldsymbol{M}\ddot{\boldsymbol{u}}(t)_{,X} + \boldsymbol{C}\dot{\boldsymbol{u}}(t)_{,X} + \boldsymbol{K}\boldsymbol{u}(t)_{,X} = -\boldsymbol{M}\boldsymbol{\delta}_{,X}\ddot{u}_g(t)_{,X} \tag{13.1}$$

$$\boldsymbol{M}\ddot{\boldsymbol{u}}(t)_{,Y} + \boldsymbol{C}\dot{\boldsymbol{u}}(t)_{,Y} + \boldsymbol{K}\boldsymbol{u}(t)_{,Y} = -\boldsymbol{M}\boldsymbol{\delta}_{,Y}\ddot{u}_g(t)_{,Y} \tag{13.2}$$

where $\boldsymbol{u}(t)_{,X}$, $\boldsymbol{u}(t)_{,Y}$ are the displacement vectors of the N degrees of freedom (translations or rotations) relative to the ground due to $\ddot{u}_g(t)_X$ and $\ddot{u}_g(t)_Y$, respectively; $\boldsymbol{M}$ is the $N \times N$ diagonal mass matrix; $\boldsymbol{C}$ and $\boldsymbol{K}$ are the $N \times N$ symmetric damping and stiffness matrices, respectively; and $\boldsymbol{\delta}_{,X}$, $\boldsymbol{\delta}_{,Y}$ are the influence vectors that describe the influence of support displacements on the structural displacements. The $\boldsymbol{u}_{,X}$ and $\boldsymbol{u}_{,Y}$ vectors (for the sake of simplicity, (t) is left out in all following expressions) are given by the following equations:

$$\boldsymbol{u}_{,X} = [\boldsymbol{u}_{X,X}\boldsymbol{u}_{Y,X}\boldsymbol{\theta}_{z,X}]^{\mathrm{T}} \tag{13.3}$$

$$\boldsymbol{u}_{,Y} = [\boldsymbol{u}_{X,Y}\boldsymbol{u}_{Y,Y}\boldsymbol{\theta}_{z,Y}]^{\mathrm{T}} \tag{13.4}$$

where $\boldsymbol{u}_{X,X}$, $\boldsymbol{u}_{Y,X}$, $\boldsymbol{\theta}_{z,X}$ are the $L \times 1$ vectors of displacements along X-axis, displacements along Y-axis, and rotations around Z (vertical)-axis, respectively, due to $\ddot{u}_{gX}$, while $\boldsymbol{u}_{X,Y}$, $\boldsymbol{u}_{Y,Y}$, $\boldsymbol{\theta}_{z,Y}$ are the $L \times 1$ vectors of displacements along X-axis, displacements along Y-axis, and rotations around Z (vertical)-axis, respectively, due to $\ddot{u}_{gY}$. The influence vectors $\boldsymbol{\delta}_{,X}$ and $\boldsymbol{\delta}_{,Y}$ are given by the following relations:

$$\boldsymbol{\delta}_{,X} = \begin{bmatrix} \boldsymbol{I} \; 0 \; 0 \end{bmatrix}^{\mathrm{T}} \tag{13.5}$$

$$\boldsymbol{\delta}_{,Y} = \begin{bmatrix} 0 \; \boldsymbol{I} \; 0 \end{bmatrix}^{\mathrm{T}} \tag{13.6}$$

where $\boldsymbol{I}$, 0 are $L \times 1$ vectors with terms equal to unity and zero, respectively. According to the superposition principle, the linear elastic response of the MDOF system to a biaxial seismic excitation $\ddot{u}_{gX}$ along X-axis and $\ddot{u}_{gY}$ along Y-axis is governed by the following equation:

$$\boldsymbol{M}\ddot{\boldsymbol{u}} + \boldsymbol{C}\dot{\boldsymbol{u}} + \boldsymbol{K}\boldsymbol{u} = -\boldsymbol{M}\boldsymbol{\delta}_{,X}\ddot{u}_{gX} - \boldsymbol{M}\boldsymbol{\delta}_{,Y}\ddot{u}_{gY} \tag{13.7}$$

where:

$$\boldsymbol{u} = \boldsymbol{u}_{,X} + \boldsymbol{u}_{,Y} \tag{13.8}$$

$$\dot{\boldsymbol{u}} = \dot{\boldsymbol{u}}_{,X} + \dot{\boldsymbol{u}}_{,Y} \tag{13.9}$$

$$\ddot{\boldsymbol{u}} = \ddot{\boldsymbol{u}}_{,X} + \ddot{\boldsymbol{u}}_{,Y} \tag{13.10}$$

Taking into account the assumption that the two components of the biaxial seismic excitation $\ddot{u}_{gX}$ and $\ddot{u}_{gY}$ are proportional to each other ($\ddot{u}_{gY} = \kappa \ddot{u}_{gX} = \kappa \ddot{u}_g$) and applying well-known principles of structural dynamics (modal analysis) N uncoupled equations describing the response of N E-SDOF systems, each one corresponding to an elastic vibration mode can be derived (Manoukas 2010):

$$M^*_{XYi}\, \ddot{D}_i + 2\, M^*_{XYi}\, \omega_i \zeta_i \dot{D}_i + V_{XYi} = -\, M^*_{XYi}\, \ddot{u}_g \tag{13.11}$$

where ω_i, ζ_i are the natural frequency and the damping ratio of mode i, respectively; D_i is the displacement of the corresponding E-SDOF system; V_{XYi} is the sum of modal base shear parallel to X-axis (V_{Xi}) plus modal base shear parallel to Y-axis multiplied by κ (κV_{Yi}); and M^*_{XYi} is given by the following equation:

$$M^*_{XYi} = M^*_{Xi} + \kappa(\nu_{Xi} L_{Yi} + \nu_{Yi} L_{Xi}) + \kappa^2 M^*_{Yi} \tag{13.12}$$

where M^*_{Xi}, M^*_{Yi} and ν_{Xi}, ν_{Yi} are the effective modal masses and the modal participation factors of vibration mode i due to independent uniaxial excitations along X- and Y-axes, respectively, while L_{Xi}, L_{Yi} are given by the following expressions ($\boldsymbol{\varphi}_i$ is the modal vector of mode i):

$$L_{Xi} = \boldsymbol{\delta}_{,X}{}^{\mathrm{T}} \boldsymbol{M} \boldsymbol{\varphi}_i \tag{13.13}$$

$$L_{Yi} = \boldsymbol{\delta}_{,Y}{}^{\mathrm{T}} \boldsymbol{M} \boldsymbol{\varphi}_i \tag{13.14}$$

Furthermore, it can be proved that D_i can be correlated to the displacement of the Nth degree of freedom u_{Ni} as follows:

$$D_i = \frac{u_{Ni}}{\nu_{XYi}\varphi_{Ni}} \tag{13.15}$$

where $\boldsymbol{\varphi}_{Ni}$ is the Nth term of $\boldsymbol{\varphi}_i$ and:

$$\nu_{XYi} = \nu_{Xi} + \kappa \nu_{Yi} \tag{13.16}$$

Equation 13.11 shows that the linear elastic response of a MDOF system with N degrees of freedom subjected to a biaxial seismic excitation $\ddot{u}_{gX}$ and

$\ddot{u}_{gY} = \kappa \ddot{u}_{gX} = \kappa \ddot{u}_g$ along X- and Y-axes, respectively, can be expressed as superposition of the responses of N SDOF systems under uniaxial excitation $\ddot{u}_g$, each one corresponding to a vibration mode having mass equal to M^*_{XYi}, displacement equal to D_i, and elastic resisting force equal to V_{XYi}.

In the inelastic range of behavior, some basic assumptions have to be made, keeping always in mind that our main intention is the development of an approximate, simplified procedure. A major assumption is that the response of a MDOF system can be expressed as superposition of the responses of appropriate SDOF systems just like in the linear range. Of course, such an assumption violates the very logic of nonlinearity, as the superposition principle does not apply in nonlinear systems. However, it must be thought as a fundamental postulate, which constitutes the basis on which many simplified pushover procedures are built. Thus, each SDOF system corresponds to a vibration "mode" i with "modal" vector $\boldsymbol{\varphi}_i$ (the quotation marks indicate that the application of the superposition principle is not strictly valid). The displacements $\boldsymbol{u}_i$ and the inelastic resisting forces $\boldsymbol{F}_{si}$ are supposed to be proportional to $\boldsymbol{\varphi}_i$ and $\boldsymbol{M\varphi}_i$, respectively. Furthermore, "modal" vectors $\boldsymbol{\varphi}_i$ are supposed to be constant, despite the successive development of plastic hinges. Also, it is supposed that Rayleigh damping is present. Obviously, Eqs. 13.8, 13.9, and 13.10 are not valid anymore; however, it is reasonable to assume that the effective earthquake forces are given by the right side of Eq. 13.7. Thus, the nonlinear response of a MDOF system with N degrees of freedom to a biaxial earthquake ground motion ($\ddot{u}_{gX}$ *and* $\ddot{u}_{gY} = \kappa \ddot{u}_{gX} = \kappa \ddot{u}_g$) may be described by the following equation:

$$\boldsymbol{M}\ddot{\boldsymbol{u}} + \boldsymbol{C}\dot{\boldsymbol{u}} + \boldsymbol{F}_s = -\boldsymbol{M}\boldsymbol{\delta}_{,X}\ddot{u}_{gX} - \boldsymbol{M}\boldsymbol{\delta}_{,Y}\ddot{u}_{gY} \tag{13.17}$$

The only difference between Eqs. 13.17 and 13.7 is that the resisting forces F_s (i.e., the forces that would have to be applied to the structure in order to obtain displacements $\boldsymbol{u}$) cannot be expressed as linear functions of the displacements $\boldsymbol{u}$, because the coefficients of stiffness matrix $\boldsymbol{K}$ do not remain constant during the loading process. However, due to the aforementioned assumptions, modal analysis is applicable even for the inelastic range of behavior, and it can be proved that Eq. 13.11 is still valid.

13.3 The Proposed Methodology

From the analysis presented above, it becomes clear that – due to some basic assumptions – the nonlinear response of an asymmetric building under biaxial seismic excitation can be decomposed into a number of responses of "modal" E-SDOF systems under uniaxial excitation, each one corresponding to a vibration mode. These assumptions are nearly identical to those of MPA. The only additional assumption introduced is that the two horizontal seismic components are proportional to each other. The proposed methodology is based on the aforementioned concept,

Table 13.1 Properties of the ith E-SDOF system

Property	Uniaxial excitation $\ddot{u}_{gX}$	Biaxial excitation $\ddot{u}_{gX} + \kappa\ddot{u}_{gY}$
Mass	M^*_{Xi}	$M^*_{XYi} = M^*_{Xi} + \kappa(\nu_{Xi}L_{Yi} + \nu_{Yi}L_{Xi}) + \kappa^2 M^*_{Yi}$
Resisting force	V_{Xi}	$V_{XYi} = V_{Xi} + \kappa\nu_{Yi}$
Displacement	$D_i = u_{Ni}/\nu_{Xi}\varphi_{Ni}$ (roof displacement u_{Ni})	$D_i = u_{Ni}/\nu_{XYi}\varphi_{Ni} = u_{Ni}/(\nu_{Xi} + \kappa\nu_{Yi})\varphi_{Ni}$ (roof displacement u_{Ni})
Damping factor	$2\,M^*_{Xi}\,\omega_i\zeta_i$	$2\,M^*_{XYi}\,\omega_i\zeta_i$

and its application process resembles the one of MPA. However, the definition of the E-SDOF systems is quite different, for the purpose of taking into account multidirectional seismic effects. In Table 13.1, the properties of the ith "modal" E-SDOF system are tabulated, along with the properties that it would have in case of uniaxial excitation (parallel to X-axis).

The proposed methodology should be implemented for all possible combinations of the seismic components. In particular, the following four combinations should be examined:

$$\ddot{u}_{gX} + \kappa\ddot{u}_{gY} \tag{13.18}$$

$$\ddot{u}_{gX} - \kappa\ddot{u}_{gY} \tag{13.19}$$

$$\ddot{u}_{gY} + \kappa\ddot{u}_{gX} \tag{13.20}$$

$$\ddot{u}_{gY} - \kappa\ddot{u}_{gX} \tag{13.21}$$

where κ may be taken equal to 0.3 in consistence with relevant code provisions (e.g., American Society of Civil Engineers 2008, Section 3.2.7.1). The equations derived by the process presented in the previous paragraphs have to be modified proportionately. It can be easily proved – by simple implementation of the process – that the consideration of the four combinations with opposite sign (e.g., $-\ddot{u}_{gX} - \kappa\ddot{u}_{gY}$ instead of $\ddot{u}_{gX} + \kappa\ddot{u}_{gY}$) leads to identical properties for the E-SDOF systems, so they can be skipped.

The steps needed for the implementation of the proposed methodology are as follows:

Step 1: Create the structural model.

Step 2: Calculate ν_{XY1} (Eq. 13.16) and M^*_{XY1} (Eq. 13.12) of the fundamental elastic vibration mode 1 for the first combination of seismic components ($\ddot{u}_{gX} + \kappa\ddot{u}_{gY}$).

Step 3: Apply to the structural model a set of lateral incremental forces (and moments) proportional to the vector $\boldsymbol{M\varphi}_1$ of the fundamental elastic vibration mode 1 and determine the (resisting force)-(displacement) curve $V_{XY1} - u_{N1}$ of the MDOF system. u_{N1} can be chosen to correspond to any degree of freedom, but usually the roof displacement parallel to X- or Y-axis is used.

Step 4: Divide the abscissas of the $V_{XY1} - u_{N1}$ diagram by the quantity $\nu_{XY1}\varphi_{N1} = u_{N1}/D_1$ and determine the (resisting force)-(displacement) curve $V_{XY1} - D_1$ of the E-SDOF system.

Step 5: Idealize $V_{XY1} - D_1$ to a bilinear curve using one of the well-known graphic procedures (e.g., American Society of Civil Engineers 2008, Section 3.3.3.2.5) and calculate the period T_1 and the yield strength reduction factor R_1 of the E-SDOF system corresponding to mode 1.

Step 6: Calculate the target displacement of mode 1 using one of the well-known procedures of displacement modification (e.g., American Society of Civil Engineers 2008, Section 3.3.3.3.2; Federal Emergency Management Agency – Applied Technology Council (ATC) 2004, Section 10.4). If the procedure is applied for research purposes using recorded earthquake ground motions, it is recommended to estimate the inelastic displacement of the E-SDOF system by means of nonlinear dynamic analysis, instead of using the relevant coefficients given in various official documents (e.g., C_1 in ASCE 41-06 and FEMA 440). This is due to the fact that the coefficient values given in such documents are based on statistical processing of data with excessive deviations and, therefore, large inaccuracies might result (Manoukas et al. 2006).

Step 7: Calculate the "modal" values of the other response quantities of interest (drifts, plastic rotations, etc.) of mode 1 by conducting pushover analysis up to the already calculated target displacement.

Step 8: Repeat steps 3–7 applying the incremental forces (and moments) in the opposite direction.

Step 9: Repeat steps 2–8 for an adequate number of modes.

Step 10: Calculate the extreme values of response parameters by utilizing one of the well-established formulas of modal superposition (SRSS or CQC).

Step 11: Repeat steps 2–10 for all possible combinations of the two horizontal components of the seismic excitation.

In general, if n is the number of modes taken into account, $2n$ pushover analyses have to be implemented (step 3) for the two possible directions of the applied lateral loads (according to step 8). Also, $2n$ target displacements (steps 4–6) and $2n$ "modal" values (step 7) of response parameters have to be calculated for each combination examined. Finally, 2^n extreme values (step 10) of response parameters are produced for each combination (2^{n+2} extreme values for all combinations). It is worth noticing that when a multimode pushover procedure (e.g., MPA) is applied independently in two orthogonal directions and the multidirectional seismic effects are taken into account by using SRSS formula or ASCE 41-06 (Sect. 3.2.7) provision, the number of resulting extreme values of the response parameters is 2^{2n} or 2^{2n+1}, respectively. It is obvious that with increasing n, ($n \geq 2$), a significant reduction of the computational cost is achieved by using the proposed procedure (2^{n+2} extreme values instead of 2^{2n} or 2^{2n+1}).

13.4 Evaluation of the Proposed Methodology

As a first step for the evaluation of the accuracy of the proposed methodology, an extensive parametric study is carried out. In particular, the methodology is applied to five single-story asymmetric R/C buildings (Fig. 13.1 and Tables 13.2 and 13.3) with different values of double normalized eccentricities $e/r = e_X/r = e_Y/r$ (where e_X, e_Y are the distances between center of mass CM and center of rigidity CR and r is the radius of gyration) ranging between 0.10 and 0.50. The modeling of the inelastic behavior of the buildings is based on the following assumptions:

1. Shear failure is precluded.
2. The inelastic deformations are concentrated at the bases of the columns (plastic hinges).
3. Strength and stiffness along the weak side of column sections are neglected.
4. Plastic hinges are modeled by bilinear elastic-perfectly plastic (moments)-(rotations) diagrams.
5. The (bending moment)-(axial force) interaction is neglected.

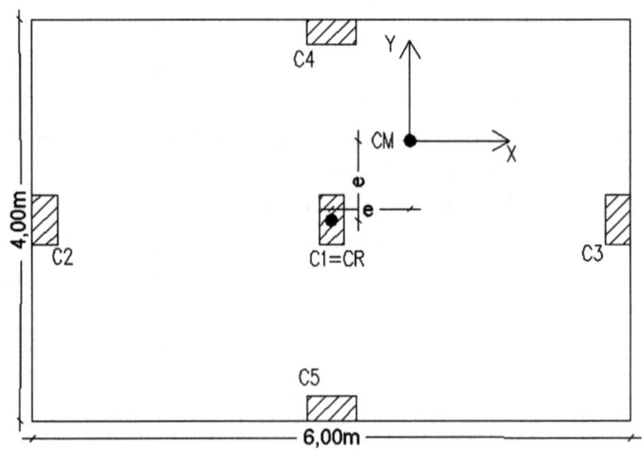

Fig. 13.1 Plan of single-story R/C asymmetric buildings

Table 13.2 Data of single-story asymmetric R/C buildings

Story height (m)	3
Column cross sections (cm)	25/50
Restraints	Columns fixed at base
Constraints	Diaphragm at roof level
Seismic mass (t)	36
Mass moment of inertia (tm^2)	156
Damping ratio (%)	5
Gravity loads	Not considered
Concrete	C16/20 ($f_{ck} = 16$ MPa)
Available plastic rotations (rad)	0.04
Yield moments (kNm)	30 (C1, C2, C3) or 50 (C4, C5)

Table 13.3 Dynamic characteristics of single-story asymmetric R/C buildings

Normalized eccentricity e/r	Periods (s) T_1 / T_2	Modal vectors $\varphi_1{}^T = [u_{x1}, u_{y1}, \theta_{z1}]$ / $\varphi_2{}^T = [u_{x2}, u_{y2}, \theta_{z2}]$	PM+ ν_{xy1} / ν_{xy2}	PM+ $M_{xy1}{}^*(t)$ / $M_{xy2}{}^*(t)$	PM− ν_{xy1} / ν_{xy2}	PM− $M_{xy1}{}^*(t)$ / $M_{xy2}{}^*(t)$
0.10	0.302	[1, −0.012, −0.019]	0.986	35.08	1.010	36.80
	0.247	[0.015, 1, 0.035]	1.009	36.87	−0.980	34.75
0.20	0.304	[1, −0.047, −0.039]	0.944	32.37	1.038	39.15
	0.250	[0.058, 1, 0.065]	1.036	39.48	−0.922	31.24
0.30	0.308	[1, −0.105, −0.060]	0.872	28.09	1.076	42.81
	0.253	[0.127, 1, 0.086]	1.076	43.66	−0.833	26.16
0.40	0.313	[1, −0.180, −0.082]	0.772	22.77	1.112	47.25
	0.257	[0.215, 1, 0.098]	1.117	48.88	−0.722	20.39
0.50	0.320	[1, −0.267, −0.104]	0.656	17.32	1.133	51.67
	0.261	[0.313, 1, 0.102]	1.148	54.26	−0.601	14.88

Table 13.4 List of seismic excitations

Excitation	Date	Magnitude (Ms)	PGA (m/s^2)	PSA (m/s^2)
Aeghio (longitudinal)	06/15/1995	6.4	4.918	12.099
Aeghio (transverse)			5.326	14.157
Thessaloniki (longitudinal)	06/20/1978	6.5	1.389	4.477
Thessaloniki (transverse)			1.430	4.809
Alkyonides (longitudinal)	02/24/1981	6.7	2.336	6.023
Alkyonides (transverse)			2.989	8.155
Kalamata (longitudinal)	09/13/1986	6.0	2.170	6.648
Kalamata (transverse)			2.913	10.125
Patras (longitudinal)	07/14/1993	5.5	1.402	4.455
Patras (transverse)			3.936	12.151
Pirgos (longitudinal)	03/26/1993	5.5	1.466	5.887
Pirgos (transverse)			4.455	7.705

The predefined values of normalized eccentricities are achieved by proper selection of the CM position. Of course, these simple buildings are not representative of all types of structural systems. However, such models are widely used in literature and seem to be appropriate for studying the response of asymmetric buildings (De la Llera and Chopra 1996). The whole investigation comprises 12 accelerograms recorded in Greece (Table 13.4). It is considered that each ground motion acts simultaneously along the two horizontal axes with the same intensity, i.e., $\kappa = 1$ and $\ddot{u}_{gX} = \ddot{u}_{gY}$. As a consequence, the possible combinations of the seismic components are only two: $\ddot{u}_{gX} + \ddot{u}_{gY}$ (PM+) and $\ddot{u}_{gX} - \ddot{u}_{gY}$ (PM−). For each building, two sets of pushover analyses are performed: (1) one based on the proposed methodology (PM) and (2) a second similar to MPA (conventional procedure – CP), which comprises pushover analyses of the examined buildings for independent uniaxial excitations along X- and Y-axes and linear addition of the response quantities. The assumptions and steps of the CP are nearly identical to those of the PM, except that

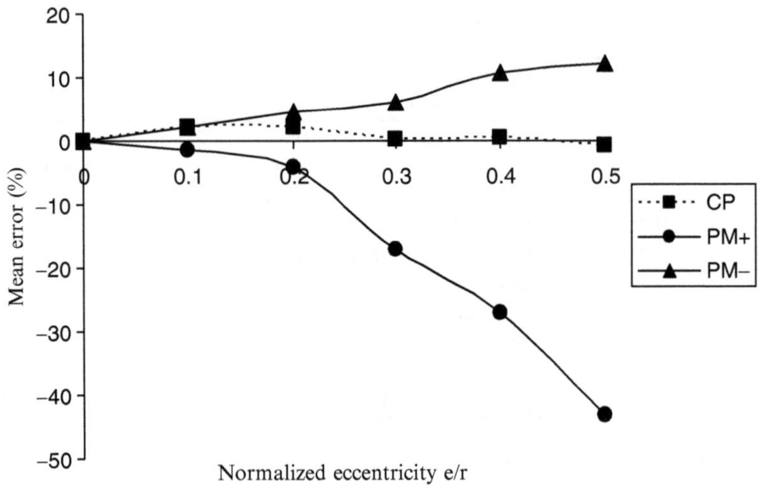

Fig. 13.2 Mean errors (%) of CM displacements along X-axis

step 11 is obviously skipped and in steps 2–4, ν_{Xi}, M^*_{Xi}, V_{Xi} or ν_{Yi}, M^*_{Yi}, V_{Yi} are used in place of ν_{XYi}, M^*_{XYi}, V_{XYi}. In both sets of pushover analyses, the first two vibration modes are taken into account and the "modal" superposition is conducted by applying the SRSS formula. It is worth noticing that the applied load pattern for each mode comprises lateral forces along X- and Y-axes as well as torsional moment around Z (vertical)-axis. The maximum "modal" response of each E-SDOF system is calculated by means of nonlinear dynamic analysis for each excitation. Then, the target displacement is estimated by multiplication of the resulting response by the quantity $\nu_{XYi}\boldsymbol{\varphi}_{Ni}$ (PM) and $\nu_{Xi}\boldsymbol{\varphi}_{Ni}$ or $\nu_{Yi}\boldsymbol{\varphi}_{Ni}$ (CP).

The response quantities of the buildings are compared to those obtained by nonlinear response history analysis (NL-RHA), which is considered as the reference solution. Each accelerogram is considered acting simultaneously along the two horizontal axes in all possible combinations ($\ddot{u}_{gX} + \ddot{u}_{gY}$, $\ddot{u}_{gX} - \ddot{u}_{gY}$, $-\ddot{u}_{gX} + \ddot{u}_{gY}$, and $-\ddot{u}_{gX} - \ddot{u}_{gY}$). In Figs. 13.2, 13.3, 13.4, 13.5, 13.6 and 13.7, the mean errors for the 12 excitations (with regard to the maximum values obtained by NL-RHA) of displacements of characteristic points (CM, C4, and C5 along X-axis and CM, C2, and C3 along Y-axis) are shown. Notice that the positive sign (+) means that response parameters obtained by PM or CP are greater than those obtained by NL-RHA. Conversely, the negative sign (−) means that the response parameters are underestimated. It is obvious that PM provides an upper-bound and a lower-bound value for each response parameter. The exact value (NL-RHA) in most cases (29 of 30) lies in this range. The mean error of the more conservative combination (PM + or PM−) does not exceed 13%. On the other hand, CP in many cases results in nonconservative response values (negative errors).

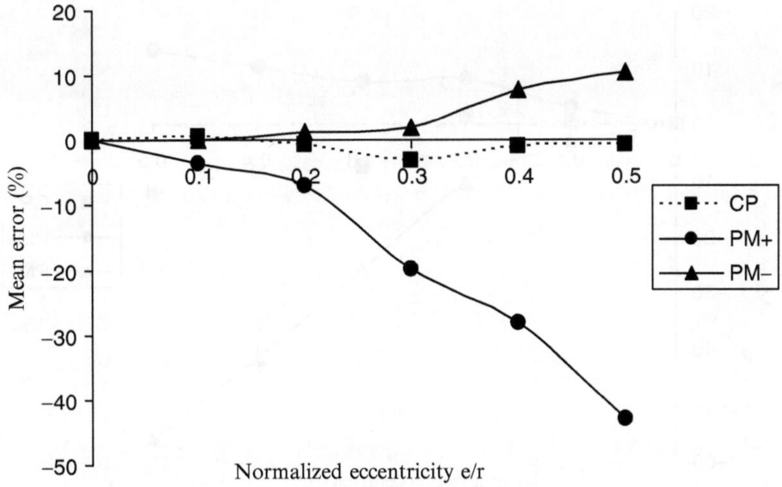

Fig. 13.3 Mean errors (%) of C4 displacements along *X*-axis

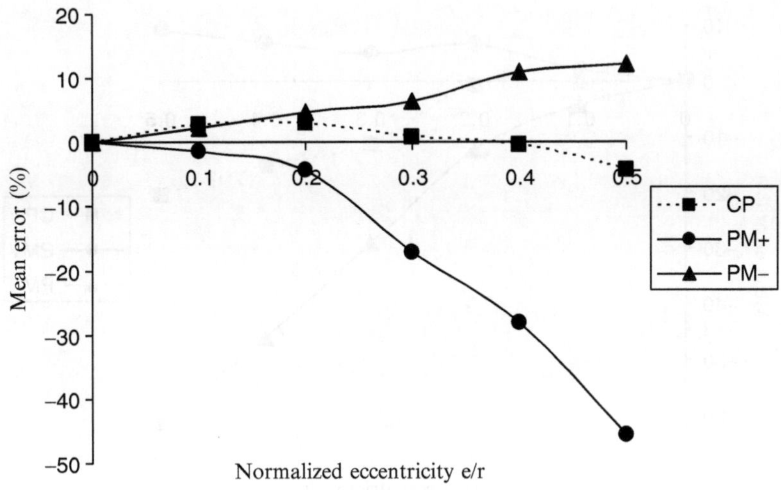

Fig. 13.4 Mean errors (%) of C5 displacements along *X*-axis

13.5 Conclusions

A new multimode pushover procedure for the approximate estimation of the seismic response of asymmetric buildings under biaxial excitation is presented in this chapter. Its main idea is that the seismic response of an asymmetric building under biaxial excitation can be related to the responses of a series of "modal" E-SDOF systems under uniaxial excitation. The whole procedure is quite similar to the

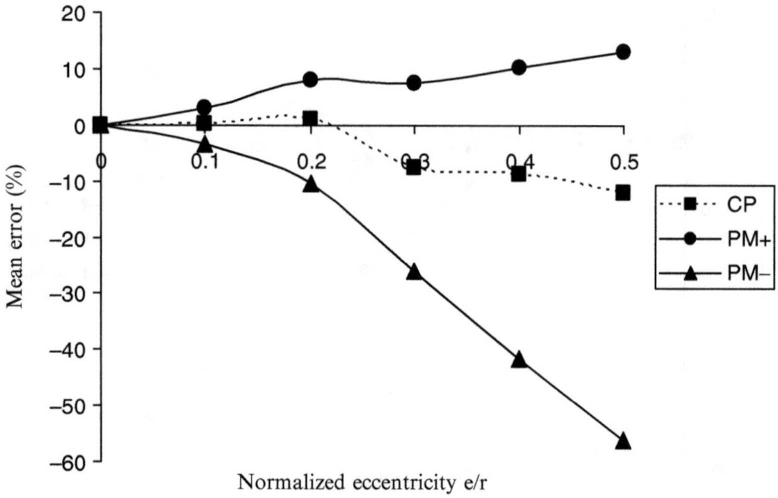

Fig. 13.5 Mean errors (%) of CM displacements along *Y*-axis

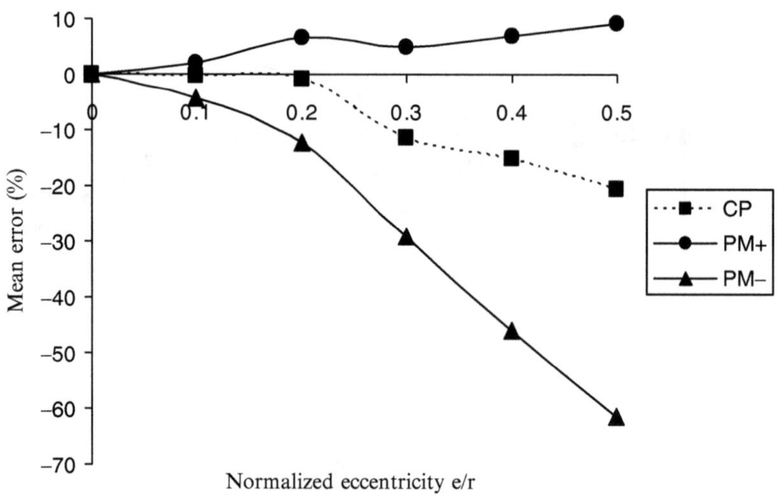

Fig. 13.6 Mean errors (%) of C2 displacements along *Y*-axis

well-known MPA. However, the properties of the E-SDOF systems are modified properly in order to take into account multidirectional seismic effects. From the presentation and the evaluation of the proposed method, the following conclusions can be derived:

1. Errors due to the application of simplified directional combination formulas to obtain bidirectional action effects are avoided.
2. A significant reduction of the computational cost as compared to other multimode pushover procedures is achieved.

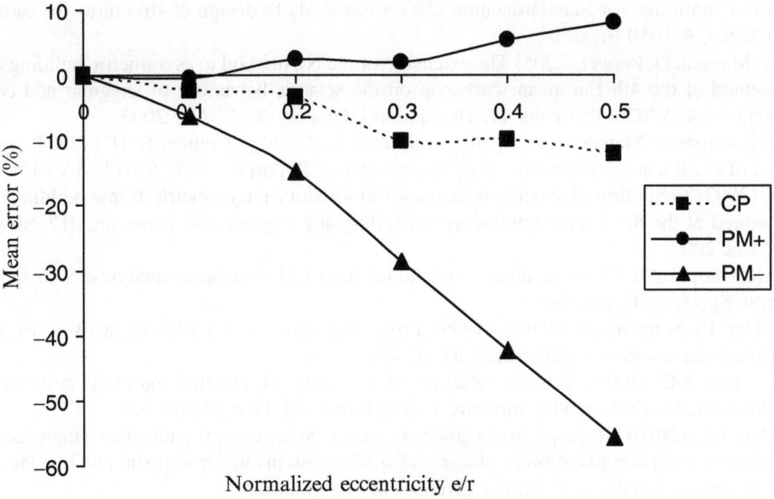

Fig. 13.7 Mean errors (%) of C3 displacements along *Y*-axis

3. The proposed methodology provides for each response parameter an upper limit and a lower limit which in most cases bound the exact value obtained by NL-RHA.
4. The results of the proposed methodology are conservative (mean error does not exceed 13%). On the contrary, the results produced by the conventional pushover procedure lead in many cases in nonconservative estimation of response values.

However, despite the fact that no restrictions are set to the development of the proposed methodology, generalization of the above conclusions to all types of asymmetric buildings requires further investigation, comprising application to a large variety of spatial structures and using an adequately high number of earthquake ground motions.

References

American Society of Civil Engineers (2008) Seismic rehabilitation of existing buildings. ASCE/SEI 41-06 standard

Applied Technology Council (ATC) (1996) Seismic evaluation and retrofit of concrete buildings vol 1. Report no. ATC-40, Redwood City, CA

Chopra AK, Goel RK (2001) A modal pushover analysis procedure to estimating seismic demands of buildings: theory and preliminary evaluation, PEER report 2001/03. Pacific Earthquake Engineering Research Center, University of California, Berkeley

Chopra AK, Goel RK (2004) A modal pushover analysis procedure to estimate seismic demands for unsymmetric-plan buildings. Earthq Eng Struct Dyn 33:903–927

De la Llera JC, Chopra AK (1996) Inelastic behavior of asymmetric multistory buildings. J Struct Eng 122:597–606

European Committee for Standardization (2004) Eurocode 8: design of structures for earthquake resistance, B-1050 Brussels

Fajfar P, Marusic D, Perus I (2005) The extension of the N2 method to asymmetric buildings. Paper presented at the 4th European workshop on the seismic behaviour of irregular and complex structures, 4EWICS, Paper no. 41, Thessaloniki, Greece, 26–27 Aug 2005

Federal Emergency Management Agency – Applied Technology Council (ATC) (2004) Improvement of nonlinear static seismic analysis procedures, Report no. ATC-55 (FEMA 440)

Fujii K (2007) Prediction of seismic response of multi-story unsymmetric frame buildings. Paper presented at the 8th Pacific conference on earthquake engineering, Paper no. 017, Singapore, 5–7 Dec 2007

Goel RK, Chopra AK (2004) Evaluation of modal and FEMA pushover analyses: SAC buildings. Earthq Spectra 20:225–254

Krawinkler H, Seneviratna GDPK (1998) Prons and cons of a pushover analysis of seismic performance evaluation. Eng Struct 20:452–464

Lin JL, Tsai KC (2008) Seismic analysis of two-way asymmetric building systems under bi-directional seismic ground motions. Earthq Eng Struct Dyn 37:305–328

Manoukas GE (2010) Static pushover analysis based on an energy equivalent single degree of freedom system – application to planar and spatial systems under uniaxial and biaxial seismic excitation. Dissertation, Aristotle University of Thessaloniki

Manoukas GE, Athanatopoulou AM, Avramidis IE (2006) Comparative evaluation of static pushover analysis' variations according to modern codes (in Greek). Paper presented at the 15th Hellenic conference on R/C structures, Alexandroupoli, Greece, 25–27 Oct 2006

Chapter 14
Three-Dimensional Modal Pushover Analysis of Unsymmetric-Plan Buildings Subjected to Two Components of Ground Motion

Juan C. Reyes and Anil K. Chopra

Abstract The original modal pushover analysis (MPA) to estimate seismic demands due to one component of ground motion is extended to consider two horizontal components simultaneously in three-dimensional analysis of buildings and to estimate internal forces and plastic hinge rotations directly from pushover analyses. Subsequently, the accuracy of the three-dimensional modal pushover analysis (MPA) procedure in estimating engineering demand parameters (EDPs) is evaluated. Eight low- and medium-rise structures were considered. Four intended to represent older buildings were designed according to the 1985 Uniform Building Code, while four other designs intended to represent newer buildings were based on the 2006 International Building Code. Median EDP values for these buildings to 39 two-component ground motions, scaled to two intensity levels, were computed by MPA and nonlinear response history analysis (RHA) and then compared. Even for these ground motions that deform the buildings significantly into the inelastic range, the MPA offers sufficient degree of accuracy. It is demonstrated that the PMPA, a variant of the MPA procedure, for nonlinear systems is almost as accurate as the well-known standard response spectrum analysis procedure is for linear systems. Thus, for practical applications, the PMPA procedure offers an attractive alternative to nonlinear RHA, whereby seismic demands can be estimated directly from the

J.C. Reyes (✉)
Civil and Environmental Engineering, Universidad de los Andes, Colombia, Carrera 1 Este No. 19A-40, Piso 6, Bogota, Colombia
e-mail: jureyes@uniandes.edu.co

A.K. Chopra
Civil and Environmental Engineering, University of California, Berkeley, 773 Davis Hall, Berkeley, CA, 94720-1710 USA
e-mail: chopra@ce.berkeley.edu

O. Lavan and M. De Stefano (eds.), *Seismic Behaviour and Design of Irregular and Complex Civil Structures*, Geotechnical, Geological and Earthquake Engineering 24, DOI 10.1007/978-94-007-5377-8_14,

(elastic) design spectrum. In contrast, the nonlinear static procedure specified in the ASCE/SEI 41-06 standard is demonstrated to grossly underestimate seismic demands for some of the unsymmetric-plan buildings considered.

14.1 Introduction

Traditional nonlinear static procedures underestimate seismic demands in intermediate and upper stories of medium- and high-rise buildings where higher-"mode" contributions to response can be significant and do not consider torsional contributions to the response for unsymmetric-plan buildings. Starting in 1997, various researchers have extended pushover analysis to unsymmetric-plan buildings. By applying a height-wise distribution of lateral forces—typical of standard planar pushover analysis—at the floor centers of mass, an approximate nonlinear static analysis procedure was developed (Kilar and Fajfar 1997). A second procedure consists of two steps: (1) three-dimensional elastic response spectrum analysis to determine roof displacement and height-wise distribution of lateral forces for each resisting element (frames, walls, etc.) and (2) planar pushover analysis of each resisting element (Moghadam and Tso 1998). Other studies have focused on the special interaction between walls and frames in pushover analysis of wall-frame structures (De Stefano and Rutenberg 1998). Another paper investigated the accuracy of applying lateral forces at different locations in the plan of unsymmetric buildings (Faella and Kilar 1998). By combining lateral displacements from nonlinear static analysis with the torsional effects from response spectrum analysis (RSA) of a three-dimensional linear model of the building, the N2 method for simplified analysis of unsymmetric-plan buildings was developed (Fajfar et al. 2005). The few comparisons of pushover analysis results with nonlinear RHA give the impression of limited success.

Because the modal pushover analysis (MPA) procedure is rooted in structural dynamics theory, this procedure can be generalized to unsymmetric-plan buildings without introducing any additional approximations (Chopra and Goel 2004). Although it was demonstrated that the MPA estimate of seismic demand is accurate for unsymmetric systems to a similar degree for a symmetric building, this conclusion was based on a very limited investigation, comparing the MPA estimate of demand (story drifts) and its exact value determined by nonlinear RHA for three mass-eccentric structural systems with different degrees of coupling between lateral and torsional motions subjected to a single ground motion. The MPA procedure has been restricted to determine seismic demands due to a single component of ground motion, although at least the two horizontal components of ground motion should be considered in computing seismic demands by three-dimensional analysis of multistory buildings. To address this issue, Reyes and Chopra (2011a) extends the MPA procedure to three-dimensional analysis of buildings—symmetric or unsymmetric in plan—subjected to two horizontal components of ground motion, simultaneously.

This investigation (Reyes 2009; Reyes and Chopra 2011b) evaluates the accuracy of the practical MPA procedure in estimating seismic demands for unsymmetric-plan buildings subjected to two horizontal components of ground motions, simultaneously, and comparatively evaluates the accuracy of the practical MPA procedure and the nonlinear static procedures specified in ASCE/SEI 41-06 (ASCE and SEI 2007), a building evaluation standard.

14.2 Structural Systems and Modeling

14.2.1 Structural Systems

The structural systems considered are three- and nine-story steel buildings with ductile frames. Four buildings intended to represent older buildings were designed according to the 1985 Uniform Building Code (UBC85), while four other designs intended to represent modern buildings were based on the 2006 International Building Code (IBC06). The plans of the selected buildings are shown in Fig. 14.1, where the moment-resisting frames are highlighted. The buildings are identified by the letters A, B, and C followed by the number of stories and the design code: plan A is rectangular with two axes of symmetry; plan B is symmetric about the y axis; and

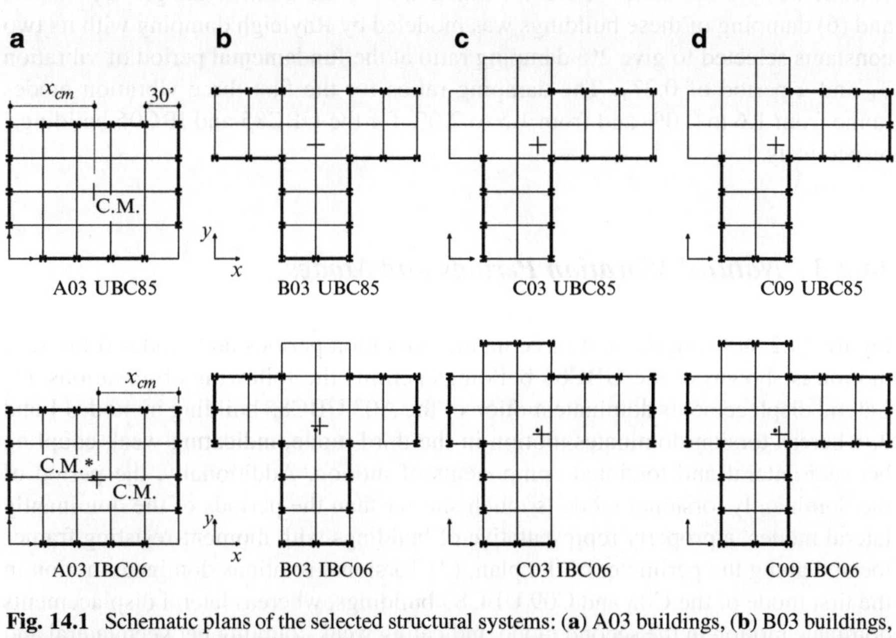

Fig. 14.1 Schematic plans of the selected structural systems: (**a**) A03 buildings, (**b**) B03 buildings, (**c**) C03 buildings, and (**d**) C09 buildings designed according to UBC85 (*upper row*) and IBC06 (*lower row*); moment-resisting frames are highlighted

plan C is unsymmetric about both x and y axes. The buildings have similar plan areas and floor weights, with a span length of 30 ft and a story height of 13 ft. Design code forces for the buildings, assumed to be located in Bell, CA (33.996 N, 118.162 W), were determined, but their member sizes were governed by drift instead of strength requirements (Reyes 2009).

14.2.2 Modeling

Nonlinear RHA, MPA, and code pushover analyses were implemented for these buildings using the PERFORM-3D computer program (Computers and Structures CSI 2006) modeled with the following features: (1) Beams and columns were modeled by a linear element with trilinear plastic hinges at the ends of the elements that include in-cycle strength deterioration, but not cyclic stiffness degradation. Axial load-moment interaction for the columns was represented by plasticity theory; (2) Panel zones were modeled as four rigid links hinged at the corners with a rotational spring that represents the strength and stiffness of the connection (Krawinkler 1978); (3) ductility capacities of girders, columns, and panel zones were specified according to the ASCE/SEI 41-06 standard (ASCE and SEI 2007); (4) columns of moment-resisting frames were assumed to be fixed at the base, whereas gravity columns were considered pinned at the base; (5) the geometric nonlinear effects were considered by a standard P-Delta formulation for both moment and gravity frames; and (6) damping of these buildings was modeled by Rayleigh damping with its two constants selected to give 2% damping ratio at the fundamental period of vibration T_1 and a period of $0.2T_1$. The damping ratios for the first three vibration modes range from 1.6 to 2.0% and from 1.5 to 2.0% for the UBC85 and IBC06 buildings, respectively.

14.2.3 Natural Vibration Periods and Modes

Figure 14.2 showing the first three natural vibration periods and modes (only roof motion is shown) of the UBC85 buildings permits the following observations. (1) Lateral displacements dominate motion of the A03 UBC85 building in modes 1 and 2, whereas torsion dominates motion in the third mode, indicating weak coupling between lateral and torsional components of motion. Additionally, the period of the dominantly torsional mode is much shorter than the periods of the dominantly lateral modes, a property representative of buildings with moment-resisting frames located along the perimeter of the plan. (2) Torsional rotations dominate motion in the first mode of the C03 and C09 UBC85 buildings, whereas lateral displacements dominate motion in the second mode, indicating weak coupling between lateral and

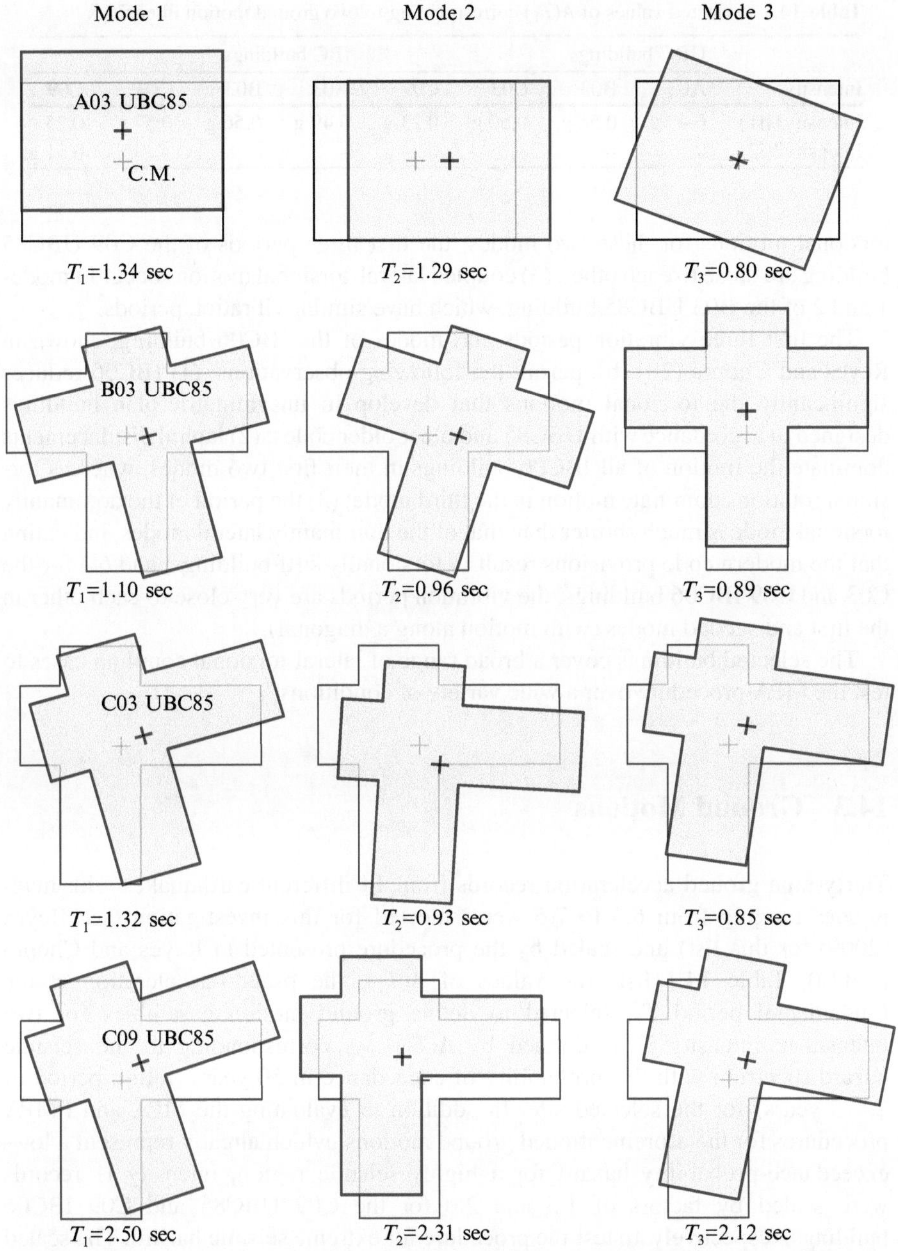

Fig. 14.2 First triplet of periods and modes of vibration of UBC85 buildings (only roof motion is shown)

Table 14.1 Selected values of $A(T_1)$ corresponding to two ground motion intensities

	UBC buildings				IBC buildings			
Intensity	A03	B03	C03	C09	A03	B03	C03	C09
Intensity1(i1)	0.49 g	0.56 g	0.50 g	0.23 g	0.49 g	0.56 g	0.57 g	0.25 g
Intensity2(i2)	–	–	–	0.34 g	–	–	–	0.51 g

torsional motions for these two modes; the first three periods of the C09 UBC85 building are close to each other. (3) coupled lateral-torsional motions occur in modes 1 and 2 of the B03 UBC85 building, which have similar vibration periods.

The first three vibration periods and modes of the IBC06 building, shown in Reyes and Chopra (2011b), permit the following observations: (1) IBC06 reduces significantly the torsional motions that develop in unsymmetric-plan buildings designed in accordance with UBC85 and other older codes; (2) lateral displacements dominate the motion of all IBC06 buildings in their first two modes, whereas torsional rotations dominate motion in the third mode; (3) the period of the dominantly torsional mode is much shorter than that of the dominantly lateral modes, indicating that the modern code provisions result in torsionally stiff buildings; and (4) for the C03 and C09 IBC06 buildings, the vibration periods are very close to each other in the first and second modes (with motion along a diagonal).

The selected buildings cover a broad range of lateral-torsional coupling cases to test the MPA procedure over a wide variety of conditions.

14.3 Ground Motions

Thirty-nine ground acceleration records from 14 different earthquakes with magnitudes ranging from 6.5 to 7.6 were selected for this investigation (see Reyes (2009) for this list) and scaled by the procedure presented in Reyes and Chopra (2011a). Table 14.1 lists the values of $A(T_1)$, the pseudo-acceleration at the fundamental period T_1, selected to define ground motion ensembles for two intensities: intensity i1 is defined by $A(T_1)_{2\%/50}$ corresponding to the seismic hazard spectrum with 2% probability of exceedance in 50 years (return period of 2,475 years) for the selected site. In addition to evaluating the MPA and PMPA procedures for the aforementioned ground motions, which already represent a low-exceedance-probability hazard for a highly seismic region, intensity i1 records were scaled by factors of 1.5 and 2.0 for the C09 UBC85 and C09 IBC06 buildings, respectively, to test the procedure for extreme seismic hazard. The scaled ground motions—identified as intensity i2 records—drive both groups of buildings significantly into the inelastic range; the median roof displacement exceeds the yield displacement (of the modal pushover curve) by a factor of about two (Reyes 2009).

14.4 Evaluation of Practical MPA

The practical MPA (PMPA) procedure described in Reyes and Chopra (2011a) permits estimation of seismic demands directly from the elastic design spectrum. The median value of the peak deformation of the *nth-mode* SDF system that is needed to obtain the reference displacement for pushover analysis is estimated from the spectrum, without nonlinear RHA, and the structure is treated as linearly elastic in estimating the seismic demand contributions of "modes" higher than the first three; this approximation has been demonstrated to be valid for the unsymmetric-plan systems considered here (Reyes 2009).

In this section, the accuracy of PMPA in estimating seismic demands is evaluated in two stages: first, the floor displacements and story drifts at the C.M. estimated in steps 1–10 of PMPA (Reyes and Chopra 2011a) are evaluated, followed by story drifts, plastic hinge rotations, and member forces at the "stiff" and "flexible" sides of the buildings estimated from the story drifts at C.M. in steps 11–12 of MPA.

14.4.1 Floor Displacements and Story Drifts at the C.M.

Figures 14.3 and 14.4 present the median values of floor displacements and story drifts at the C.M., determined by the nonlinear RHA, MPA, and PMPA procedures for the UBC85 and IBC06 buildings, respectively. Results for two different intensities of ground motion are included for the C09 buildings. In general, PMPA provided a larger estimate of seismic demands compared to MPA because it overestimated the median value of the peak deformation of the nth-mode inelastic SDF system, as shown in Table 14.2, where the ratio of the empirical value of $\hat{D}_{\mathrm{n}}$ and the median of the D_{n} values determined by nonlinear RHA for the first "mode" of the UBC85 buildings associated with *x*- and *y*-components of ground motion is shown in Table 14.2. This is to be expected for the systems considered because the empirical equation for the inelastic deformation ratio C_{Rn} does not permit values below 1.0, whereas the exact data does fall below 1.0 (Chopra and Chintanapakdee 2004). The PMPA generally provided a more accurate estimate of seismic demand for those cases where MPA underestimated demand (relative to nonlinear RHA), but a less accurate estimate for cases where MPA already overestimated the demand. For most of the analyzed cases, the PMPA procedure tends to overestimate floor displacements and story drifts, especially for the A03 UBC85 and A03 IBC06 buildings subjected to i1-intensity ground motions and the C09 IBC06 building subjected to i2-intensity ground motions. The height-wise average discrepancy in story drifts is 27, 13, 13, 5, and 7% for the A03, B03, C03, and C09 with i1-intensity and C09 with i2-intensity UBC85 buildings, respectively, and 21, 25, 10, 12, and 20% for the A03, B03, C03, and C09 with i1-intensity and CO9 with i2-intensity IBC06 buildings, respectively. Such overestimation is

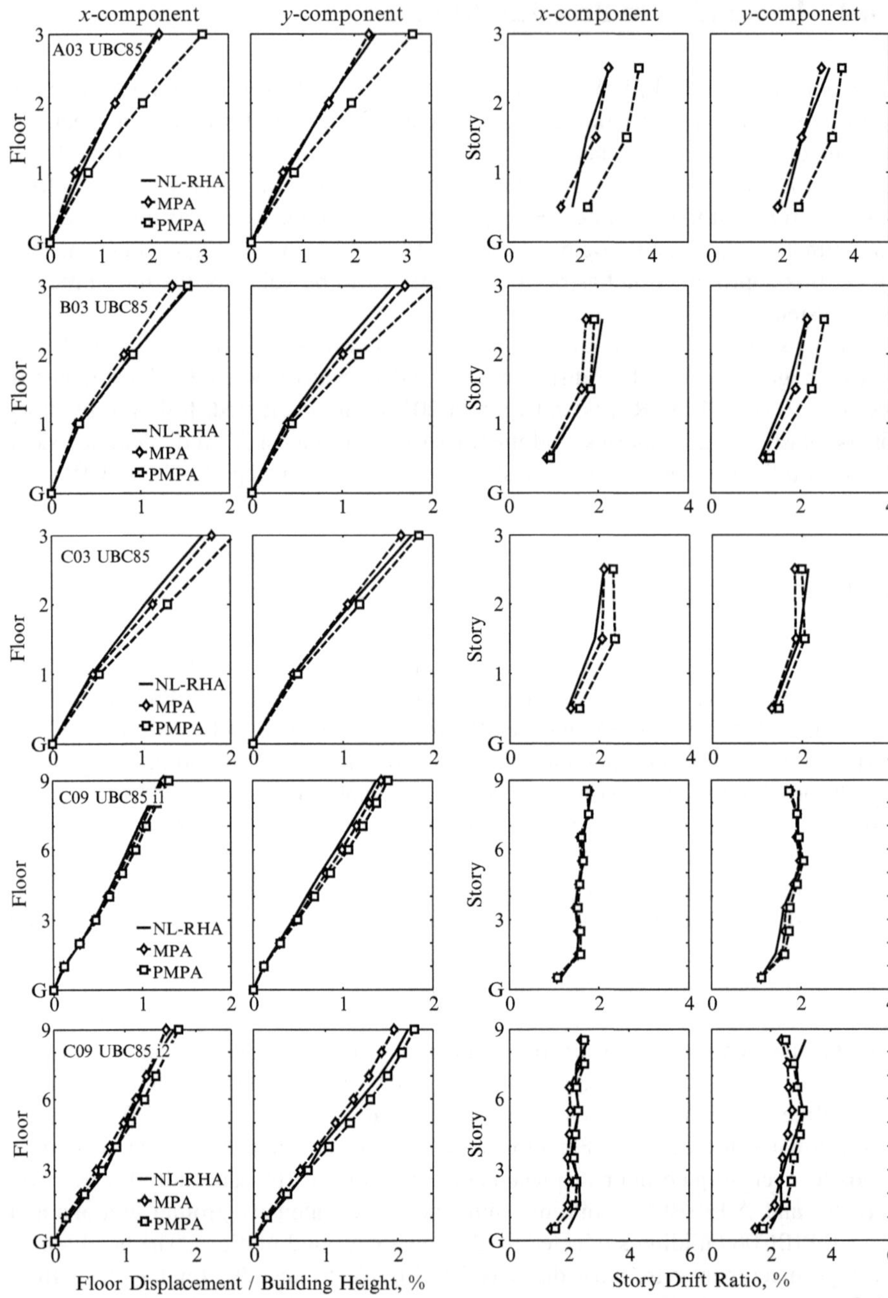

Fig. 14.3 Median floor displacements (columns 1–2) and story drifts (columns 3–4) at the C.M. of the UBC85 buildings determined by nonlinear RHA, MPA, and PMPA. Two ground motion intensities (i1 and i2) are included for the C09 buildings

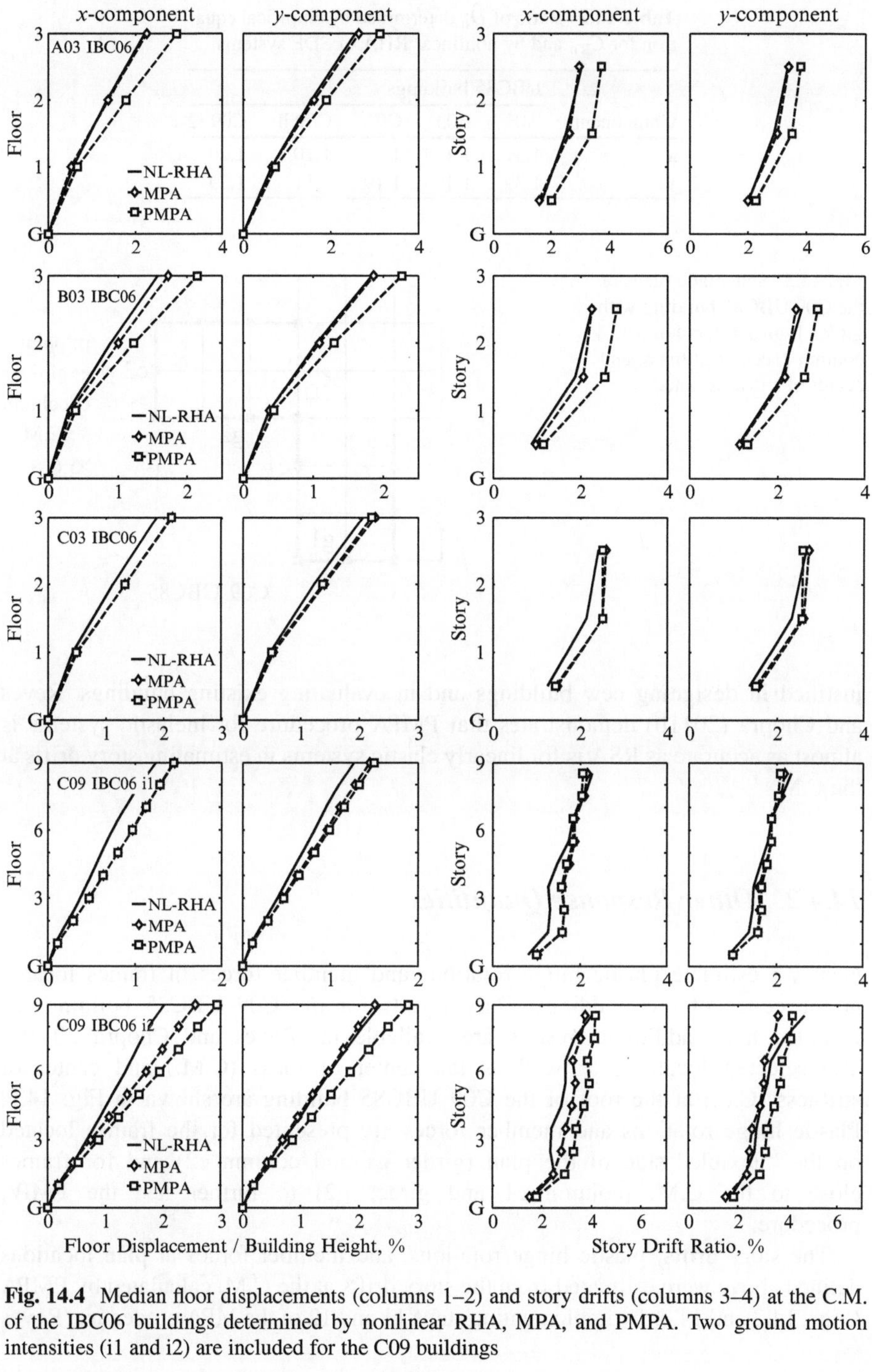

Fig. 14.4 Median floor displacements (columns 1–2) and story drifts (columns 3–4) at the C.M. of the IBC06 buildings determined by nonlinear RHA, MPA, and PMPA. Two ground motion intensities (i1 and i2) are included for the C09 buildings

Table 14.2 Ratio of $\hat{D}_n$ determined by empirical equation for C_{Rn} and by nonlinear RHA of SDF systems

	UBC85 buildings				
Component	A03	B03	C03	C09 i1	C09 i2
x	1.37	1.16	1.24	1.10	1.20
y	1.39	1.19	1.19	1.11	1.29

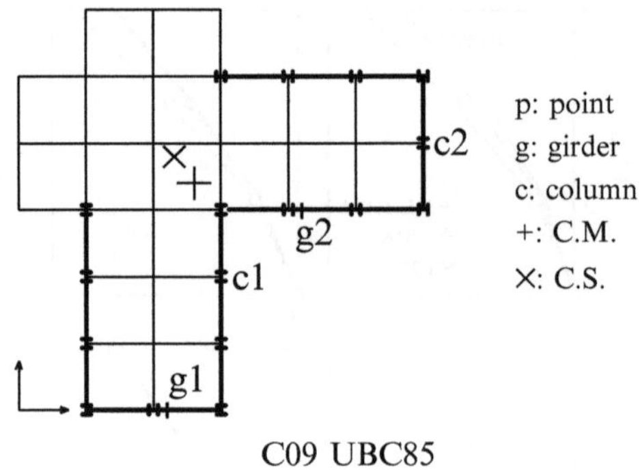

Fig. 14.5 Schematic plans of the C09 UBC85 building with girder 1, girder 2, column 1, column 2, center of mass, and center of stiffness noted

justified in designing new buildings and in evaluating existing buildings. Reyes and Chopra (2011b) demonstrates that PMPA procedure for inelastic systems is almost as accurate as RSA is for linearly elastic systems in estimating story drifts at the C.M.

14.4.2 Other Response Quantities

Next we estimate plastic hinge rotations and member forces in frames located at the edges of the buildings. Only results for the C09 UBC85 building are included here; additional results are available in (Reyes and Chopra 2011b). The selected locations as well as the center of mass (C.M.) and center of stiffness (C.S.) at the roof of the C09 UBC85 building are shown in Fig. 14.5. Plastic hinge rotations and member forces are presented for the frames located on the "flexible" side of the plan (girder g1 and column c2) and for frames close to the C.M. (column c1 and girder g2) to further test the PMPA procedure.

The story drifts, plastic hinge rotations, and member forces at plan locations defined above were estimated from the story drifts at the C.M. calculated by PMPA (Figs. 14.3 and 14.4) by implementing steps 11 and 12 of the MPA procedure (Reyes

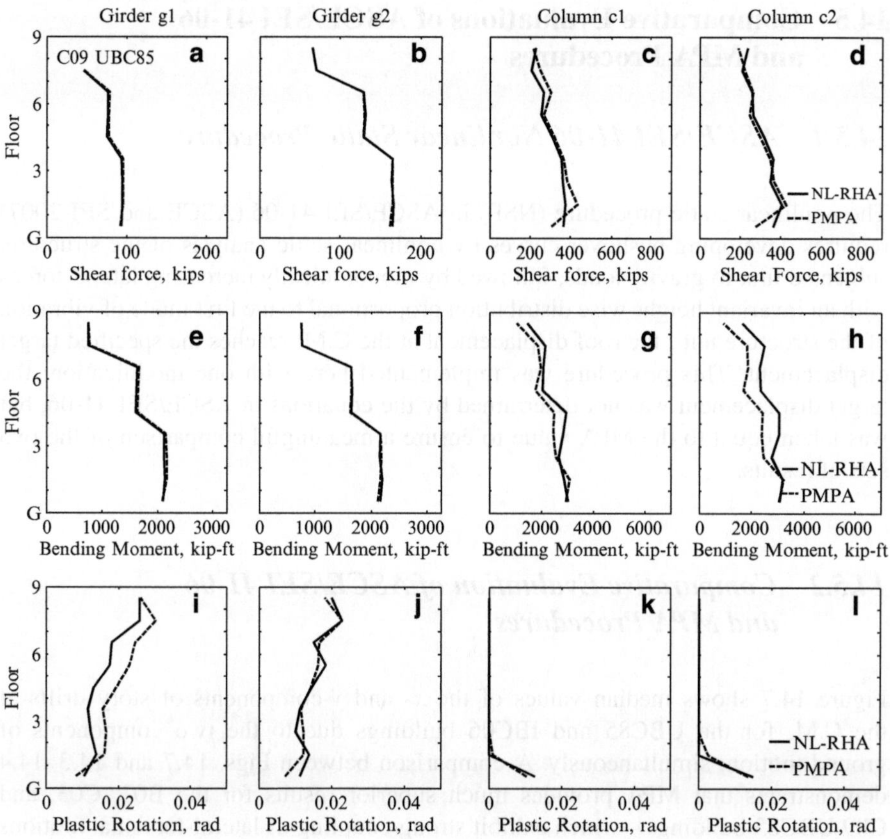

Fig. 14.6 Median shear force, bending moment, and plastic hinge rotation for girders 1 and 2 and columns 1 and 2 (identified in Fig. 14.6) of the C09 UBC85 building determined by nonlinear RHA and PMPA

and Chopra 2011a) and compared with the "exact" results from nonlinear RHA; see Fig. 14.6. The shear forces, bending moments, and plastic hinge rotations in the girders and columns, identified in Fig. 14.5, were estimated conservatively in most cases (Fig. 14.6). Plastic hinge rotations are generally overestimated in frames located on the "flexible" side of the building and underestimated in those located at the "stiff" side, e.g., compare Fig. 14.6i with 14.6j. In most cases, PMPA provides accurate estimates of shear forces and bending moments in frames located on both the "stiff" and "flexible" sides (Fig. 14.6a–f). The error in internal forces is generally smaller than the error in hinge rotations because internal forces increase slowly with hinge rotation for members that deform beyond the elastic limit at both ends. As a result, even a large error in the hinge rotation leads to only small error in the computed internal forces.

14.5 Comparative Evaluations of ASCE/SEI 41-06 and MPA Procedures

14.5.1 ASCE/SEI 41-06 Nonlinear Static Procedure

The nonlinear static procedure (NSP) in ASCE/SEI 41-06 (ASCE and SEI 2007) requires developing pushover curves by nonlinear static analysis of the structure, subjected first to gravity loads, followed by monotonically increasing lateral forces with an invariant height-wise distribution proportional to the first mode of vibration of the structure until the roof displacement at the C.M. reaches the specified target displacement. This procedure was implemented here with one modification: the target displacement was not determined by the equations in ASCE/SEI 41-06, but was taken equal to the MPA value to ensure a meaningful comparison of the two sets of results.

14.5.2 Comparative Evaluation of ASCE/SEI 41-06 and MPA Procedures

Figure 14.7 shows median values of the *x*- and *y*-components of story drifts at the C.M. for the UBC85 and IBC06 buildings due to the two components of ground motion, simultaneously. A comparison between Figs. 14.7 and 14.3–14.4 demonstrates that MPA provides much superior results for the B03, C03, and C09 UBC85 buildings—which exhibit strong coupling of lateral-torsional motions and closely spaced periods (Fig. 14.2)—compared to the code. The ASCE/SEI 41-06 force distributions grossly underestimated the story drifts for the C03 and C09 UBC85 buildings (Fig. 14.7a, rows 3, 4, and 5) and for the C09 IBC06 building, especially in the upper stories (Fig. 14.7b, rows 4 and 5). In contrast, the MPA procedure provided a much better estimate of story drift demands in the upper stories of these buildings (Figs. 14.3 and 14.4), because it includes all three components of forces in pushover analysis and higher-"mode" contributions to the response. Note that these higher-"mode" contributions are especially significant for the nine-story building (see Figures 5.21 and 5.22 in Reyes (2009)) and the MPA procedure accounted for them, providing excellent estimates of story drifts (Fig. 14.5, rows 4 and 5). Because the response of the A03 UBC85, A03 IBC06, and B03 IBC06 buildings is dominated by the first "mode" (see Figure 5.12 in Reyes (2009)), in which the motion is dominantly translational, the ASCE 41-06 force distribution is adequate (Fig. 14.7a, row 1; and Fig. 14.7b, rows 1 and 2).

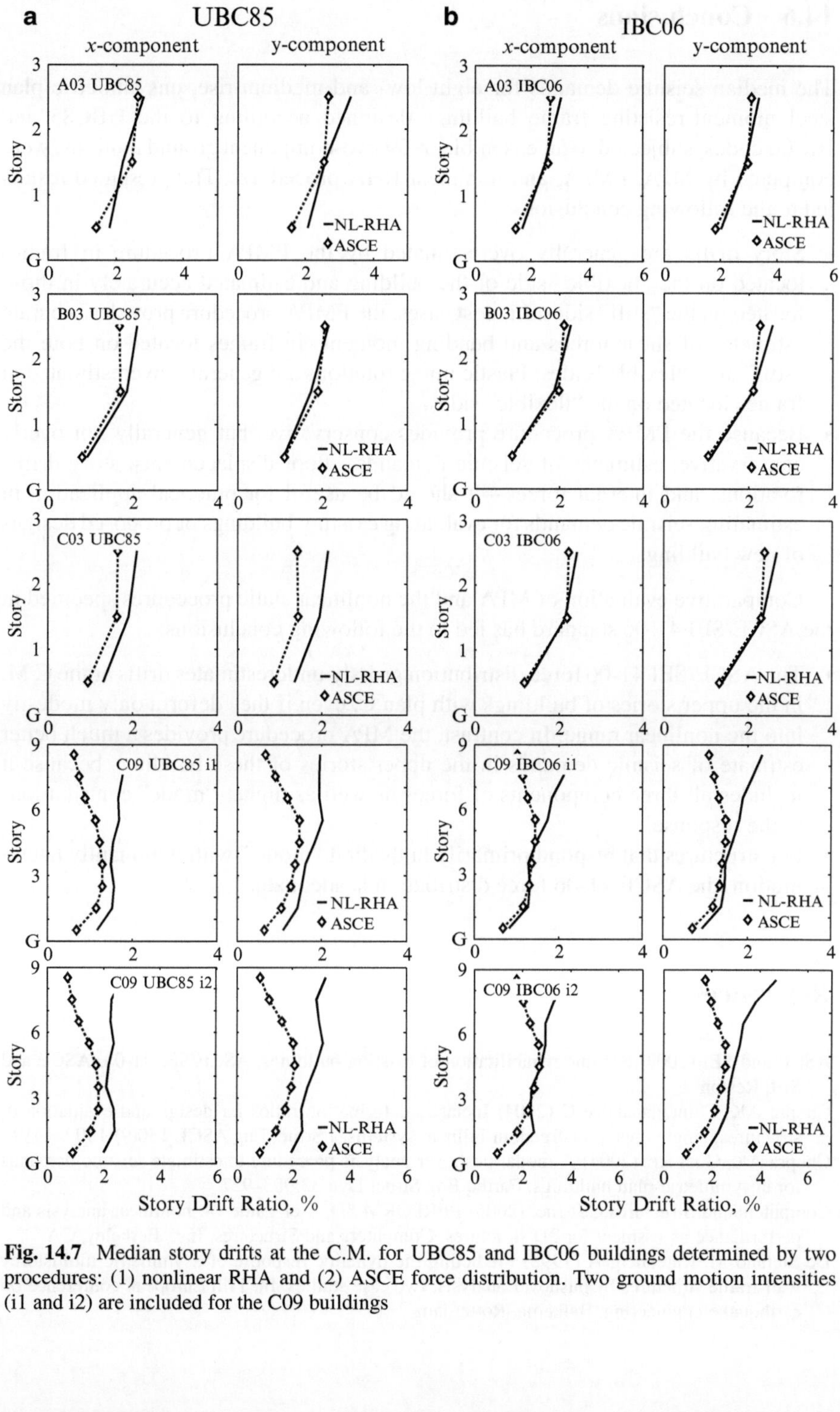

Fig. 14.7 Median story drifts at the C.M. for UBC85 and IBC06 buildings determined by two procedures: (1) nonlinear RHA and (2) ASCE force distribution. Two ground motion intensities (i1 and i2) are included for the C09 buildings

14.6 Conclusions

The median seismic demands for eight low- and medium-rise, unsymmetric-plan steel moment-resisting frame buildings designed according to the UBC85 and IBC06 codes, subjected to an ensemble of 39 two-component ground motions, were computed by MPA, PMPA, and nonlinear RHA procedures. The presented results led to the following conclusions:

- Story drifts are generally overestimated by the PMPA procedure in frames located on the "flexible" side of the building and estimated accurately in those located on the "stiff" sides. In most cases, the PMPA procedure provides accurate estimates of shear forces and bending moments in frames located on both the "stiff" and "flexible" sides. Plastic hinge rotations are generally overestimated in frames located on the "flexible" sides.
- Because the PMPA procedure provides conservative, but generally not overly conservative, estimates of seismic demands—floor displacements, story drifts, rotations, and internal forces—it should be useful for practical application in estimating seismic demands for evaluating existing buildings or proposed designs of new buildings.

Comparative evaluation of MPA and the nonlinear static procedures specified in the ASCE/SEI 41-06 standard has led to the following conclusions:

- The ASCE/SEI 41-06 force distribution grossly underestimates drifts at the C.M. in the upper stories of buildings with plan C, even if they deform only modestly into the nonlinear range. In contrast, the MPA procedure provides a much better estimate of seismic demands in the upper stories of these buildings because it includes all three components of forces as well as higher-"mode" contributions to the response.
- For structures that respond primarily in the first "mode," with dominantly lateral motion, the ASCE 41-06 force distribution is adequate.

References

ASCE and SEI (2007) Seismic rehabilitation of existing buildings, ASCE/SEI 41-06. ASCE and SEI, Reston

Chopra AK, Chintanapakdee C (2004) Inelastic deformation ratios for design and evaluation of structures: single-degree-of-freedom bilinear systems. J Struct Eng ASCE 130(9):1309–1319

Chopra AK, Goel RK (2004) A modal pushover analysis procedure to estimate seismic demands for unsymmetric-plan buildings. Earthq Eng Struct Dyn 33:903–927

Computers and Structures CSI Inc. (2006) PERFORM 3D, user guide v4, nonlinear analysis and performance assessment for 3D structures. Computers and Structures, Inc., Berkeley, CA

De Stefano M, Rutenberg A (1998) Predicting the dynamic response of asymmetric multistorey wall-frame structures by pushover analysis: two case studies. In: 11th European conference on earthquake engineering, Balkema, Rotterdam

Faella G, Kilar V (1998) Asymmetric multistorey R/C frame structures: push-over versus non-linear dynamic analysis. In: 11th European conference on earthquake engineering, Balkema, Rotterdam

Fajfar P, Marusic D, Perus I (2005) Torsional effects in the pushover-based seismic analysis of buildings. J Earthq Eng 9(6):831–854

Kilar V, Fajfar P (1997) Simple push-over analysis of asymmetric buildings. Earthq Eng Struct Dyn 26:233–249

Krawinkler H (1978) Shear in beam-column joints in seismic design of frames. Eng J AISC 15(3):82–91

Moghadam AS, Tso WK (1998) Pushover analysis for asymmetrical multistorey buildings. In: 6th U.S. national conference on earthquake engineering. EERI, Oakland, CA

Reyes JC (2009) Estimating seismic demands for performance-based engineering of buildings. Ph.D. dissertation, Department of Civil and Environmental Engineering, University of California, Berkeley, CA

Reyes JC, Chopra AK (2011a) Three-dimensional modal pushover analysis of buildings subjected to two components of ground motion, including its evaluation for tall buildings. Earthq Eng Struct Dyn 40(7):789–806

Reyes JC, Chopra AK (2011b) Evaluation of three-dimensional modal pushover analysis for unsymmetric-plan buildings subjected to two components of ground motion. Earthq Eng Struct Dyn 40(13):1475–1494

Chapter 15
Estimating Torsional Demands in Plan Irregular Buildings Using Pushover Procedures Coupled with Linear Dynamic Response Spectrum Analysis

Carlos Bhatt and Rita Bento

Abstract The limitation of the commonly used nonlinear static procedures (NSPs), including the ones recommended by the seismic codes (Eurocode 8 – N2 method, ATC40 and FEMA440 – capacity spectrum method, CSM), is their inability to capture the torsional behaviour of plan-asymmetric buildings. Fajfar and his team have extended the N2 method to this kind of structures through the application of correction factors which depend on both linear dynamic response spectrum analysis and pushover analysis. In this chapter, the proposed correction factors are applied to the N2 method and to the CSM with the features recommended in FEMA440 in order to assess the seismic response of three existing plan irregular buildings. The torsional demands estimated by the aforementioned NSPs are duly compared with the ones defined by means of the most precise nonlinear dynamic analyses for several levels of seismic intensity. The torsional correction factors used seem to improve the performance of existing code pushover methodologies in estimating the seismic response of plan irregular structures.

15.1 Introduction

The use of NSPs on the seismic assessment or design of structures has gained considerable popularity in recent years, backed by a large number of extensive verification studies that have demonstrated their relatively good accuracy in estimating the seismic response of regular structures (planar frames and bridges).

However, the extension of such use to the case of 3D irregular structures has been the object of a limited number of scientific studies, which effectively ends up

---

C. Bhatt • R. Bento (✉)
Department of Civil Engineering and Architecture, Instituto Superior Técnico,
Technical University of Lisbon, Avenida Rovisco Pais, 1049-001 Lisbon, Portugal
e-mail: cbhatt@civil.ist.utl.pt; rbento@civil.ist.utl.pt

O. Lavan and M. De Stefano (eds.), *Seismic Behaviour and Design of Irregular and Complex Civil Structures*, Geotechnical, Geological and Earthquake Engineering 24, DOI 10.1007/978-94-007-5377-8_15,

by limiting significantly the use of NSPs to assess actual existing structures, the majority of which do tend to be non-regular (Fajfar et al. 2005a; Chopra and Goel 2004; D'Ambrisi et al. 2009; Erduran and Ryan 2011).

The major limitations of the existing NSPs, including the ones recommended by the seismic codes (e.g. the N2 method (Fajfar and Fischinger 1988; Fajfar 2000) proposed in Eurocode 8 (CEN 2004) and the CSM (Freeman et al. 1975; Freeman 1998) with the features presented in ATC40 (ATC 1996) and in FEMA440 (ATC 2005)), is their inability to capture the torsional behaviour of plan irregular buildings. Generally they cannot capture the torsional effects distorting the real structural response.

Fajfar and his team have developed the extended N2 method (Fajfar et al. 2005a, b) which is able to capture the torsional behaviour of plan-asymmetric buildings. This procedure is based on the application of correction factors to the pushover results obtained with the N2 method. The correction factors depend on the results of a dynamic elastic analysis and of a pushover analysis.

Bhatt and Bento have also extended the CSM-FEMA440 to plan-asymmetric buildings (Bhatt and Bento 2011a) using the same correction factors definition proposed by Fajfar.

In this chapter, the results obtained in three existing plan irregular buildings, using the extended N2 method and the extended CSM-FEMA440 to plan-asymmetric structures, are compared with the nonlinear dynamic median results and with the linear response spectrum analysis. The application of torsional correction factors to improve the performance of existing code pushover methodologies in estimating the seismic response of such structures is therefore evaluated.

In the first part of the work, the extension of both code procedures to the 3D case is described. Afterwards, the three case studies analysed, their modelling options as well as the seismic action considered are depicted. The results obtained with the evaluated procedures are presented in terms of normalised top displacements in order to better understand the torsional response of the buildings. Final conclusions are pointed out in the end.

15.2 Torsional Correction Factors to Use in Pushover Methods

Extensive parametric studies have been performed by Fajfar and his co-workers (2005a) in order to investigate the parameters that influence the inelastic torsional response of building structures. Several conclusions were drawn and are herein presented:

1. The inelastic torsional response is qualitatively similar to the elastic torsional response. Quantitatively, the torsional effects depend on the ductility demand and, therefore, on the ground motion intensity.

2. An upper bound of the torsional amplifications can be estimated with a linear dynamic response spectrum analysis.
3. The torsional effects decrease with the increase of plastic deformations. This trend is clear with the smaller amplification of displacements on the flexible side. However, if the structure is subjected to small plastic deformations, characterised by ductility less than 2.0, the amplification on the flexible edge may be slightly higher than in the elastic structure.
4. The response on the stiff edge depends on the influence of different modes of vibration and on the ground motion in the transverse direction. This response depends on the structural and ground motion characteristics in both directions. It is difficult to make general conclusions about the response on the stiff side. De-amplification of displacements due to torsion on the stiff side decreases with increasing plastic deformations in elastic torsionally stiff structures. Sometimes, a transition from de-amplification to amplification could happen. In elastic torsionally flexible structures, the amplification due to torsion decreases with increasing plastic deformations.
5. For large plastic deformations, the smaller torsional effects in the inelastic range when compared with the elastic range are usually illustrated by a flattening of the displacement envelopes in the horizontal plane.
6. The dispersion of results is larger in the inelastic range than in the elastic regime.

Based on the results obtained, the following conclusions were taken (Fajfar et al. 2005a). They are important for the development of simplified analysis methods and code guidelines:

1. A conservative estimation of the amplification of displacements due to torsion in the inelastic range can be determined by a dynamic elastic analysis.
2. Any reduction of displacements on the stiff side compared to the counterpart symmetric building, obtained from elastic analysis, will decrease or even disappear in the inelastic range.

These conclusions were used by Fajfar and his team (2005a, b) to develop an extension of the N2 method to plan-asymmetric building structures. The entire procedure can be summarised in the following steps:

1. Perform pushover analyses with positive and negative signs for each X and Y direction of a 3D numerical model. Compute the target displacement – displacement demand at the CM at roof level – for each direction as the larger value of the + and – sign pushover, using the original N2 method proposed in Eurocode 8.
2. Perform a linear modal response spectrum analysis in two X and Y directions combining the results according to the SRSS rule.
3. Determine the torsional correction factors. This factor is computed by the ratio between the normalised roof displacements obtained by the elastic response spectrum analysis and by the pushover analysis. The normalised roof displacement is obtained by normalising the displacement value at a specific location with respect to those of the centre of mass (CM). If the normalised roof displacement obtained

from the elastic response spectrum analysis is smaller than 1.0, one should consider 1.0 to avoid any favourable torsional effect (reduction of displacements) given by the elastic analysis.

4. Multiply the quantity under study at a certain location by the correction factor calculated for that location.

As one can conclude from the previous steps, the extended N2 method uses both nonlinear static pushover and elastic dynamic analysis. The displacement demand and its distribution along the height at the centre of mass of each storey are determined using the original N2 method. The amplification of displacements due to torsion is calculated by elastic dynamic analysis. The reduction of displacements due to torsion is not taken into account. The results obtained by Fajfar and his team show that this extended procedure leads to conservative estimations of the torsional response of plan-asymmetric buildings.

In (2011a), Bhatt and Bento have extended the CSM-FEMA440 to plan-asymmetric buildings following the aforementioned procedure suggested by Fajfar. The difference lies in step (1), where the target displacement is calculated using the CSM with the features proposed in FEMA440.

15.3 Case Studies

Three real plan-asymmetric RC buildings were analysed in this endeavour. The first case study is the three-storey SPEAR building. It represents typical existing three-storey buildings in the Mediterranean region following Greece's concrete design code in force between 1954 and 1995. This structure was designed only for gravity loads based on the construction practice applied in the early 1970s that included the use of smooth rebars. A prototype was tested in full scale at Ispra within the European framework project SPEAR. Further details on the structure and its pseudo-dynamic testing can be found in Fardis (2002) and Fardis and Negro (2006). The SPEAR building is plan asymmetric in both *X* and *Y* directions, but it is regular in elevation (Fig. 15.1).

The second case study is a five-storey building. It is a real Turkish reinforced concrete structure which experienced the 1999 Golcuk earthquake without any damage. The building is asymmetric along the *X*-axis (Fig. 15.2a), and all the floors keep the same height (Fig. 15.2b). There are potential weak connections due to the existence of beams framing into beams. There are also walls and elongated columns (wall-like column), as presented in Fig. 15.2a. For more details on the building's characteristics, see Vuran et al. (2008).

The last case study is a real eight-storey Turkish reinforced concrete building (Bhatt et al. 2010). It is a plan-asymmetric structure in both *X*- and *Y*-axes; see Fig. 15.3a. The first storey is 5.0 m high and the upper floors are 2.7 m high (Fig. 15.3b). There are beams framing into beams leading to possible weak connections in the structure. There are also walls and elongated columns (wall-like

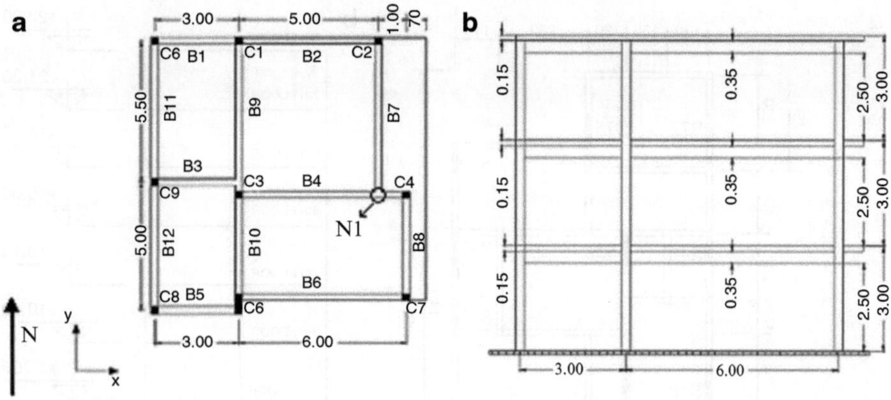

Fig. 15.1 Three-storey building configuration. (**a**) In plan and (**b**) at the southwest facade (units in metres)

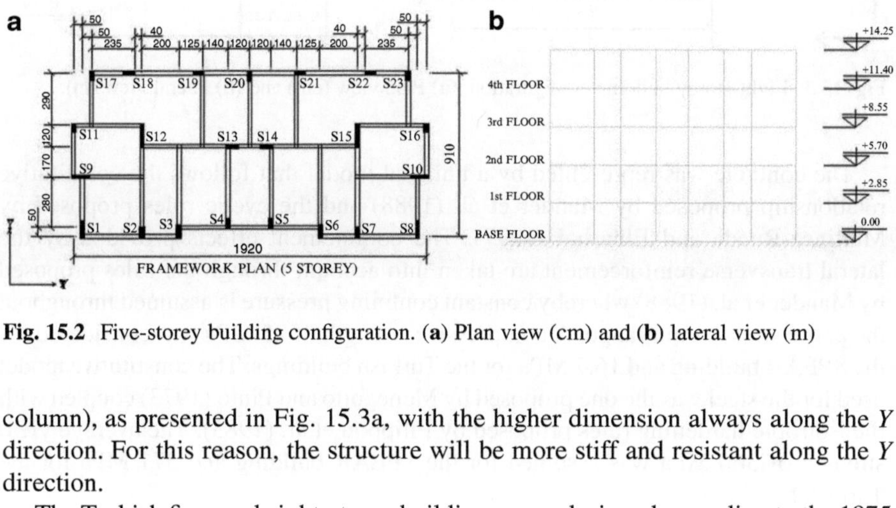

Fig. 15.2 Five-storey building configuration. (**a**) Plan view (cm) and (**b**) lateral view (m)

column), as presented in Fig. 15.3a, with the higher dimension always along the Y direction. For this reason, the structure will be more stiff and resistant along the Y direction.

The Turkish five- and eight-storey buildings were designed according to the 1975 Seismic Code of Turkey.

15.4 Modelling Options

The 3D models representing the buildings under analysis were developed with the fibre element-based finite element program SeismoStruct (Seismosoft 2006). They were built using space frames assuming the centreline dimensions. The inelastic behaviour of the structural elements was modelled using a fibre element model, with each fibre being characterised by the material relationships described below.

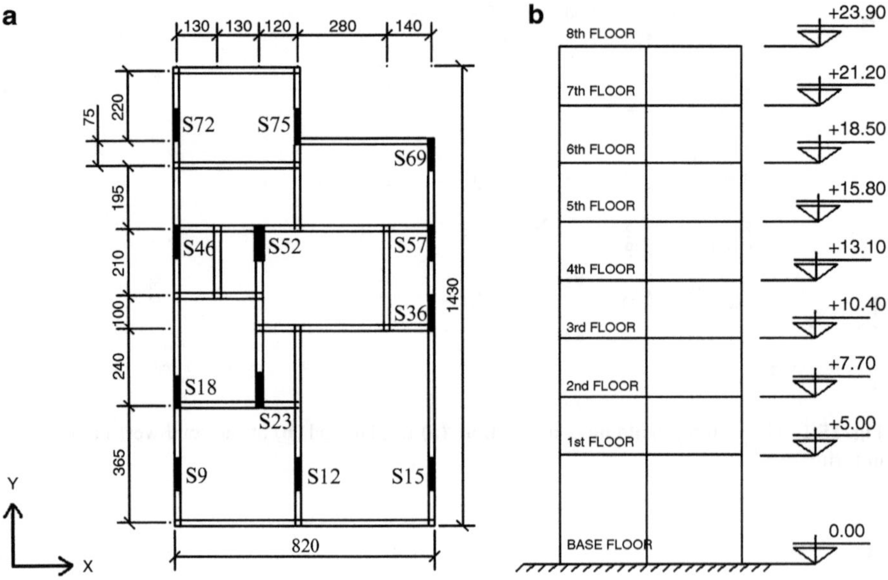

Fig. 15.3 Eight-storey building configuration. (**a**) Plan view (cm) and (**b**) lateral view (m)

The concrete was represented by a uniaxial model that follows the constitutive relationship proposed by Mander et al. (1988) and the cyclic rules proposed by Martinez-Rueda and Elnashai (1997). The confinement effects provided by the lateral transverse reinforcement are taken into account through the rules proposed by Mander et al. (1988) whereby constant confining pressure is assumed throughout the entire stress-strain range. A compressive strength of 25 MPa was considered for the SPEAR building and 16.7 MPa for the Turkish buildings. The constitutive model used for the steel was the one proposed by Menegotto and Pinto (1973) coupled with the isotropic hardening rules proposed by Filippou et al. (1983). The average yield strength of 360 MPa was assumed for the SPEAR building and 371 MPa for the Turkish buildings.

The rigid diaphragm modelling calibration of the SPEAR building can be found in Pinho et al. (2008). The comparisons between the analytical results and the experimental tests for this structure are described in Bento et al. (2010).

15.5 Seismic Assessment

In this section, the parametric study is described as well as the seismic action definition and the numerical model used in the performed structural analyses.

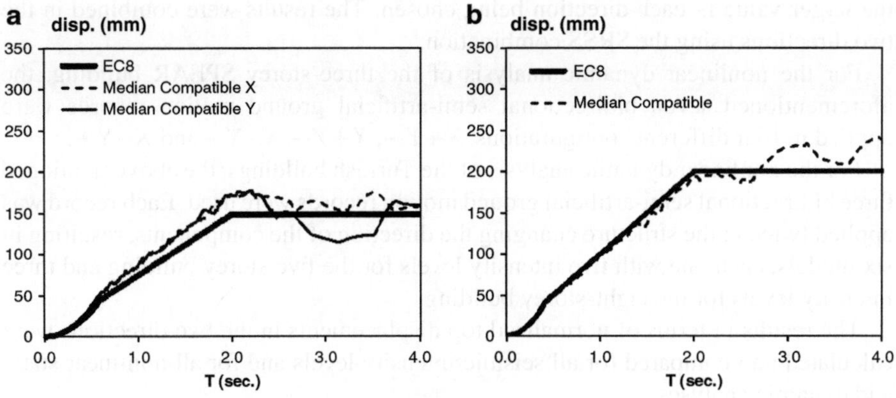

Fig. 15.4 Displacement response spectra. (**a**) Three-storey and (**b**) five- and eight-storey buildings

15.5.1 Seismic Action

Seven bidirectional semi-artificial ground motion records from the SPEAR project fitted to the EC8 (CEN 2004) elastic design spectrum (type 1 soil C) were used in the three-storey building case. For the five- and eight-storey buildings, combinations of three bidirectional semi-artificial ground motion records were applied. The three considered ground motions are real records from the PEER's database website (PEER 2009). They were fitted to the Eurocode 8 elastic design spectrum (with the Turkish code features – type 1 soil A) using the software RSPMatch2005 (Hancock et al. 2006).

The ground motions were scaled and applied for a wide range of peak ground intensities in order to assess the performance of the NSPs throughout different levels of structural inelasticity. The accelerograms were scaled for peak ground accelerations of 0.05, 0.1, 0.2 and 0.3 g for the three-storey building and of 0.1, 0.2, 0.4, 0.6 and 0.8 g for the Turkish buildings. The median displacement response spectra of each set of ground motions were used to compute the nonlinear static procedures response. They are represented in Fig. 15.4a, b as defined for the three-storey building and for the Turkish buildings, respectively. In these figures are also plotted the EC8's response spectra with which the real accelerograms were matched.

15.5.2 Structural Analyses Performed

In the pushover analyses, lateral forces were applied to the structure in the form of modal load pattern (for the N2 method, a uniform distribution was also applied). The loads were applied independently in the two horizontal positive/negative directions, resulting in four analyses. For each one, the target displacement was computed with the larger value in each direction being chosen. The results were combined in the two directions using the SRSS combination.

For the nonlinear dynamic analysis of the three-storey SPEAR building, the aforementioned seven bidirectional semi-artificial ground motion records were applied in four different configurations: $X+Y+$, $X+Y-$, $X-Y-$ and $X-Y+$.

For the nonlinear dynamic analysis of the Turkish buildings, the abovementioned three bidirectional semi-artificial ground motion records were used. Each record was applied twice in the structure changing the direction of the components, resulting in six models, each one with five intensity levels for the five-storey building and three intensity levels for the eight-storey building.

The results in terms of normalised top displacements in the two directions were calculated and compared for all seismic intensity levels and for all nonlinear static and dynamic analyses.

15.6 Dynamic Properties of the Case Studies

The modal properties of the three analysed buildings are herein presented. The three-storey building has a fundamental mode of 0.617 s characterised by translation along the *X* direction, a second mode of 0.527 s with torsional motion and a third mode of 0.441 s with translation along the *Y* direction. It is mentioning that, in this case study, both translational modes are coupled with torsion. The five-storey building presents a first mode of 0.617 s with translation along the *X* direction, a second mode of 0.593 s with translation along the *Y* direction and a third mode of 0.509 s with torsional motion. Finally, the eight-storey building has a first mode of 1.445 s with translation along the *X* direction, a second mode of 0.636 s with torsional motion and a third mode of 0.482 s with translation along the *Y* direction. Further details on the modal results can be found in Bhatt and Bento (2011b).

The analysed buildings present torsional features due to their irregularities in plan. According to Fajfar et al. (2005a), the period ratios of a structure have an important influence on its torsional behaviour. The period ratios Ωx and Ωy are defined as the uncoupled translational period divided by the uncoupled torsional period in the *X* and *Y* directions. The influence of the predominantly torsional mode of vibration on the response in the direction considered when compared with the predominantly translational mode increases if the period ratio decreases. Structures with period ratios larger than 1 are usually classified as torsionally stiff and structures with period ratios smaller than 1 as torsionally flexible. A structure can be torsionally stiff in one direction and torsionally flexible in the other. In Table 15.1 are represented the period ratios in the *X* and *Y* directions for the three analysed buildings. One can conclude that the three-storey building is classified as torsionally stiff in the *X* direction and torsionally flexible in the *Y* direction. The five-storey building is torsionally stiff in both *X* and *Y* directions. The eight-storey building is torsionally stiff in the *X* direction and torsionally flexible in the *Y* direction.

Table 15.1 Period ratios

	Ωx	Ωy
3-storey building	1.2	0.8
5-storey building	1.2	1.2
8-storey building	2.3	0.8

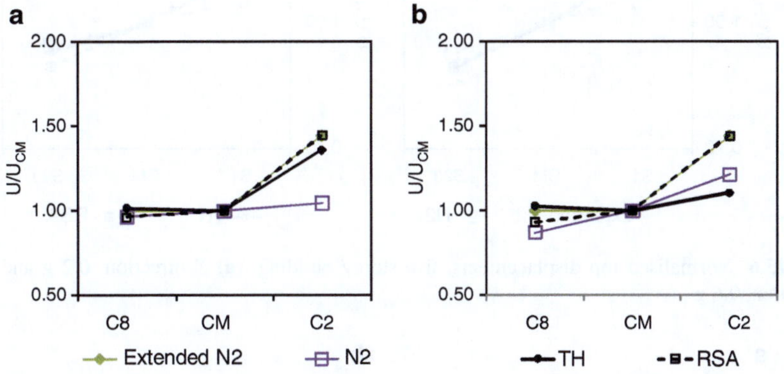

Fig. 15.5 Normalised top displacements, three-storey building. (**a**) *X* direction, 0.1 g and (**b**) *Y* direction, 0.2 g

15.7 Results of the Parametric Studies

The results obtained with the extension of the N2 method and of the CSM-FEMA440 for plan-asymmetric buildings are plotted against the median time history nonlinear dynamic results (TH), the response spectrum analysis (RSA) and the original procedures, in terms of normalised top displacements. When dealing with plan-asymmetric buildings, the normalised top displacements is the measure one should analyse in order to understand the torsional behaviour of the structure (Fajfar et al. 2005a). This measure is obtained by normalising the edge displacement values with respect to those of the centre of mass. Several plots are presented showing the performance of the analysed procedures in estimating the torsional motion of the evaluated buildings.

In Figs. 15.5, 15.6 and 15.7 are depicted the results of the extended N2 method and in Figs. 15.8, 15.9 and 15.10 are presented the results of the extended CSM-FEMA440, for the three analysed buildings and for different seismic intensities. More results on the performance of the extended N2 method and of the extended CSM-FEMA440 can be found in Bhatt and Bento (2011b) and in Bhatt and Bento (2011a), respectively.

From the obtained results, one can observe that the two original N2 and CSM-FEMA440 methods lead to similar normalised top displacements in the three analysed buildings. The same happens with both extended procedures.

One can conclude that torsional effects are generally higher for lower ground motion intensities. For increasing seismic intensities, one can understand a flattening

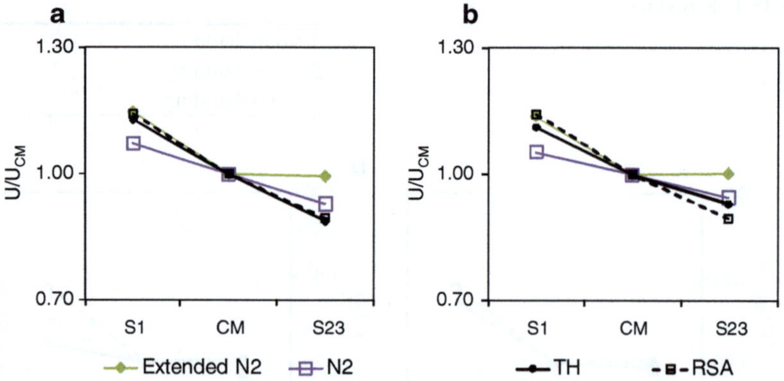

Fig. 15.6 Normalised top displacements, five-storey building. (**a**) *X* direction, 0.2 g and (**b**) *X* direction, 0.6 g

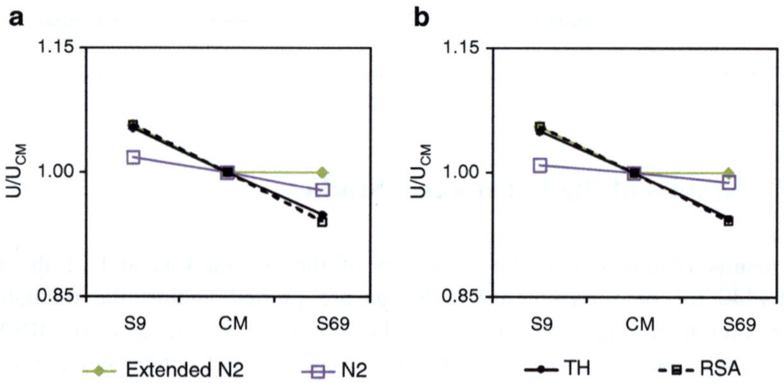

Fig. 15.7 Normalised top displacements, eight-storey building. (**a**) *X* direction, 0.2 g and (**b**) *X* direction, 0.4 g

on the normalised top displacements. This can be observed in all the analysed buildings. This conclusion confirms the idea that torsional effects are generally smaller in the inelastic range compared to what happens in the elastic one.

The plots clearly show that the RSA estimates an upper bound of the torsional amplification on the flexible side of the buildings, in both elastic and inelastic ranges.

The extended procedures reproduce in a very good fashion the nonlinear dynamic results for all the buildings analysed and through all the seismic intensities tested. These methods show, for these case studies, a much better performance in estimating the torsional behaviour of the buildings than the original methods. Generally the last ones are not capable to reproduce the torsional response of the buildings.

The abovementioned plots show that the extended methods reproduce in a very accurate way the torsional amplification on the flexible edge in all the buildings

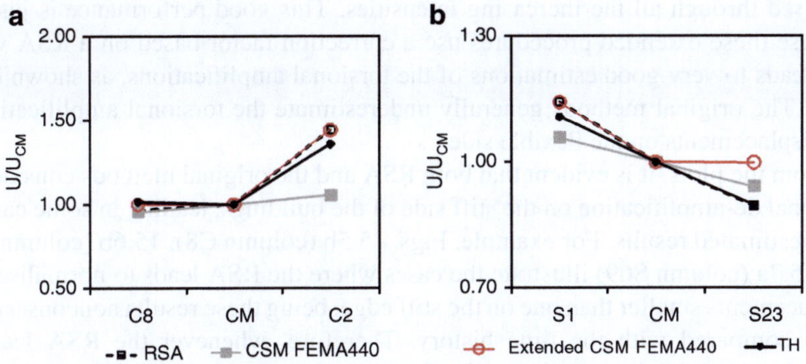

Fig. 15.8 Normalised top displacements. (**a**) Three-storey building, *X* direction, 0.1 g and (**b**) five-storey building, *X* direction, 0.1 g

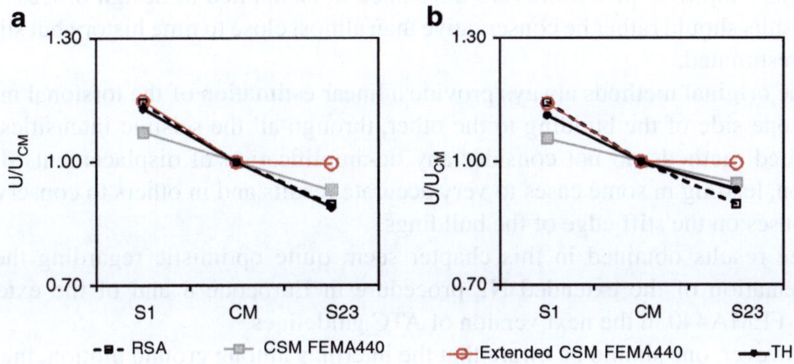

Fig. 15.9 Normalised top displacements, five-storey building. (**a**) *X* direction, 0.2 g and (**b**) *X* direction, 0.6 g

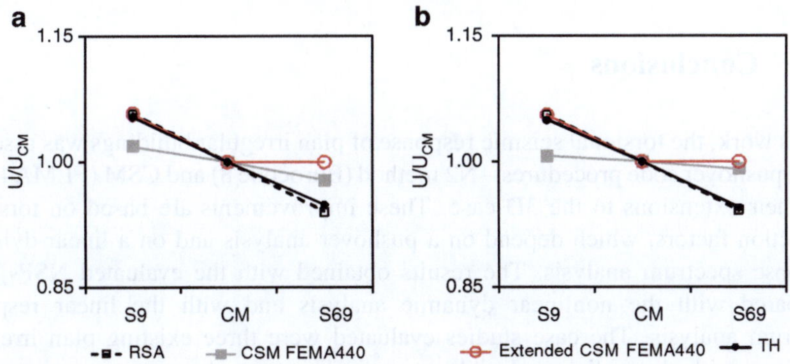

Fig. 15.10 Normalised top displacements, eight-storey building. (**a**) *X* direction, 0.2 g and (**b**) *X* direction, 0.4 g

analysed through all the increasing intensities. This good performance is justified because these extended procedures use a correction factor based on a RSA which also leads to very good estimations of the torsional amplifications, as shown in the plots. The original methods generally underestimate the torsional amplification of the displacements on the flexible side.

From the plots, it is evident that both RSA and the original methods consider the torsional de-amplification on the stiff side of the buildings, leading in some cases to underestimated results. For example, Figs. 15.5b (column C8), 15.6b (column S23) and 15.7a (column S69) illustrate the cases where the RSA leads to normalised top displacements smaller than one on the stiff edge, being these results nonconservative when compared with the time history. Therefore, whenever the RSA leads to normalised top displacements smaller than one, the extended methods consider this value to be equal to one. This recommendation avoids the extended methods to produce nonconservative results on this stiff edge. One can say that this is a safe criterion for designing and assessment of structures. In fact it must not be forgotten that this simplified procedures are developed to be applied in design offices where the results should rather be conservative than almost close to time history but slightly underestimated.

The original methods always provide a linear estimation of the torsional motion from one side of the building to the other, through all the seismic intensities. The extended methods do not consider any de-amplification of displacements due to torsion, leading in some cases to very accurate results and in others to conservative responses on the stiff edge of the buildings.

The results obtained in this chapter seem quite optimistic regarding the implementation of the extended N2 procedure in Eurocode 8 and of the extended CSM-FEMA440 in the next version of ATC guidelines.

However, one should be aware that the interplay among ground motion, inelastic amplification or de-amplification of displacements and structural system is complex. Therefore, more studies in different buildings should be developed in order to consolidate these nonlinear static approaches.

15.8 Conclusions

In this work, the torsional seismic response of plan irregular buildings was assessed using pushover code procedures – N2 method (Eurocode 8) and CSM (FEMA440) – and their extensions to the 3D case. These improvements are based on torsional correction factors, which depend on a pushover analysis and on a linear dynamic response spectrum analysis. The results obtained with the evaluated NSPs were compared with the nonlinear dynamic analysis and with the linear response spectrum analysis. The case studies evaluated were three existing plan irregular buildings with three, five and eight storeys. The procedures were evaluated by comparing the results in terms of normalised top displacements of the three case studies for different ground motion intensities. This measure gives an idea about the torsional behaviour of the structures.

The results obtained from this study showed that torsional effects are in general higher for lower ground motion intensities. In fact, for increasing seismic intensities, one can notice a flattening on the normalised top displacements of each building. This confirms the idea that torsional effects are generally smaller in the inelastic range than in the elastic stage.

The extended methods performed in a much more accurate way than their original counterparts in estimating the torsional behaviour of all buildings analysed through all the seismic intensities tested. They generally captured in a very precise way the torsional amplification in terms of displacements on the flexible side of the buildings.

The extended procedures do not take into account any de-amplification of displacements due to torsion. Therefore, the response on the stiff side of the buildings was in some cases estimated in a very precise way by the methods and overestimated in others.

The original methods are not capable in general to reproduce the torsional motion of the buildings, usually leading to a linear estimation of the torsional motion from one side of the building to the other.

The original procedures considered the de-amplification on the stiff side of the buildings, underestimating in some cases the response. On the flexible side, the normalised top displacements were generally nonconservative in respect to the time history results.

Recently, several procedures have been proposed taking into account torsion in simplified nonlinear static procedures; however, definitive answers have not yet been reached. The results obtained herein added to the ones already published (Fajfar et al. 2005a, b; D'Ambrisi et al. 2009; Bhatt and Bento 2011b; Koren and Kilar 2011), confirm the idea that the extended N2 method has a potential to be implemented in the next version of Eurocode 8 in order to correctly estimate the torsional response in real plan-asymmetric RC buildings through the use of pushover analysis.

The extension of the CSM-FEMA440 also presents potential to be incorporated in future codes, namely, the ATC guideline. However, this procedure should be further tested in order to consolidate definitive conclusions.

Acknowledgements The authors would like to thank Dr Ihsan Bal for providing valuable data related to research work on the topic of this chapter. A special acknowledgement to Dr Rui Pinho for the valuable discussions that certainly enriched the work herein presented. The authors would like to acknowledge the financial support of the Portuguese Foundation for Science and Technology (Ministry of Science and Technology of the Republic of Portugal) through the research project PTDC/ECM/100299/2008 and through the PhD scholarship SFRH/BD/28447/2006 granted to Carlos Bhatt.

References

ATC, Applied Technology Council (1996) Seismic evaluation and retrofit of concrete buildings, vols 1 and 2, Report no. ATC-40. Redwood City, CA

ATC (2005) Improvement of nonlinear static seismic analysis procedures, FEMA440 report. Redwood City, CA

Bento R, Bhatt C, Pinho R (2010) Using nonlinear static procedures for seismic assessment of the 3D irregular SPEAR building. Earthq Struct 1(2):177–195

Bhatt C, Bento R (2011a) Extension of the CSM-FEMA440 to plan-asymmetric real building structures. Earthq Eng Struct Dyn 40(11):1263–1282

Bhatt C, Bento R (2011b) Assessing the seismic response of existing RC buildings using the extended N2 method. Bull Earthq Eng 9(4):1183–1201

Bhatt C, Bento R, Pinho R (2010) Verification of nonlinear static procedures for a 3D plan-irregular building in Turkey. In: Proceedings of the 14th European conference on earthquake engineering, Ohrid, Macedonia

CEN, Comité Européen de Normalisation (2004) Eurocode 8: design of structures for earthquake resistance. Part 1: general rules, seismic actions and rules for buildings. EN 1998-1:2004. Brussels, Belgium

Chopra AK, Goel RK (2004) A modal pushover analysis procedure to estimate seismic demands for unsymmetric-plan buildings. Earthq Eng Struct Dyn 33:903–927

D'Ambrisi A, Stefano M, Tanganelli M (2009) Use of pushover analysis for predicting seismic response of irregular buildings: a case study. J Earthq Eng 13:1089–1100

Erduran E, Ryan K (2011) Effects of torsion on the behaviour of peripheral steel-braced frame systems. Earthq Eng Struct Dyn 40(5):491–507

Fajfar P (2000) A nonlinear analysis method for performance-based seismic design. Earthq Spectra 16(3):573–592

Fajfar P, Fischinger M (1988) N2 – a method for non-linear seismic analysis of regular buildings. In: Proceedings of the ninth world conference in earthquake engineering, Tokyo-Kyoto, Japan, pp 111–116

Fajfar P, Marusic D, Perus I (2005a) Torsional effects in the pushover-based seismic analysis of buildings. J Earthq Eng 9(6):831–854

Fajfar P, Marušic D, Perus I (2005b) The extension of the N2 method to asymmetric buildings. In: Proceedings of the 4th European workshop on the seismic behaviour of irregular and complex structures, Thessaloniki

Fardis MN (2002) Design of an irregular building for the SPEAR project – description of the 3-storey structure. Research report. University of Patras, Greece

Fardis MN, Negro P (2006) SPEAR – Seismic performance assessment and rehabilitation of existing buildings. In: Proceedings of the international workshop on the SPEAR project, Ispra, Italy

Filippou FC, Popov EP, Bertero VV (1983) Modelling of R/C joints under cyclic excitations. J Struct Eng 109(11):2666–2684

Freeman SA (1998) Development and use of capacity spectrum method. In: Proceedings of the sixth U.S. national conference on earthquake engineering, Seattle, Oakland, USA

Freeman SA, Nicoletti JP, Tyrell JV (1975) Evaluation of existing buildings for seismic risk – a case study of Puget Sound Naval Shipyard, Bremerton, Washington. In: Proceedings of U.S. national conference on earthquake engineering, Berkeley, USA, pp 113–122

Hancock J, Watson-Lamprey J, Abrahamson NA, Bommer JJ, Markatis A, McCoy E, Mendis R (2006) An improved method of matching response spectra of recorded earthquake ground motion using wavelets. J Earthq Eng 10(S1):67–89

Koren D, Kilar V (2011) The applicability of the N2 method to the estimation of torsional effects in asymmetric base-isolated buildings. Earthq Eng Struct Dyn 40(8):867–886

Mander JB, Priestley MJN, Park R (1988) Theoretical stress-strain model for confined concrete. ASCE J Struct Eng 114(8):1804–1826

Martinez-Rueda JE, Elnashai AS (1997) Confined concrete model under cyclic load. Mater Struct 30(197):139–147

Menegotto M, Pinto PE (1973) Method of analysis for cyclically loaded RC plane frames including changes in geometry and non-elastic behaviour of elements under combined normal force and bending. In: Symposium on the resistance and ultimate deformability of structures anted on by well defined loads, International Association for Bridges and Structural Engineering, Zurich, Switzerland, pp 15–22

PEER (2009) Strong Ground Motion Database, Available from: http://peer.berkeley.edu/nga/

Pinho R, Bhatt C, Antoniou S, Bento R (2008) Modelling of the horizontal slab of a 3D irregular building for nonlinear static assessment. In: Proceedings of the 14th world conference on earthquake engineering, Beijing, China, pp 05-01-0159

Seismosoft (2006) Seismostruct – a computer program for static and dynamic nonlinear analysis of framed structures. Available online from URL: www.seismosoft.com

Vuran E, Bal YE, Crowley H, Pinho R (2008) Determination of equivalent SDOF characteristics of 3D dual structures. In: Proceedings of the 14th world conference on earthquake engineering, Beijing, China, pp S15–031

Chapter 16
Analysis of Collapse of Irregular Tall Structures Using Mixed Lagrangian Formulation

Georgios Apostolakis, Andrei M. Reinhorn, Gary Dargush, Oren Lavan, and Mettupalayam Sivaselvan

Abstract Recent extreme events, such as earthquakes, hurricanes, terrorist attacks, and fire, showed that the best designed tall structures can be vulnerable to collapse. A new approach was developed by the authors using the mixed Lagrangian formulation for analysis of tall structures. The formulation attempts to solve problems using a force-based approach in which momentum appears explicitly and can be used to deal with structures where deterioration and fracture occur before collapse. In this chapter, a time integral of functions of the response over the duration of the response is considered. The kernel of the integral consists of two functions – the Lagrangian and the dissipation function – of the response variables that describe the configuration of the structure and their rates. The Lagrangian function is energy-like and describes the conservative characteristics of the system, while the dissipation function similar to a flow potential describes the dissipative characteristics. Irregular structures are extremely sensitive to loss of supporting column, which can lead to amplified out-of-plane and torsional motions. This chapter presents the formulation using the mixed Lagrangian formulation (MLF) which allows considering the three-dimensional effects during collapse. An example, using a fifteen-story building, is presented, to illustrate the performance of the above approach. The structure is

---

G. Apostolakis • A.M. Reinhorn (✉) • M. Sivaselvan
Department of Civil, Structural and Environmental Engineering, University at Buffalo, The State University of New York, Buffalo, NY 14260, USA
e-mail: ga59@buffalo.edu; reinhorn@buffalo.edu; siva@colorado.edu

G. Dargush
Department of Mechanical and Aerospace Engineering, University at Buffalo, The State University of New York, Buffalo, NY 14260, USA
e-mail: gdargush@buffalo.edu

O. Lavan
Faculty of Civil and Environmental Engineering, Technion – Israel Institute of Technology, Haifa 32000, Israel
e-mail: lavan@technion.ac.il

O. Lavan and M. De Stefano (eds.), *Seismic Behaviour and Design of Irregular and Complex Civil Structures*, Geotechnical, Geological and Earthquake Engineering 24, DOI 10.1007/978-94-007-5377-8_16,

irregular and the loss of either a single or several columns on the periphery leads to an excessive sway and loss of capability to sustain the gravity loads (i.e., collapse). The failure of eccentrically located columns in an otherwise symmetric building leads to the coupled lateral-torsional collapse, well captured by the formulation presented herein.

16.1 Introduction

Recent extreme events, such as earthquakes, hurricanes, terrorist attacks, and fire, showed that the best designed tall structures can be vulnerable to collapse. The current state of the art in analysis of tall structures offers very few solutions for post elastic behavior and structure instability leading to collapse. Most advanced analytical solutions are developed for planar structural systems which are used to analyze the entire structure in each lateral direction separately. Although several analytical tools are capable to analyze structures with both geometric and material nonlinearities in three dimensions, these computational platforms are cost ineffective when a collapse analysis is desired.

A new approach was developed by the authors using the mixed Lagrangian formulation for analysis of tall structures (Sivaselvan et al. 2009; Sivaselvan and Reinhorn 2006). The formulation attempts to solve problems using a force-based approach in which momentum appears explicitly and can be used to deal with structures where deterioration (Sivaselvan et al. 2009), fracture (Lavan et al. 2009), and contact (Lavan 2010) occur before collapse. In conventional formulations, the response of the structure is considered as the solution of a set of differential equations in time. In this chapter, a time integral of functions of the response over the duration of the response is considered. The kernel of the integral consists of two functions – the Lagrangian and the dissipation function – of the response variables that describe the configuration of the structure and their rates. The integral is called the action integral. It is shown here that in considering elastic-plastic systems, it is natural to also include the time integrals of internal forces in the structure as configuration variables. The Lagrangian function is energy-like and describes the conservative characteristics of the system, while the dissipation function similar to a flow potential describes the dissipative characteristics. In a conservative system, the action integral is rendered stationary (maximum, minimum, or saddle point) by the response. For nonconservative systems such as elastic-plastic systems, such a variational statement is not possible, and only a weak form which is not a total integral is possible. It is shown however that the form of the Lagrangian is invariant under finite deformations. Such a formulation enables the construction of numerical integration schemes, applicable to three-dimensional structures.

An implementation of the method is made for tall structures considering sparsity of resulting matrices. The formulation includes both geometric and material nonlinearities. The formulation is suitable for analyses of structures in which columns are in jeopardy in case of extreme loading.

Irregular structures are extremely sensitive to loss of a single gravity-load supporting columns, which can lead to amplified out-of-plane and torsional motions. Such failure is intolerable and in most cases may lead to a complete collapse. This chapter presents the formulation using the mixed Lagrangian formulation (MLF) which allows considering the three-dimensional effects during collapse. An example, using a fifteen-story building, is presented, to illustrate the performance of the above approach. The structure is irregular and the loss of either a single or several columns on the periphery leads to an excessive sway and loss of capability to sustain the gravity loads (i.e., collapse).

16.2 Mixed Lagrangian Formalism

The Hamiltonian approach in this chapter considers both conservative and dissipative processes. An overview of the mixed Lagrangian formulation for spatially discrete multiple-degree-of-freedom (MDOF) systems is presented below. For more information on the Lagrangian formulation, the reader is referred to (Sivaselvan and Reinhorn 2006; Sivaselvan et al. 2009). The weak formulation starts with the proper selection of state variables and the construction of a Lagrangian L and dissipation Φ functions. In principle, the Lagrangian function includes the conservative characteristics of the system, while the dissipation function incorporates the nonconservative characteristics of the system. Both are functions of the generalized coordinates q_k of the system and their first-order time derivatives for $k = 1, \ldots n$. The action integral is introduced, and by applying Hamilton's action principle (Hamilton 1834), one arrives at the Euler-Lagrange equations, which are the governing equations of the system.

Partitioning the degrees of freedom (DOF) of the structure into those that have associated mass of damping, those that do not have mass or damping, and those with prescribed displacement or velocity, the governing equations of the structure are

$$\mathbf{M}\dot{\mathbf{v}} + \mathbf{C}\mathbf{v} + \mathbf{B}_1\mathbf{F} = \mathbf{P} \tag{16.1a}$$

$$\mathbf{A}\dot{\mathbf{F}} + \partial_{\mathbf{F}}\Phi(\mathbf{F},\boldsymbol{\zeta}) - \mathbf{B}_1^{\mathbf{T}}\mathbf{v} - \mathbf{B}_2^{\mathbf{T}}\mathbf{v}^{\mathbf{o}} - \mathbf{B}_3^{\mathbf{T}}\mathbf{v}^{\mathbf{p}} = \mathbf{P} \tag{16.1b}$$

$$\mathbf{B}_2\mathbf{F} = \boldsymbol{\Pi} \tag{16.1c}$$

$$\mathbf{G}\dot{\boldsymbol{\zeta}} + \partial_{\boldsymbol{\zeta}}\Phi(\mathbf{F},\boldsymbol{\zeta}) = \mathbf{0} \tag{16.1d}$$

where $\mathbf{M}$ is the mass matrix, $\mathbf{C}$ is the damping matrix, $\mathbf{B}$ is the equilibrium matrix, $\mathbf{A}$ is the block diagonal matrix of element flexibilities, $\mathbf{G}$ is the block diagonal matrix of inverse hardening moduli, $\mathbf{F}$ is the vector of internal forces in the structure, $\mathbf{v}$ is the vector of velocities at DOF with associated mass or damping, $\mathbf{v}^{\mathbf{o}}$ is the vector of velocities at DOF with neither mass nor damping (quasi-static), $\mathbf{v}^{\mathbf{p}}$ is the vector of velocities at DOF with prescribed displacement/velocity, $\boldsymbol{\zeta}$ is the vector of internal variables (plastic strains and time derivatives), $\mathbf{P}$ is the vector of external forces on DOF with mass or damping, and $\boldsymbol{\pi}$ is the vector of external forces at quasi-static DOF. When geometric nonlinearities are considered, the equilibrium matrix $\mathbf{B}$ is a function of displacement and is partitioned as $\mathbf{B}=[\mathbf{B_1}^{\mathrm{T}}, \mathbf{B_2}^{\mathrm{T}}, \mathbf{B_3}^{\mathrm{T}}]^{\mathrm{T}}$ corresponding to DOF with mass or damping, quasi-static DOF, and DOF with prescribed displacement or velocity, respectively. The governing equations include the linear momentum equation, the deformation compatibility in the elements equation, the equilibrium equation at quasi-static DOF, and the evolution of constitutive internal variables equation, respectively.

16.3 Discrete Lagrangian Formalism

The governing equations of a structural system, obtained by the Lagrangian formalism presented in the previous section, are to be solved numerically. The proposed scheme is a discrete variational integrator based on Cadzow's discrete calculus of variations (Cadzow 1970). Starting by constructing a discrete Lagrangian L_{d}, which is an approximation of the continuous Lagrangian L, the discrete action sum S_{d} that corresponds to L_{d} is constructed, where

$$S_{\mathrm{d}} = \sum_{k=1}^{N} L_{\mathrm{d}}\left(q_{k-1}, q_k\right) \tag{16.2}$$

while N is the total number of time-steps and q_k represents the discrete value of a state variable at step k.

Following Cadzow's discrete variational calculus, the solution of the discrete system governed by the Lagrangian L_{d} is the one that produces the extreme discrete action sum S_{d}. The action integral is discretized using the midpoint rule with a time-step h. The resulting discrete set of equations in terms of the state variables introduced earlier is

$$\mathbf{M}\left(\frac{\mathbf{v}_{n+1}-\mathbf{v}_n}{h}\right)+\mathbf{C}\left(\frac{\mathbf{v}_{n+1}+\mathbf{v}_n}{2}\right)+\mathbf{B}_1\left(\frac{\mathbf{F}_{n+1}+\mathbf{F}_n}{2}\right)=\left(\frac{\mathbf{P}_{n+1}+\mathbf{P}_n}{2}\right) \tag{16.3a}$$

$$\left(\mathbf{A}+\frac{h^2}{4}\mathbf{B}_1^{\mathrm{T}}\tilde{\mathbf{M}}^{-1}\mathbf{B}_1\right)\mathbf{F}_{n+1}+\frac{h}{2}\partial_{\mathbf{F}}\Phi\left(\mathbf{F}_{n+1},\zeta_{n+1}\right)-\mathbf{B}_2^{\mathrm{T}}\boldsymbol{\mu}_{n+1}-\mathbf{B}_3^{\mathrm{T}}\boldsymbol{\mu}_{n+1}^{\mathbf{p}}=\boldsymbol{\beta} \tag{16.3b}$$

$$\mathbf{B}_2\mathbf{F}_{n+1}=\boldsymbol{\Pi}_{n+1} \tag{16.3c}$$

$$\mathbf{G}\boldsymbol{\zeta}_{n+1} + \frac{h}{2}\partial_{\zeta}\Phi\left(\mathbf{F}_{n+1}, \zeta_{n+1}\right) = \boldsymbol{\gamma} \tag{16.3d}$$

where

$$\tilde{\mathbf{M}} = \mathbf{M} + \frac{h}{2}\mathbf{C}$$

$$\boldsymbol{\beta} = \mathbf{A}\mathbf{F}_n + \mathbf{B}_3^{\mathrm{T}}\boldsymbol{\mu}_{n+1}^{\mathbf{P}} + \frac{h^2}{4}\mathbf{B}_1^{\mathrm{T}}\tilde{\mathbf{M}}^{-1}\left(\mathbf{P}_{n+1} + \mathbf{P}_n - \mathbf{B}_1\mathbf{F}_n + \frac{4}{h}\mathbf{M}\mathbf{v}_n\right) - \frac{h}{2}\partial_{\mathbf{F}}\Phi\left(\mathbf{F}_n, \zeta_n\right)$$

$$\boldsymbol{\mu}_{n+1} = \frac{h}{2}\left(\mathbf{v}_{n+1} + \mathbf{v}_n\right) = \text{displacement increment in quasi} - \text{static DOF}$$

$$\boldsymbol{\mu}_{n+1}^{\mathbf{P}} = \frac{h}{2}\left(\mathbf{v}_{n+1}^{\mathbf{P}} + \mathbf{v}_n^{\mathbf{P}}\right) = \text{prescribed displacement increment}$$

$$\boldsymbol{\gamma} = \mathbf{G}\boldsymbol{\zeta}_n - \frac{h}{2}\partial_{\zeta}\Phi\left(\mathbf{F}_n, \zeta_n\right)$$

It can be shown that if either internal force **F** or velocity **v** state variable is eliminated from the above set of discrete equations, the remaining set of equations can be defined as a minimization problem. The minimization problem, in terms of internal force vector **F** as state variable, can be restated as

$$\left(\mathbf{F}_{n+1}, \zeta_{n+1}\right) = \arg\min \frac{1}{2}\mathbf{F}^{\mathrm{T}}\left(\mathbf{A} + \frac{h^2}{4}\mathbf{B}_1^{\mathrm{T}}\tilde{\mathbf{M}}^{-1}\mathbf{B}_1\right)\mathbf{F} + \frac{1}{2}\boldsymbol{\zeta}^{\mathrm{T}}\mathbf{G}\boldsymbol{\zeta} - \boldsymbol{\beta}^{\mathrm{T}}\mathbf{F} - \boldsymbol{\gamma}^{\mathrm{T}}\boldsymbol{\zeta}$$

$$\text{subject to } \mathbf{B}_2\mathbf{F} = \boldsymbol{\Pi}_{n+1} \text{ and } \frac{h}{2}\varphi\left(\mathbf{F}\right) \leq 0 \tag{16.4}$$

where the yield function of the structure $\varphi(.)$ appears as a constraint. The minimization problem (16.4) is solved as a sequence of linear equality-constraint problems, where the constraints are represented by an augmented Lagrangian approach. For a detailed presentation of the minimization approach, the reader is referred to (Sivaselvan and Reinhorn 2006; Sivaselvan et al. 2009).

16.4 Numerical Applications

Numerical examples, using a fifteen-story building, are presented, to illustrate the performance of the above approach. The structure is irregular and the loss of either a single or several columns on the periphery leads to an excessive sway and loss of capability to sustain the gravity loads (i.e., collapse). In addition, the applicability of the algorithm to collapse simulation of large-scale structures is demonstrated and collapse mechanism modes are identified.

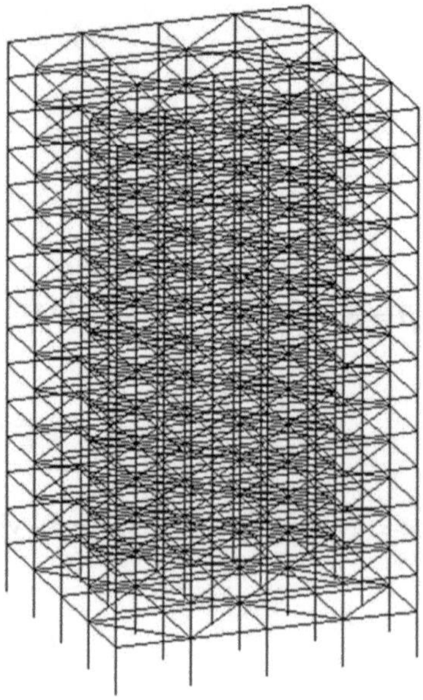

Fig. 16.1 Fifteen-story building example

The fifteen-story building is illustrated in Fig. 16.1. The building has four bays in each direction. In plan view the building is roughly 30 m × 30 m and the total height is 55 m. In all analyses geometric nonlinearities are included, and the elastic-plastic behavior of the members is modeled. The damping ratio is assumed 5%. For each example a convergence study was performed using time-steps of 0.02, 0.01, and 0.005 s. The results presented here are for the smaller time-step of 0.005 s.

16.4.1 Example 1: Symmetric Collapse Mechanism Under Gravity Loads

In this simulation, a symmetric collapse mechanism of the fifteen-story building is demonstrated. The approach used for initiating collapse is weakening of base story columns. For the analyses of the building, two analysis steps are employed. The first step is a nonlinear static analysis where 1.5 times the gravity loads are applied in ten increments. The above factor on the gravity loads is selected in order to speed up the collapse of the building. The second step is nonlinear dynamic analysis where the instantaneous weakening of base story columns occurs.

A plan view of the first story with the position of the weakened columns is shown in Fig. 16.2. From Fig. 16.2, it can be seen that the first two rows of base columns are weakened (columns 1–10). The deformed shape of the building as collapse progresses in bird view (azimuth=20° and elevation=20°) and plan view (azimuth=20° and elevation=90°) is shown at different time instances in Figs. 16.3 and 16.4, respectively.

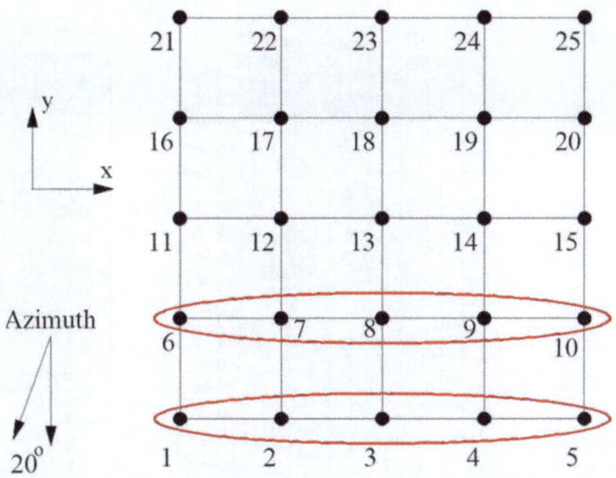

Fig. 16.2 Example 1 – Plan view of building with weakened base columns

The weakening of the front two base story column rows results in loss of the building's capability to sustain the gravity loads which leads to collapse of the building. The collapse mechanism of the building is a symmetric one and progresses as collapse of all the bays directly above the weakened base story columns as a single block. The rest of the building is leaning toward the collapsed part of the building (Figs. 16.3 and 16.4).

In Fig. 16.5, the number of iterations of the simulation is plotted at each time increment. It is seen that the number of iterations is small and does not change during the progression of collapse.

The displacement response of the building at different story levels (nodes directly above node 3, see Fig. 16.2) in the *y*-direction and the *z*-direction is presented in Fig. 16.6, respectively. From Fig. 16.6, it is seen that all the stories are moving as a block in the vertical direction.

16.4.2 Example 2: Torsional Collapse Mechanism – Gravity Loads

In this simulation, a torsional collapse mechanism of the fifteen-story building is demonstrated. The approach used for initiating collapse is weakening of the base story columns. For the analyses of the building, two analysis steps are employed. The first step is a nonlinear static analysis where 1.5 times the gravity loads are applied in ten increments. The second step is nonlinear dynamic analysis where the instantaneous weakening of base story columns occurs in two stages.

A plan view of the first story with the position of the weakened columns is shown in Fig. 16.7. The same number of columns as example 1 is weakened, but in the present simulation, the weakening of the columns takes place in two stages. The first

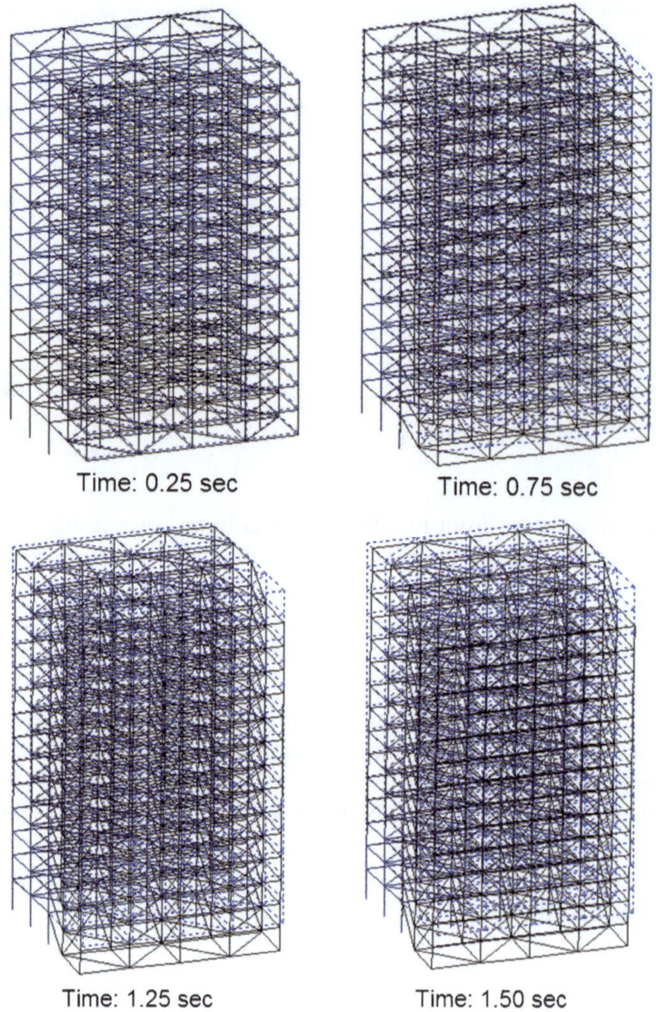

Fig. 16.3 Deformed shape of the building as collapse progresses in bird view

instantaneous weakening of columns 1–4 and 6–9 occurs at time $t = 0$ s. The second instantaneous weakening of columns 11–12 occurs at time $t = 2$ s. The deformed shape of the building as collapse progresses plan view is shown at different time instances in Fig. 16.8.

The weakening of the first set of columns at the base story results in loss of the building's capability to sustain the gravity loads which leads to significant sway of the building and initiation of collapse. The weakening of the second set of columns results to excessive sway of the building and progression of the collapse mechanism. The collapse mechanism of the building is torsional and progresses as collapse of all

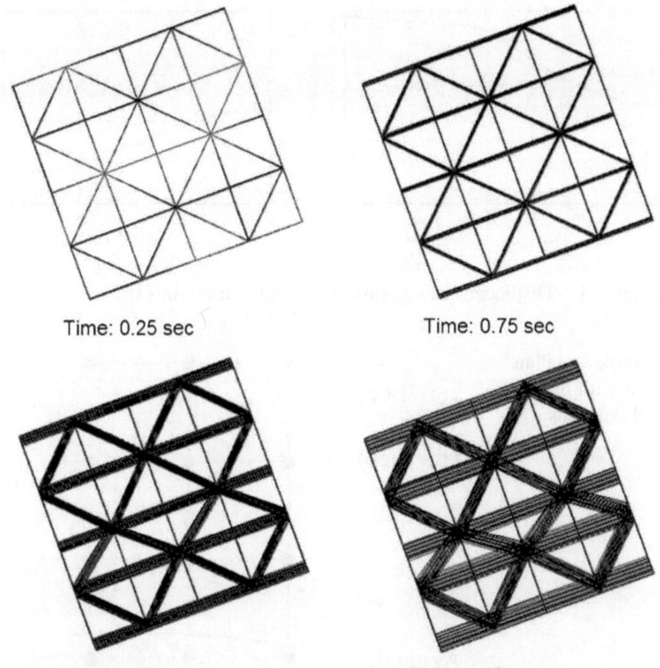

Fig. 16.4 Deformed shape of the building as collapse progresses in plan view

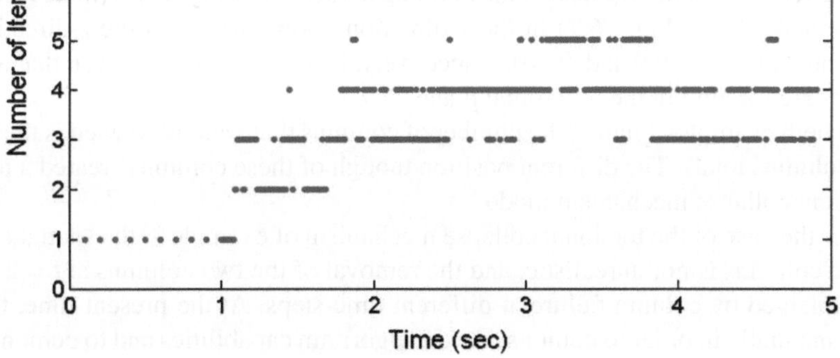

Fig. 16.5 Example 1 – Number of iterations

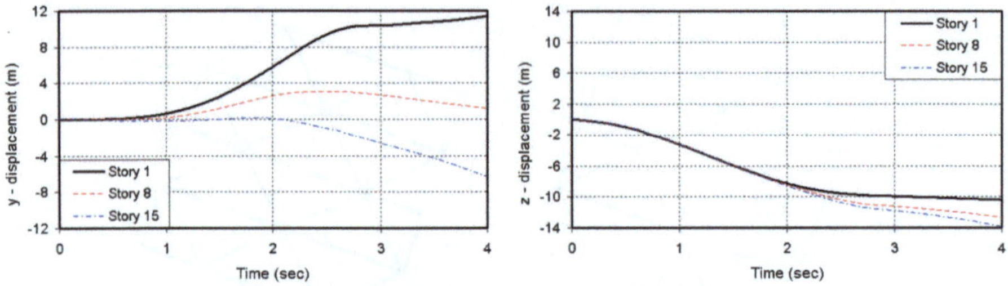

Fig. 16.6 Example 1 – Displacement response in *y*- and *z*-directions (m)

Fig. 16.7 Example 2 – Plan view of building with base story weakened columns

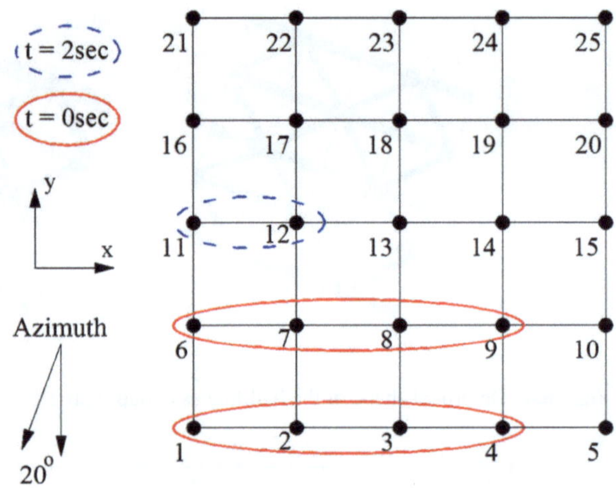

the structure as can be seen in Fig. 16.8. Significant rotation of the whole structure about a point in the upper left corner is seen in Fig. 16.8. Like in example 1, the number of iterations is small and does not change during the progression of collapse.

The displacement response of the building at different story levels (nodes directly above node 3, see Fig. 16.7) in the *x*-direction, *y*-direction, and the *z*-direction is presented in Figs. 16.9 and 16.10, respectively. From Fig. 16.9 it is seen that all the stories are rotating in the horizontal plane.

In both examples 1 and 2, the number of columns that were weakened is the same (10 columns total). The different position though of these columns created a totally different collapse mechanism mode.

For the case of the torsional collapse mechanism of example 2, the spatial choice of the columns is not unrealistic, and the removal of the two columns at $t = 2$ s can be explained by column failure at different time-steps. At the present time, this is done manually in order to demonstrate the algorithm capabilities and to demonstrate that torsional is a possible collapse mechanism even initially symmetric structures. In the future, when the algorithm has the capability to detect and model column failure, this will be done automatically.

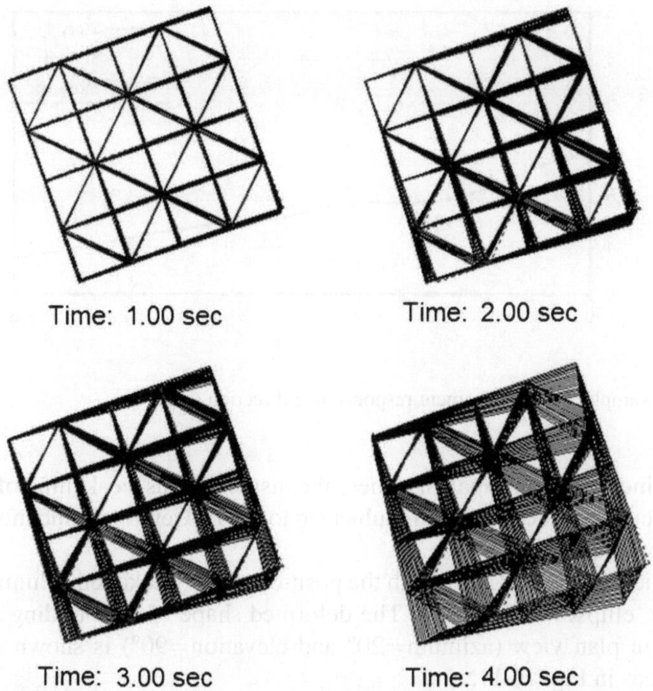

Fig. 16.8 Deformed shape of the building as collapse progresses in plan view

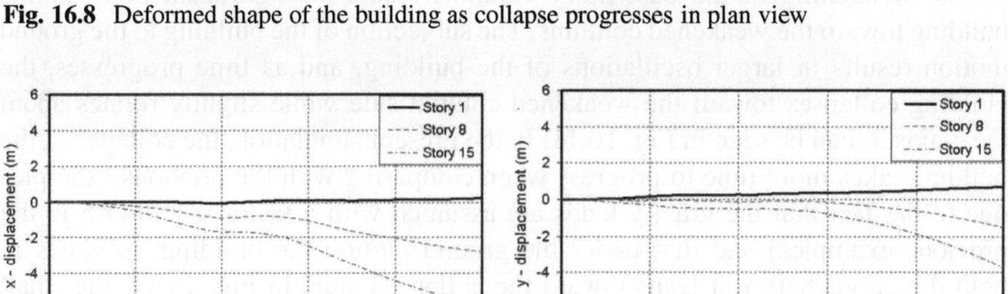

Fig. 16.9 Example 2 – Displacement response in *x*- and *y*-directions (m)

16.4.3 Example 3: Collapse Mechanism – Earthquake Excitation

In this simulation, a collapse mechanism of the fifteen-story building under the 1940 El-Centro ground motion in the *y*-direction is demonstrated. The approach used for initiating collapse is weakening of the base story columns. For the analyses of the building, two analysis steps are employed. The first step is a nonlinear static analysis where 1.0 times the gravity loads are applied in ten increments. The second

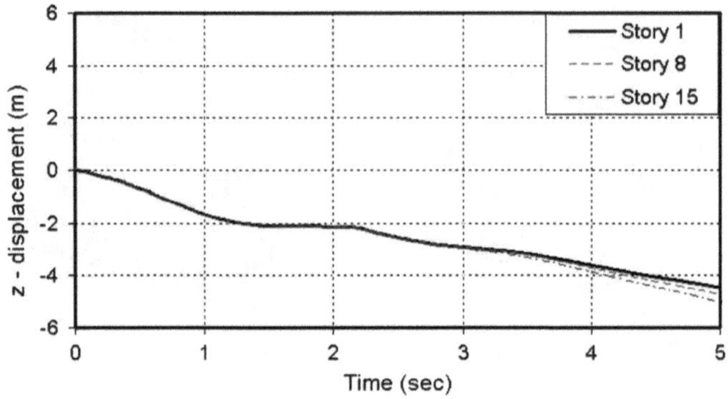

Fig. 16.10 Example 2 – Displacement response in *z*-direction (m)

step is nonlinear dynamic analysis where the instantaneous weakening of base story columns occurs and the structure is subjected to the El-Centro ground motion scaled with a factor of 5.0.

A plan view of the first story with the position of the weakened columns is shown in solid line ellipses in Fig. 16.7. The deformed shape of the building as collapse progresses in plan view (azimuth=20° and elevation=90°) is shown at different time instances in Fig. 16.11.

The weakening of the base story columns results in significant sway of the building toward the weakened columns. The subjection of the building to the ground motion results in larger oscillations of the building, and as time progresses, the building collapses toward the weakened column side while slightly rotates about the *z*-axis as can be seen in Fig. 16.11. In the present simulation, the collapse of the building takes more time to progress when comparing with the previous examples due to the fact that the gravity loads are assumed with a factor of 1.0 (1.5 in the previous examples) and that under the ground motion the building oscillates in both directions before it leans toward the collapsed side. In Fig. 16.11, the small rotation of the structure while collapsing is seen. Similarly to the previous examples, it is seen that the number of iterations is small and does not change during the progression of collapse.

The displacement response of the building at different story levels (nodes directly above node 3, see Fig. 16.7) in the *x*-direction, *y*-direction, and the *z*-direction is presented in Figs. 16.12 and 16.13, respectively. From the above figures the oscillations of the building due the ground motion, while collapsing, are seen.

16.4.4 Example 4: Collapse Mechanism – Earthquake Excitation – Twist

In this simulation, a collapse mechanism of the fifteen-story building under the 1940 El-Centro ground motion in the *y*-direction is demonstrated.

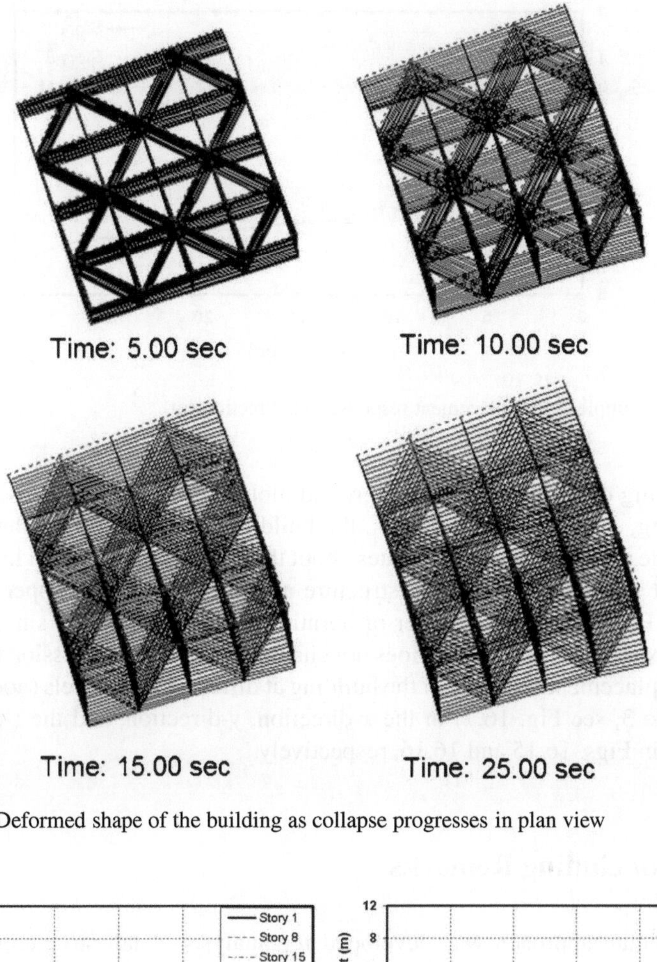

Fig. 16.11 Deformed shape of the building as collapse progresses in plan view

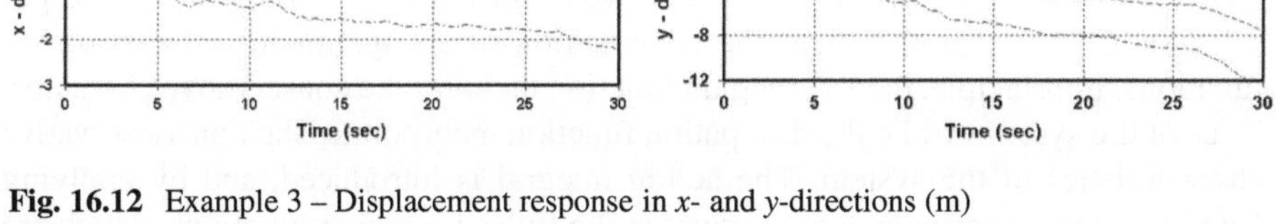

Fig. 16.12 Example 3 – Displacement response in *x*- and *y*-directions (m)

The present simulation is the same as example 3, with the difference that here we assume 1.5 times the gravity loads and the ground motion is scaled with a factor of 4.0. The objective is a collapse mechanism with significant rotation of the building in the *z*-direction. The weakening of the base story columns results in significant sway of the building toward the weakened columns.

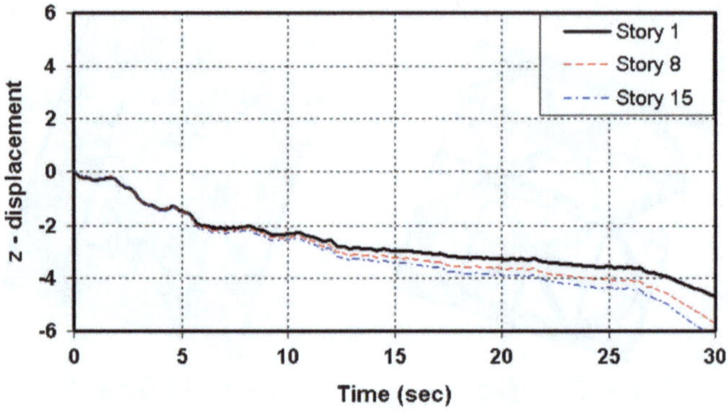

Fig. 16.13 Example 3 – Displacement response in *z*-direction (m)

Subjecting of the building at the ground motion results in larger oscillations of the building, and as time progresses, the building collapses toward the weakened column side while significantly rotates about the *z*-axis as can be seen in Fig. 16.14. Significant rotation of the whole structure about a point in the upper left corner is seen in Fig. 16.14. The number of iterations of the simulation, similarly to the previous examples, is small and does not change during the progression of collapse.

The displacement response of the building at different story levels (nodes directly above node 3, see Fig. 16.7) in the *x*-direction, *y*-direction, and the *z*-direction is presented in Figs. 16.15 and 16.16, respectively.

16.5 Concluding Remarks

A Hamiltonian approach was developed for analysis of tall structures using the mixed Lagrangian formulation (MLF), a force-based approach that considers both conservative and dissipative processes, to deal with structures where deterioration and fracture occur before collapse. The weak formulation starts with the proper selection of state variables and the construction of a Lagrangian and dissipation functions. In principle, the Lagrangian function includes the conservative characteristics of the system, while the dissipation function incorporates the nonconservative characteristics of the system. The action integral is introduced, and by applying Hamilton's action principle, one arrives at the Euler-Lagrange equations, which are the governing equations of the system. A minimization problem is solved at every time-step (after discretization of the action integral using a central difference half step) as a sequence of linear equality-constraint problems.

The formulation is capable to solve large deformations (with geometric nonlinearities) and inelastic deteriorations (caused by material nonlinearities) which occur during gradual and progressive collapse. Four numerical examples are provided to illustrate the solution capability using static and dynamic analyses, tracing the large

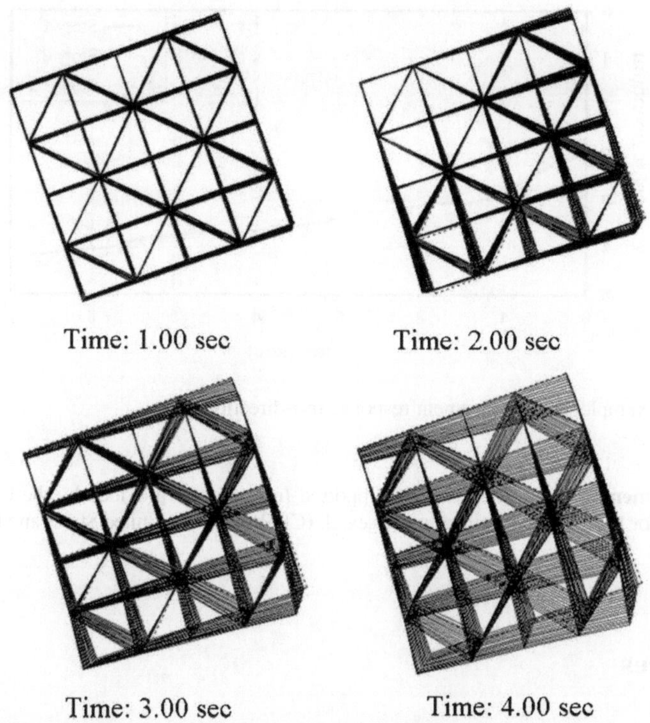

Fig. 16.14 Deformed shape of the building as collapse progresses in plan view

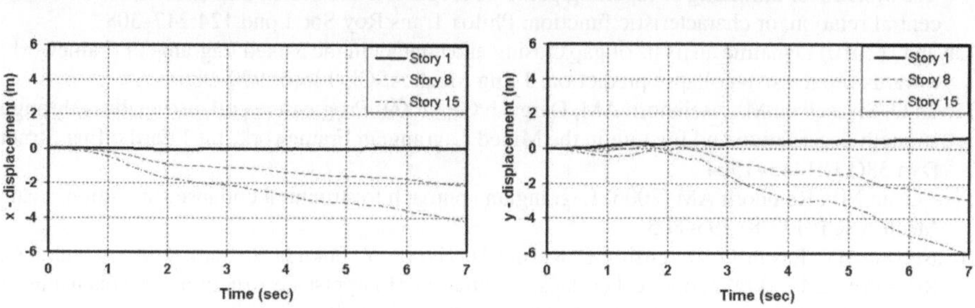

Fig. 16.15 Example 4 – Displacement response in x-direction (m)

displacements causing coupled sway and torsion before collapse. It should be noted that failure of eccentrically located columns in an otherwise symmetrical structure leads to either pure torsional or coupled torsional-lateral collapse. In the numerical cases, presented herein, the solution converges within four iterations, in average. The structure in the numerical examples was well designed for seismic motions. Therefore, gravity loads and earthquake intensities used in the examples were magnified, in order to produce relatively quick collapse for illustration. With the unscaled loads, the structure will need much more time and repetitive earthquakes in order to collapse.

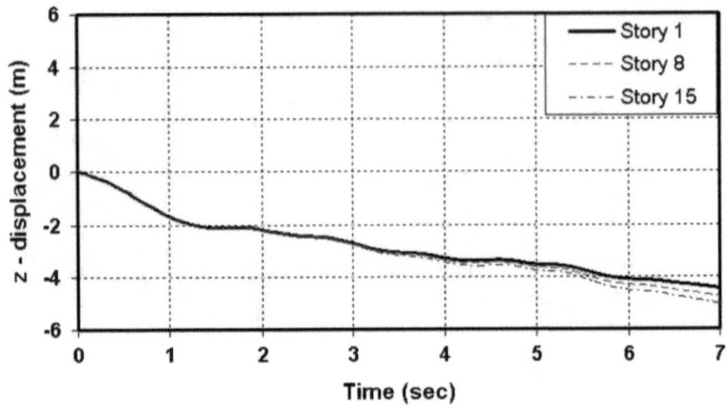

Fig. 16.16 Example 4 – Displacement response in *z*-direction (m)

Acknowledgments This work has been supported from a grant provided by the Consortium of Universities for Earthquake Engineering Research (CUREE) in the United States and Kajima Corp. of Japan.

References

Cadzow JA (1970) Discrete calculus of variations. Int J Control 11(3):393–407

Hamilton WR (1834) On a general method in dynamics: by which the study of the motions of all free systems of attracting or repelling points is reduced to the search and differentiation of one central relation, or characteristic function. Philos Trans Roy Soc Lond 124:247–308

Lavan O (2010) Dynamic analysis of gap closing and contact in the Mixed Lagrangian Framework: towards progressive collapse prediction. J Eng Mech ASCE 136(8):979–986

Lavan O, Sivaselvan MV, Reinhorn AM, Dargush GF (2009) Progressive collapse analysis through strength degradation and fracture in the Mixed Lagrangian Framework. Int J Earthq Eng Struct Dyn 38(13):1483–1504

Sivaselvan MV, Reinhorn AM (2006) Lagrangian approach to structural collapse simulation. J Eng Mech ASCE 132(8):795–805

Sivaselvan MV, Lavan O, Dargush GF, Kurino H, Hyodo Y, Fukuda R, Sato K, Apostolakis G, Reinhorn AM (2009) Numerical collapse simulation of large-scale structural systems using an optimization-based algorithm. Int J Earthq Eng Struct Dyn 38:655–677

Part IV
Seismic Design of Irregular Structures

Chapter 17
Improved Earthquake-Resistant Design of Irregular Steel Buildings

Miltiadis T. Kyrkos and Stavros A. Anagnostopoulos

Abstract In the past several years, the seismic behavior of eccentric buildings has been studied with detailed models of the plastic hinge type, first for reinforced concrete buildings and then for steel, braced frame type buildings, all of them designed in accordance with the appropriate new Eurocodes. In all cases, it was found that the distribution of ductility demands is not as uniform throughout the structure as one might have expected and desired for a well-designed structure. Such an uneven distribution indicates suboptimal material use and a potential for premature failure of certain members. In this chapter, a design modification that has been proposed earlier and improved substantially the inelastic earthquake behavior of buildings with biaxial eccentricity but with rectangular layouts is applied to eccentric, L-shaped buildings. Both a torsionally stiff and a torsionally flexible building are examined, and it is found that the modification gives also good results for such buildings, especially the torsionally stiff one. The improvement of the behavior of the torsionally flexible building may be considered marginal, but this is probably associated with characteristics of the specific building. In any case, modern codes suggest avoidance of torsionally flexible buildings whose seismic behavior is more difficult to control, and for this reason, stricter design requirements are specified.

17.1 Introduction

Earthquake response of irregular buildings in the inelastic range is an open research area that includes also assessment of code provisions pertaining to the design of such buildings (Rutenberg 1998, 2002; De Stefano and Pintucchi 2008). Most of

---

M.T. Kyrkos • S.A. Anagnostopoulos (✉)
Department of Civil Engineering, University of Patras, Patras 26500, Greece
e-mail: mtkyrk@gmail.com; saa@upatras.gr

O. Lavan and M. De Stefano (eds.), *Seismic Behaviour and Design of Irregular and Complex Civil Structures*, Geotechnical, Geological and Earthquake Engineering 24, DOI 10.1007/978-94-007-5377-8_17,

the pertinent studies have been based on elastic analyses of idealized multistory buildings or on inelastic analyses of highly simplified, one-story, inelastic models of the shear beam type with 3 degrees of freedom (Rutenberg 1998, 2002; De Stefano and Pintucchi 2008; Chopra and Goel 1991; De La Colina 2003; Stathopoulos and Anagnostopoulos 2000, 2003; Humar and Kumar 2000). In the past decade, however, research on earthquake-induced torsion started using more sophisticated, multistory, multi-degree-of-freedom inelastic models of the plastic hinge type (e.g., Ghersi et al. 2000; Fajfar et al. 2004; Stathopoulos and Anagnostopoulos 2005; Kyrkos and Anagnostopoulos 2011a, b; Kyrkos 2011; Anagnostopoulos et al. 2010). This research revealed that the widely used simplified, one-story, 3-degree-of-freedom models used in most of the past studies could lead to erroneous conclusions, unless model properties were very carefully selected to closely match key properties of the multistory building (Anagnostopoulos et al. 2010). Hence, code provisions based, in part at least, on such results might be questionable. Moreover, a number of controversies had been generated from such studies, and a few publications were devoted to them (Rutenberg 1998, 2002; De Stefano and Pintucchi 2008). One such controversy that lingered for some time was the question whether in a code-designed building the critical edge, where the term critical is used to mean "having the highest ductility demands," is the so-called "flexible" or "stiff" edge. A convincing answer to this has been presented recently, based on analyses using both detailed and simplified models (Anagnostopoulos et al. 2010). The same studies of torsion with the detailed plastic hinge models also showed that the ductility demand differences between the two edges, "flexible" and "stiff," were often very large, with the demands in the "flexible" edge being always substantially greater than the demands in the "stiff" edge of the building. This was initially found for concrete buildings, where both rotational ductility factors and damage indices were used as measures of inelastic deformations. Subsequently, the same was confirmed for eccentric, braced frame type, steel buildings, and a design modification was proposed to alleviate this problem (Kyrkos and Anagnostopoulos 2011a, b; Kyrkos 2011).

The work reported by Kyrkos and Anagnostopoulos was for steel, eccentric braced frame buildings with one, three, and five stories, all with biaxial eccentricities and rectangular layouts (Kyrkos and Anagnostopoulos 2011a, b; Kyrkos 2011). This chapter addresses the same problem for two L-shaped, three-story eccentric buildings, one planned and designed as torsionally stiff and the other as torsionally flexible. The results indicate a similar problem of the as-designed buildings for ductility demand distribution, as observed in the buildings with the rectangular layouts, and further that the same modification proposed before works well also for these two buildings, especially for the torsionally stiff one.

17.2 Methodology

The present investigation was carried out using two three-story, steel, braced frame buildings, the first torsionally stiff and the second torsionally flexible. Both buildings are L-shaped, and their typical layouts can be seen in Figs. 17.1 and 17.2. Note that

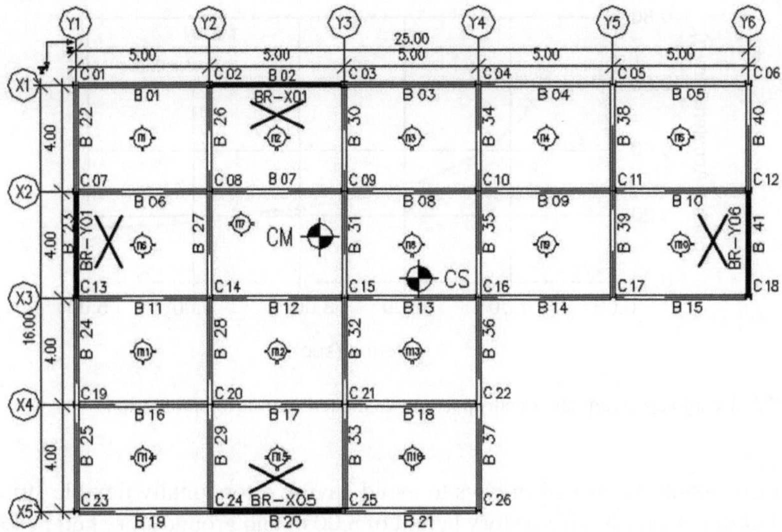

Fig. 17.1 Layout of the three-story torsionally stiff steel building

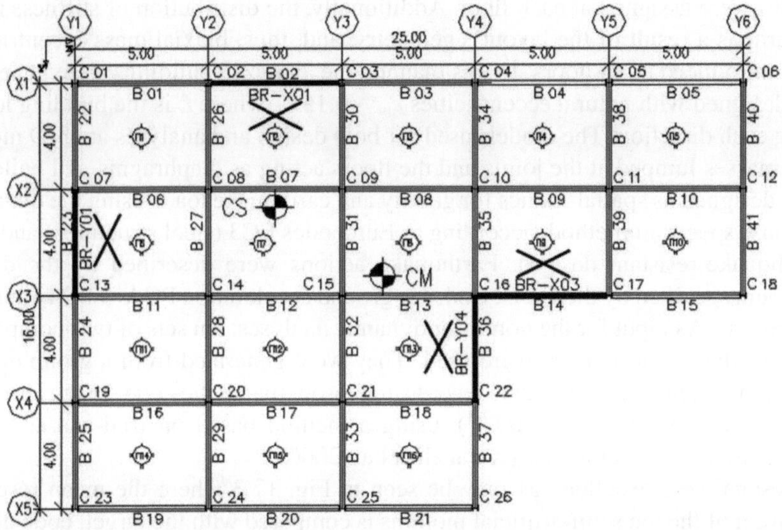

Fig. 17.2 Layout of the three-story torsionally flexible steel building

each building is formed by 5 frames along the *x*-axis (FR-X1 to FR-X5) and 6 frames along the *y*-axis (FR-Y1 to FR-Y6). In order to have a torsionally stiff and a torsionally flexible building just for the purpose of our work, braces were used to stiffen specific bays as shown in Figs. 17.1 and 17.2, respectively. In a real design,

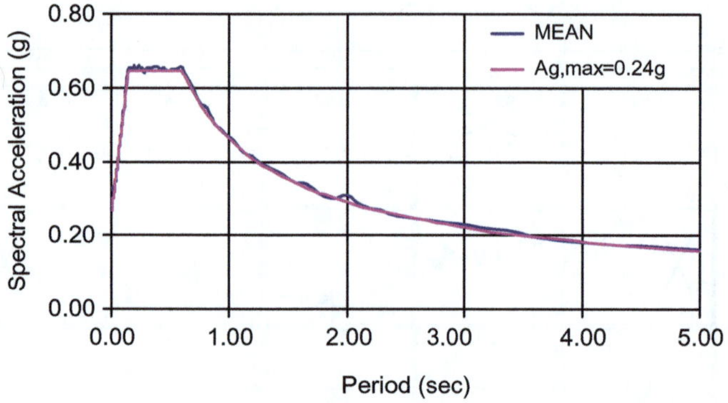

Fig. 17.3 Design spectrum and mean spectrum of the ten semi-artificial motions

the braces would be placed in bays to avoid having a torsionally flexible structure. Both buildings have a typical story height of 3.00 m and ground story height 4.00 m.

Using appropriate distributions of the floor loads, e.g. through non-symmetric live load distribution, non-symmetric balconies (common causes of mass eccentricity in typical buildings, not shown in the given layout), nonsymmetric joint masses were assigned at each floor. Additionally, the distribution of stiffness is not uniform as a result of the layout's geometry, and, thus, biaxial mass eccentricities were introduced in all floors. In this manner, the analyzed buildings were generated and designed with natural eccentricities e_m ~0.15 L, where L is the building length along each direction. The models used for both design and analyses are 3-D models with masses lumped at the joints and the floors acting as diaphragms. All buildings were designed as spatial frames for gravity and earthquake loads using the dynamic, response spectrum method, according to Eurocodes EC3 (steel structures) and EC8 (earthquake-resistant design). Earthquake actions were described by the design spectrum specified by the Greek Code for ground acceleration PGA = 0.24g and soil category II. As input for the nonlinear dynamic analyses, ten sets of two-component semi-artificial motion pairs were used. They were generated from a group of five, two-component, real earthquake records, to closely match the code design spectrum (with a descending branch $1/T^{2/3}$), using a method based on trial-and-error and Fourier transform techniques (Karabalis et al. 2000).

Results were excellent, as may be seen in Fig. 17.3 where the mean response spectrum of the ten semi-artificial motions is compared with the target, code design spectrum. Each synthetic motion pair, derived from the two horizontal components of each historical record, was applied twice by mutually changing the components along the x and y system axes. Thus, each design case was analyzed for ten sets of two-component motions, and mean values of peak response indices were computed. In this manner, the effects of individual motions are smoothed, and the conclusions become less dependent on specific motion characteristics.

The three lowest periods of the three-story torsionally stiff building are $T_x = T_y$=0.58 s and $T_\theta = 0.31$ s, while the mean natural eccentricities in each horizontal direction are $\varepsilon_x = 0.15$ and $\varepsilon_y = 0.09$. Notice that the first torsional period is lower than the two translational periods, as it is expected for a torsionally stiff building. The torsionally flexible building has a fundamental torsional mode with period T_θ=0.67 and two translational modes with periods T_y=0.61 s and T_x=0.59 s, while its average natural eccentricities are ε_x=0.15 and ε_y=0.12. It is noted that in multistory buildings, the CR cannot be really defined, except under very restrictive conditions. Thus, an approximate CR was computed herein for reference purposes, on a floor-by-floor basis as follows

$$e_{sx} = \frac{\sum_{i=1}^{m} K_{f-iy} x_i}{\sum_1^m K_{f-iy}} \qquad e_{sy} = \frac{\sum_{i=1}^{n} K_{f-ix} y_i}{\sum_1^n K_{f-ix}} \tag{17.1}$$

$$K_{f-i} = \frac{24E}{h^2}\left[\frac{2}{\sum K_c} + \frac{1}{\sum K_{ba}} + \frac{1}{\sum K_{bb}}\right]^{-1} + \sum \frac{AE}{L}\cos^2\varphi \tag{17.2}$$

where e_{sx} and e_{sy} are the x and y coordinates of the approximate stiffness center CR; K_{f-i} designates the approximate story stiffness of frame i, x and y the directions of the frame axis, and m and n the number of frames along the y- and x-axes, respectively; E = modulus of elasticity; $K_c = I_c/h$; $K_b = I_b/\ell$; I_c, I_b = section moment of inertia of columns and beams, respectively; h = story height and ℓ = beam length; A = area of brace section; L = brace length; and φ = angle of brace member and the horizontal plane. The second indices, a and b, in K_{ba} and K_{bb}, designate the upper (above) and lower (below) floor beams of the frame in the considered story.

17.3 Nonlinear Dynamic Analyses

The nonlinear analyses were carried out using the program RUAUMOKO (Carr 2005). Frame beams and columns were modeled with the well-known plastic hinge model, in which yielding at member ends is idealized with plastic hinges of finite length having bilinear moment-curvature relationship and strain hardening ratio equal to 0.05. A moment-axial force interaction diagram was also employed for columns, giving the yield moment as a function of the applicable axial force on the column section. Bracing members, yielding in tension and buckling in compression, were modeled with a nonsymmetric bilinear force-axial deformation relationship (Fig. 17.4).

The basic measure used to assess the severity of inelastic response is the ductility factor of the various members.

For bracing members, the ductility factor is defined as

$$\mu_u = 1 + \left(\frac{u_p}{u_y}\right) \tag{17.3}$$

where u_p is the maximum plastic member elongation and u_y the elongation at first yield.

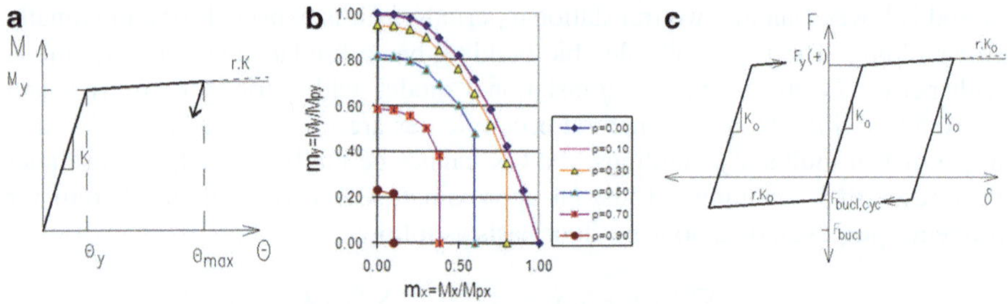

Fig. 17.4 (**a**) Nonlinear moment-rotation relations for beam-columns, (**b**) column M-N interaction diagram, and (**c**) nonlinear force deformation diagram for braces

For beams and beam-columns, the rotational ductility factor has traditionally been defined as

$$\mu_\theta = 1 + \left(\frac{\theta_p}{\theta_y}\right) \tag{17.4}$$

where θ_p is the maximum plastic hinge rotation at either end of a member (beam or column) and θ_y is a normalizing "yield" rotation, typically set equal to $\theta_y = M_y \ell / 6EI$. For columns, the yield moment M_y is usually taken to correspond to the yield moment under the action of gravity loads. In the present study, an alternative definition of the rotational ductility factor, based on the post-yield plastic moment, has been used (Anagnostopoulos 1981)

$$\mu = 1 + \left(\frac{\Delta M}{p \cdot M_y}\right) \tag{17.5}$$

where $\Delta M = M_{\max} - M_y$, M_y = yield bending moment, and $p = 0.05$, the strain hardening ratio.

In addition to the above measures, peak floor displacements and interstory drifts are used to assess the inelastic behavior of the buildings.

17.4 Results from Nonlinear Analyses of "As-Designed" Buildings

Results from time history analyses of the two buildings are presented in terms of mean values of the peak response parameters over the ten pairs of applied motions. In the case of the beam ductility factors, the response parameter averaged over the ten pairs of motion is the maximum rotational ductility demand in any of the beams in the considered frame and floor. Following standard terminology based on static application of the lateral load in torsionally stiff buildings, the edge

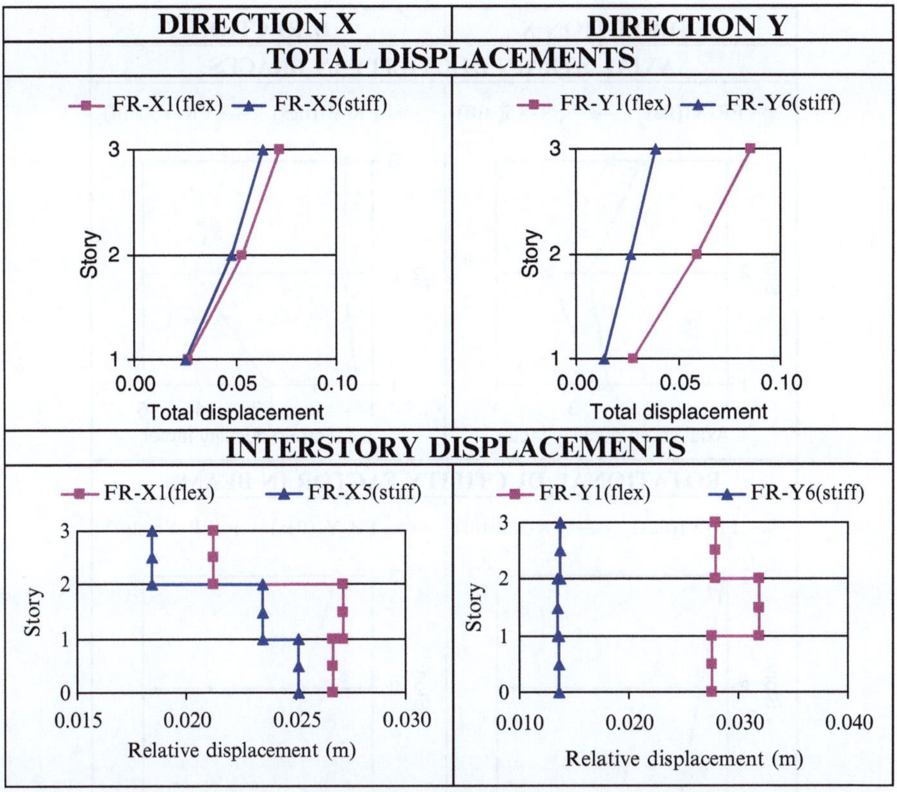

Fig. 17.5 Total displacements and interstory drifts of three-story torsionally stiff building (FR-X1 and FR-Y1: "flexible" edges, FR-X5 and FR-Y6: "stiff" edges)

where the displacement from rotation is added to the pure floor translation is called "flexible" edge, while the opposite edge where the displacement due to rotation is subtracted from the pure translation is called "stiff" edge. Since the examined buildings have biaxial eccentricity, the edge distinction just mentioned applies to both the *x* and *y* horizontal directions of the buildings. Thus, results are presented for each edge frame and each direction. In torsionally flexible buildings, however, it is not necessarily the "flexible" edge that experiences the largest translation, but it could well be the "stiff" edge, depending on the relative values of the torsional and translational periods and on the input characteristics.

17.4.1 Three-Story Torsionally Stiff Building

Displacement results and ductility demands for braces and beams of the "flexible" and "stiff" edges are presented in Figs. 17.5 and 17.6, respectively. Ductility

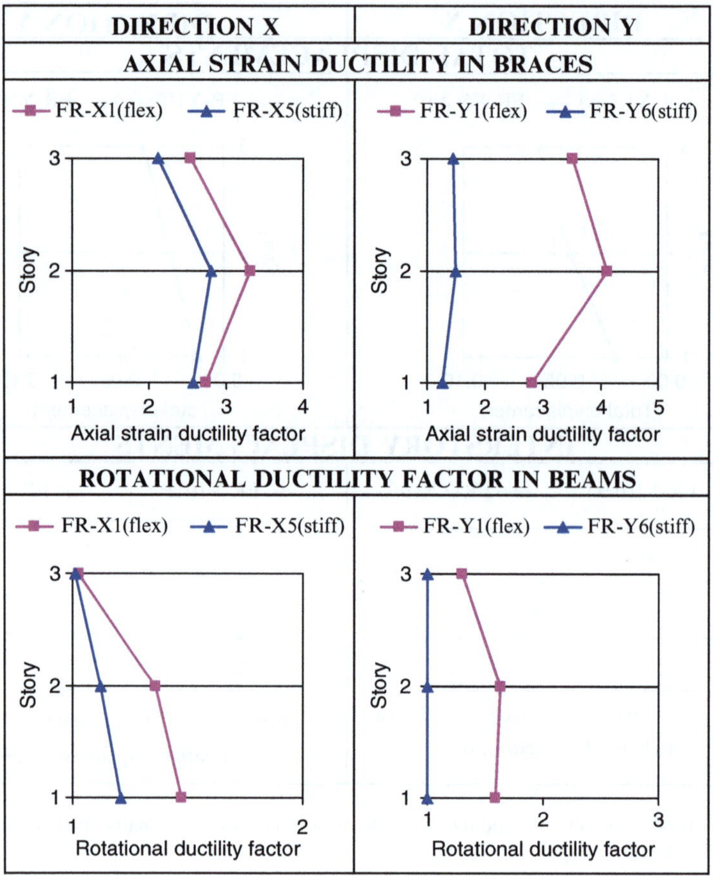

Fig. 17.6 Member ductility demands of three-story torsionally stiff building (FR-X1 and FR-Y1: "flexible" edges, FR-X5 and FR-Y6: "stiff" edges)

demands are presented only for beams and brace members because the columns remained essentially elastic. Looking into Fig. 17.5, we can see that displacements at the "flexible" edges of the torsionally stiff eccentric building are substantially greater than those at the "stiff" edges due to the induced earthquake rotations, especially in the y direction. The same is true for ductility demands of the braces and the beams (Fig. 17.6).

17.4.2 Three-Story Torsionally Flexible Building

Displacement results for the torsionally flexible building are shown in Fig. 17.7 and ductility demands in Fig. 17.8. Compared to the results of the torsionally

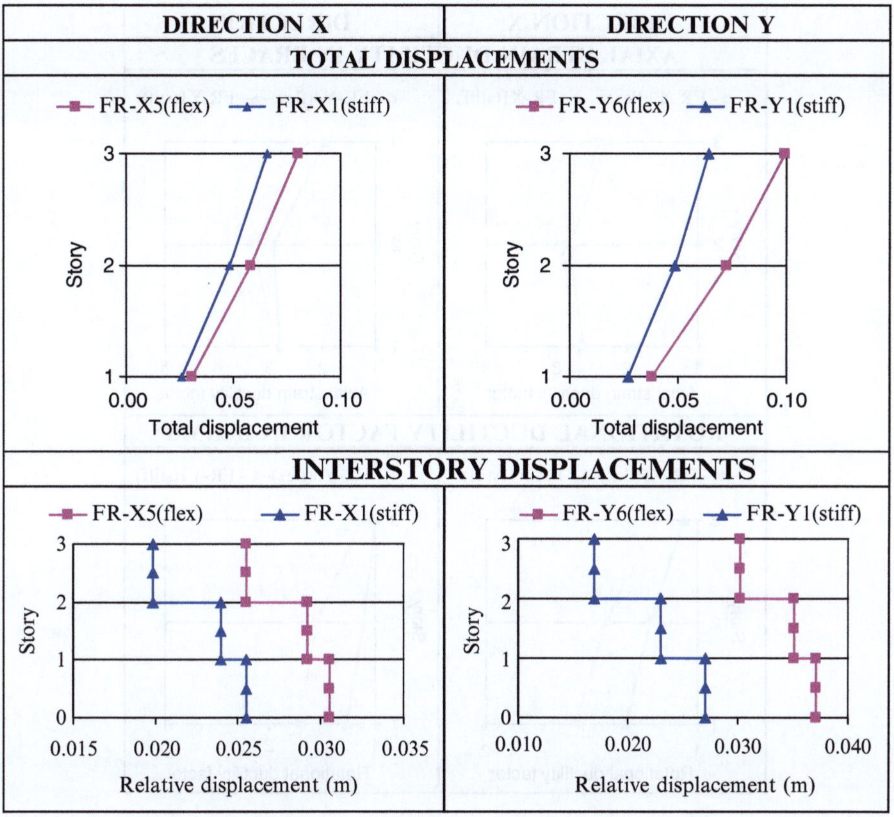

Fig. 17.7 Total displacements and interstory drifts of three-story torsionally flexible building (FR-X5 and FR-Y6: "flexible" edges, FR-X1 and FR-Y1: "stiff" edges)

stiff building, quantitative differences aside, the behavior pattern is similar, with displacement and ductility demands being larger in the frames at the "flexible" edges or sides. It is further noticed that the differences in ductility demands in the braces here are much smaller than in the torsionally stiff building, since the braces in this case are placed near the core of the building.

17.5 Modification Procedure

A structural design can be characterized as satisfactory when the limiting values of the controlling response parameters do not have wide variations within the groups of structural members to which they apply. In the opposite case, suboptimal use of material may be present as well as a potentially higher risk of failure in cases of unexpected overloads. Thus, the observed substantial differences in

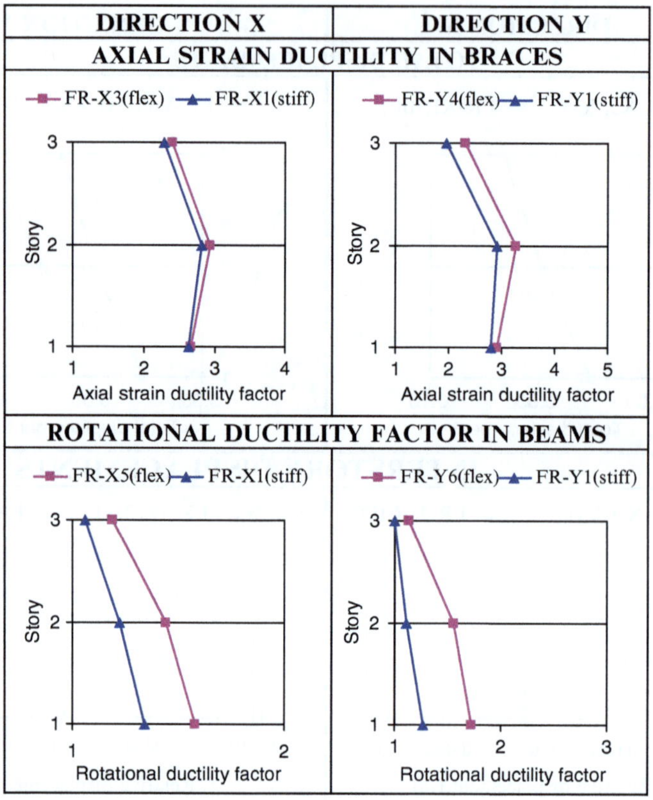

Fig. 17.8 Member ductility demands of three-story torsionally flexible building (FR-X5,X3 and FR-Y4,Y6: "flexible" edges, FR-X1 and FR-Y1: "stiff" edges)

ductility demands between the opposite edges of the examined buildings point to the need for a design modification that would eliminate or reduce these differences. The modifications for the torsionally stiff building are the same as in Kyrkos and Anagnostopoulos (2011b). A modified version of this modification will be applied to the torsionally flexible building (Kyrkos 2011).

17.5.1 Modification Procedure and Results for the Torsionally Stiff Building

The modification procedure for the torsionally stiff building aims at increasing the strength of the bracing members at the "flexible" edges and reducing the strength of the braces at the "stiff" edges without affecting the strength of the other structural elements (columns, beams). The first step for application of this modification is to obtain the top-story displacements at the "flexible" and "stiff" edges of the building in both horizontal directions due to the earthquake loading considered and then compute the following factors in each horizontal direction:

$$f_{i,\text{flex}} = 2 \cdot \frac{u_{i,\text{flex}}}{(u_{i,\text{flex}} + u_{i,\text{stiff}})} \qquad (17.6)$$

$$f_{i,\text{stiff}} = 2 \cdot \frac{u_{i,\text{stiff}}}{(u_{i,\text{flex}} + u_{i,\text{stiff}})}$$

where $u_{i,\text{flex}}$ is the top-story displacement of the "flexible" edge in the i-direction and $u_{i,\text{stiff}}$ the top-story displacement of the "stiff" edge also in the i-direction. These displacements are obtained by the dynamic response spectrum method for the seismic combinations considered. The factors are ratios of the top-story displacements at the "flexible" and "stiff" edges in a given direction (x *or* y) to their mean values. The design modification that was subsequently applied was to multiply the axial areas of the bracing members in both the "stiff" and "flexible" edges by the corresponding factors in each direction. The values of these factors, for modifying frames in direction x, are 0.89 for the "stiff" edge and 1.11 for the "flexible" edge. Similarly, the values for modifying frames in direction y are 0.78 for the "stiff" edge and 1.22 for the "flexible" edge. After this modification, each structure was checked again for full compliance with the applicable codes. The new, modified structures were again subjected to the same two-component earthquake set, and their responses were again computed as before.

Figures 17.9 and 17.10 show displacements and ductility demands for the torsionally stiff building for the initial and the modified design. If we compare the results obtained from the modified design with that of the original design, we see a substantial improvement of response in all cases: the overall maximum ductility demand factor in each group is reduced and so are the differences between "flexible" and "stiff" edges, producing more uniform distribution of such demands.

17.5.2 Modification Procedure and Results for the Torsionally Flexible Building

The modification procedure for the torsionally flexible building aims at increasing the strength of the structural members (columns, beams, and braces) at the "flexible" edges and reducing the strength of the braces at the "stiff" edges without affecting the strength of the other structural elements (columns, beams). The ratios of the top-story displacements $f_{i,\text{flex}}$ and $f_{i,\text{stiff}}$ (Eq. 17.6) are also used, but now the displacements are obtained by the equivalent static method for the seismic combinations considered. The design modification that was subsequently applied was to multiply the axial areas of the bracing members in both the "stiff" and "flexible" edges by the corresponding factors in each direction and to do the same

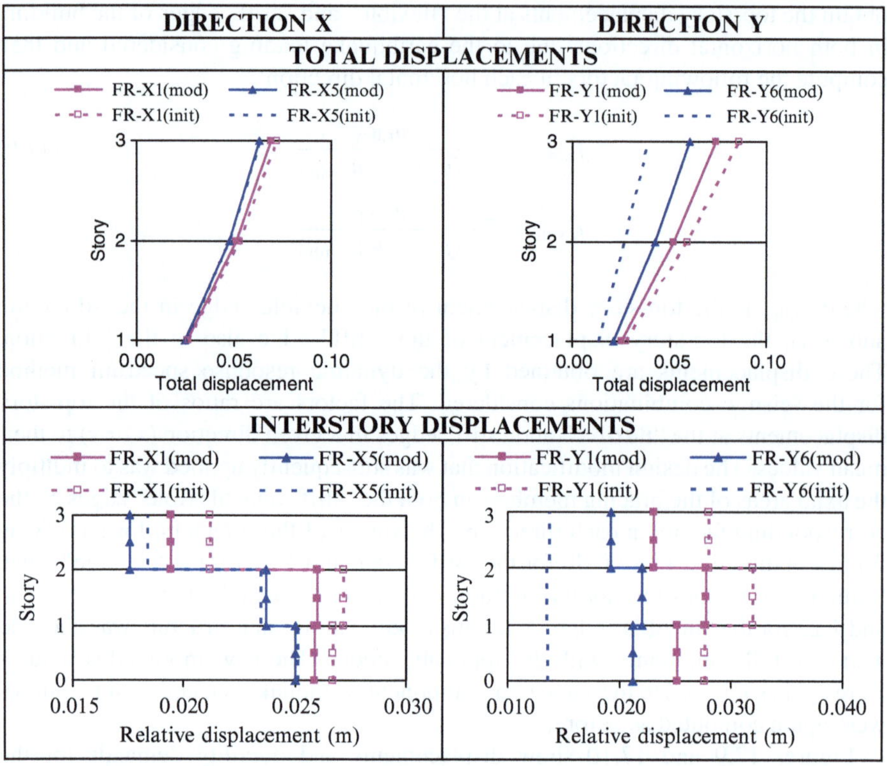

Fig. 17.9 Comparison of total displacements and interstory drifts of three-story torsionally stiff building, for the initial and modified design (FR-X1 and FR-Y1: "flexible" edges, FR-X5 and FR-Y6: "stiff" edges)

for the beam and column sections, but *only* in the "flexible" edges. The cross sections of columns and beams of the "stiff" edges are not reduced, as their strength is controlled mainly by gravity loads (loading combinations without earthquake). After this modification, each structure was checked again for full compliance with the applicable codes. This procedure gave modification factors of 0.88 and 0.66 for the braces in the stiff x and y sides, respectively (i.e., for braces in frames $X1$ and $Y1$), and factors 1.12 and 1.34 for beams and columns in the "flexible" edges along the x and y directions, respectively. However, contrary to what happened in the torsionally stiff building, now the reduction in the brace sections at the "stiff" side proved excessive, and these two factors were increased from 0.88 and 0.66 to 0.94 and 0.92, respectively, to satisfy all the code-required checks. The new, modified structures were again subjected to the same two-component motion earthquake set, and their responses were again computed as before.

Figures 17.11 and 17.12 show displacements and ductility demands for the torsionally flexible building for the initial and the modified design. If we compare the results obtained from the modified design with that of the original design,

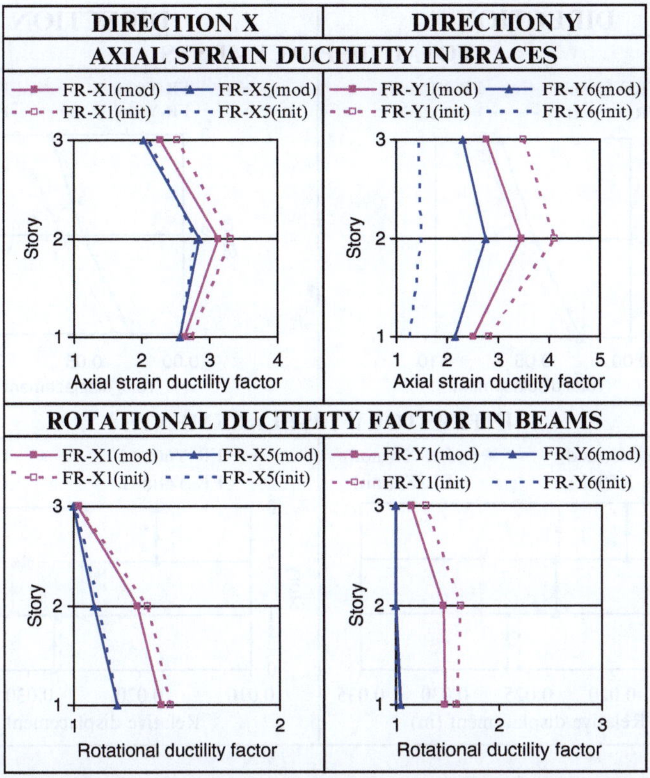

Fig. 17.10 Comparison of ductility demands of three-story torsionally stiff building, for the initial and modified design (FR-X1 and FR-Y1: "flexible" edges, FR-X5 and FR-Y6: "stiff" edges)

we see again some noticeable improvement of response: the overall maximum ductility demand factor in each group is reduced and so are the differences between "flexible" and "stiff" edges, producing more uniform distribution of such demands. The improvements, however, are not as significant as in the torsionally stiff building.

17.6 Some Comments on the Proposed Modification

To get more insight about the consequences of the proposed modification, the new stiffness centers of the modified buildings were computed using Eqs. (17.1) and (17.2) and compared with those of the original designs. Results are presented in Table 17.1. We can see that the proposed modification brings the approximate stiffness center of each story closer to the mass center, and, thus, the torsional motions are reduced. This reduction is obviously greater for the "flexible" edge and hence the reduction in the observed differences of ductility demands between

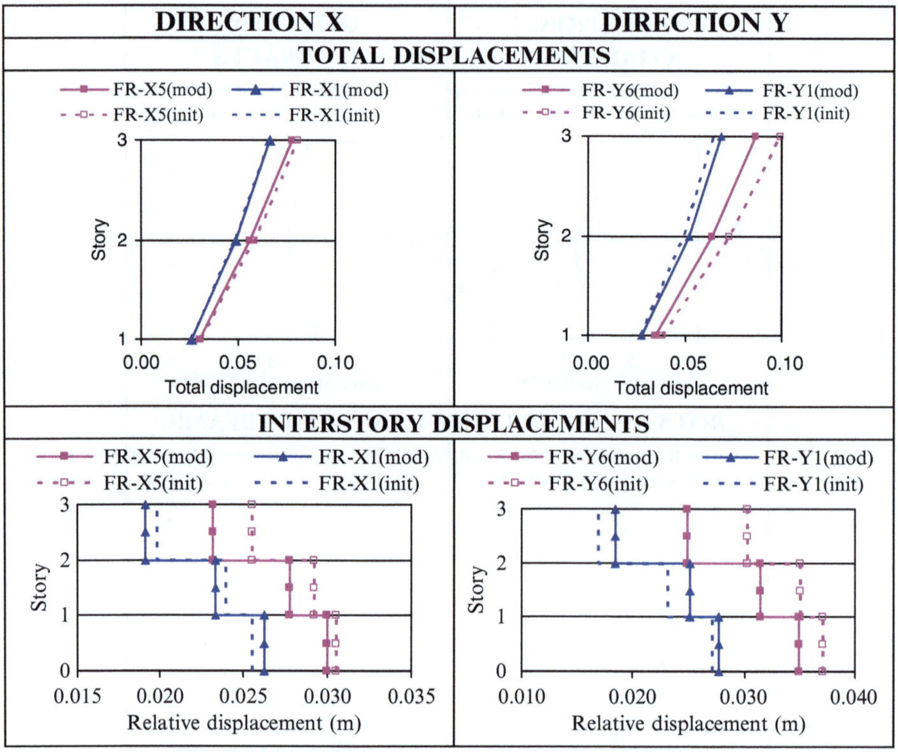

Fig. 17.11 Comparison of total displacements and interstory drifts of three-story torsionally flexible building, for the initial and modified design (FR-X5 and FR-Y6: "flexible" edges, FR-X1 and FR-Y1: "stiff" edges)

"flexible" and "stiff" edges. We must note here that bringing the stiffness center as close as possible to the mass center, in other words trying to minimize the physical eccentricity, is a well-known design objective in earthquake engineering as it minimizes torsional motion. The proposed modification is thus a "blind" way of achieving this without significant extra effort.

17.7 Conclusions

In this chapter, the earthquake response of two irregular L-shaped steel, braced frame buildings, one torsionally stiff and the other torsionally flexible, both designed in accordance with Eurocodes EC3 and EC8 were examined, and similar, overall, results were obtained compared to earlier findings for eccentric rectangular, steel, and reinforced concrete frame buildings. More specifically, it was found that under the action of two-horizontal-component earthquake loadings, compatible with the

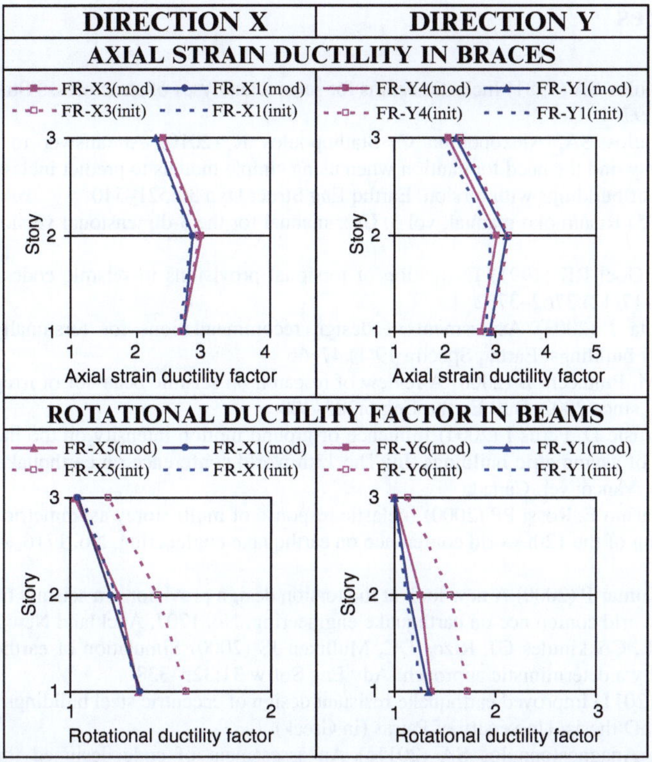

Fig. 17.12 Comparison of ductility demands of three-story torsionally flexible building, for initial and modified design (FR-X3,X5 and FR-Y4,Y6:"flexible" edges, FR-X1 and FR-Y1:"stiff" edges)

Table 17.1 Mean natural eccentricities for the initial and modified designs of the three-story buildings

	Mean natural eccentricity			
	Initial design		Modified design	
	ε_x	ε_y	ε_x	ε_y
3-st torsionally stiff	0.15	0.09	0.06	0.03
3-st torsionally flexible	0.15	0.12	0.06	0.09

design spectra, ductility demands at the "flexible" edges were significantly greater than ductility demands at the "stiff" edges. Subsequently, the original designs were modified using two slightly different procedures for each building, and it was found that the response of the new designs was improved: ductility demands at the "flexible" edges generally decreased and the differences between the two sides diminished so that a more uniform distribution of ductility demands was achieved.

On the basis of these findings, a code modification may appear desirable. However, additional studies covering other types of irregular buildings and a wider spectrum of parameters will be required, before any firm recommendation is put forward.

References

Anagnostopoulos SA (1981) Inelastic beams for seismic analyses of structures. J Struct Div ASCE 107(7):1297–1311
Anagnostopoulos SA, Alexopoulou C, Stathopoulos K (2010) An answer to an important controversy and the need for caution when using simple models to predict inelastic earthquake response of buildings with torsion. Earthq Eng Struct Dyn 39:521–540
Carr AJ (2005) Ruaumoko manual, vol 3. User manual for the 3-dimensional version- Ruaumoko 3D
Chopra AK, Goel RK (1991) Evaluation of torsional provisions in seismic codes. J Struct Div (ASCE) 117(12):3762–3782
De La Colina J (2003) Assessment of design recommendations for torsionally unbalanced multistory buildings. Earthq Spectra 19(1):47–66
De Stefano M, Pintucchi B (2008) A review of research on seismic behavior of irregular building structures since 2002. Bull Earthq Eng 6:285–308
Fajfar P, Marusic D, Perus I (2004) Influence of ground motion intensity on the inelastic torsion response of asymmetric buildings. In: The 13th world conference on earthquake engineering, No. 3496, Vancouver, Canada
Ghersi A, Marino E, Rossi PP (2000) Inelastic response of multi-storey asymmetric buildings. In: Proceeding of the 12th world conference on earthquake engineering, No. 1716, Auckland New Zealand
Humar JL, Kumar P (2000) A new look at the torsion design provisions in seismic building codes. In: 12th world conference on earthquake engineering, No. 1707, Auckland New Zealand
Karabalis DL, Cokkinides GJ, Rizos DC, Mulliken JS (2000) Simulation of earthquake ground motions by a deterministic approach. Adv Eng Softw 31:329–338
Kyrkos MT (2011) Improved earthquake resistant design of eccentric steel buildings with diagonal braces. PhD thesis, University of Patras (in Greek)
Kyrkos MT, Anagnostopoulos SA (2011a) An assessment of code designed torsionally stiff asymmetric steel buildings under strong earthquake excitations. Earthq Struct 2:109–126
Kyrkos MT, Anagnostopoulos SA (2011b) Improved earthquake resistant design of torsionally stiff asymmetric steel buildings. Earthq Struct 2:127–147
Rutenberg A (1998) EAEE Task Group (TG)8: behavior of irregular and complex structures – state of the art report: seismic nonlinear response of code-designed asymmetric structures. In: 11th European conference on earthquake engineering, Paris
Rutenberg A (2002). EAEE Task Group (TG)8: behavior of irregular and complex structures – asymmetric structures – progress since 1998. In: 12th European conference on earthquake engineering, London
Stathopoulos KG, Anagnostopoulos SA (2000) Inelastic earthquake response of buildings subjected to torsion. In: 12th world conference on earthquake engineering no. 781, Auckland New Zealand
Stathopoulos KG, Anagnostopoulos SA (2003) Inelastic earthquake response of single-story asymmetric buildings: an assessment of simplified shear- beam models. Earthq Eng Struct Dyn 32:1813–1831
Stathopoulos KG, Anagnostopoulos SA (2005) Inelastic torsion of multistory buildings under earthquake excitations. Earthq Eng Struct Dyn 34:1449–1465

Chapter 18
Design of a Plan Irregular RC Frame Building by Direct Displacement-Based Design Method

Vittorio Capozzi, Gennaro Magliulo, and Roberto Ramasco

Abstract In this chapter, an application of the direct displacement-based design (DDBD) to multistorey irregular in-plan RC frame buildings is made. A case study is carried out in order to extend and validate the methodology to this type of structures. The design of a torsionally flexible system is carried out according to DDBD, and its seismic performance is compared, through nonlinear dynamic analyses, to the performance of the same building designed according to elastic modal response spectrum analysis. Lumped plasticity models are implemented for nonlinear dynamic analyses, which are carried out according to EC8 provisions: seven real earthquakes, selected in order to fit on average the elastic design spectrum, are used as input. The two different design methods provide very different reinforcement ratios: the DDBD allows a reinforcement saving of about 70% for beams and 50% for columns. In spite of this, the verification at the ultimate limit state, performed by nonlinear dynamic analyses according to EC8, is satisfied. Furthermore, nonlinear analyses show a better response of the structure designed by DDBD: the torsional twist is reduced and the damage is better distributed.

18.1 Introduction

The design method provided by the most important seismic codes, based on the acceleration spectrum and identified as force-based design method (FBD), is characterised by the application on the structure of reduced seismic forces, admitting, in the case of strong seismic event, an excursion in plastic range based on structural ductility. The ductility, also depending on structural typology, is assured

---

V. Capozzi • G. Magliulo (✉) • R. Ramasco
Department of Structural Engineering (DIST), University of Naples Federico II, via Claudio 21, I-80125 Naples, Italy
e-mail: vittorio.capozzi@unina.it; gmagliul@unina.it; ramasco@unina.it

O. Lavan and M. De Stefano (eds.), *Seismic Behaviour and Design of Irregular and Complex Civil Structures*, Geotechnical, Geological and Earthquake Engineering 24, DOI 10.1007/978-94-007-5377-8_18,

by detailing in sections; indeed, overall ductile behaviour is ensured if the ductility demand globally involves a large volume of the structure, spread to different elements and locations of all its storeys. The design also involves the application of the capacity design rules that aim to avoid plastic hinges in columns and brittle shear failures. However, for several years, drawbacks and limits of FBD method (Priestley 1993) have been underlined. First, the elastic stiffness of the elements is proportional to strength, while the yielding curvature does not essentially depend on strength; consequently, it is not possible to perform an accurate analysis of either the elastic structural periods or the elastic distribution of required strength throughout the structure until the member strengths have been determined. Furthermore, the distribution of seismic forces based on initial stiffness cannot provide in general adequate estimates of force distribution and displacements in the inelastic range (Priestley 1993; Priestley et al. 2007). Finally, the code force reduction for the structure is not able to assure the requirement of uniform safety.

Considering the limits of FBD methods and in order to control the damage for different levels of seismic intensity, researchers are developing design methods based on displacements that are better correlated to damage. These new approaches based on displacements are known as displacement-based design procedures. The most developed and effective one is the "direct displacement-based design" proposed by Priestley (1998) and disseminated through the draft model code (Priestley et al. 2007).

The method considers a substitute structure (SDOF system) with linear behaviour (secant stiffness to the maximum displacement) that approximates the nonlinear behaviour of a complex structure (MDOF system). The design departs from a design displacement, which assures an acceptable damage for the considered seismic intensity.

The method has been recently applied to a lot of categories of regular in-plan structures, with very interesting results (Belleri and Riva 2009; Benedetti et al. 2008; Garcia et al. 2010; Maley et al. 2010; Pettinga et al. 2007; Rizzato et al. 2009).

The purpose of this chapter is to contribute to the extension and validation of the DDBD method for the design of multistorey RC irregular in-plan framed structures. The design of a torsionally flexible system is carried out according to DDBD and its seismic performance is compared with that of the same building designed according to EC8 FBD rules (CEN 2003). The two designed buildings, named "FBD structure" and "DDBD structure", present same materials and geometry, also in terms of cross-section dimensions. The seismic performances of designed frames are finally evaluated through nonlinear dynamic analyses using the computer program CANNY99 (Li 1996).

18.2 Torsional Response of Non-regular In-Plan Buildings

Structures with asymmetry in-plan are subjected to torsional rotations as well as direct translation under seismic response. In traditional elastic analysis of torsional effects in buildings, only centre of mass (C_M) and stiffness (C_R) are considered, and a structure is assumed to have plan eccentricity when the two points do not coincide. In inelastic response range, the centre of shear strength (C_V) is very important as C_R (Paulay 2001); the structure rotates around it in inelastic field. Generally C_V and C_R do not coincide.

The eccentricity of the centre of stiffness with respect to the centre of mass is found from (Fig. 18.1)

$$e_{RX} = \frac{\sum_{i=1}^{n} k_{Zi} x_i}{\sum_{i=1}^{n} k_{Zi}}; \quad e_{RZ} = \frac{\sum_{j=1}^{m} k_{Xj} z_j}{\sum_{j=1}^{m} k_{Xj}} \tag{18.1}$$

where k_{Zj} and k_{Xj} are the element stiffness in the Z and X directions, respectively, and x_j and z_j are measured from the centre of mass.

The eccentricity of the centre of strength, C_V, is defined by

$$e_{VX} = \frac{\sum_{i=1}^{n} V_{Zi} x_i}{\sum_{i=1}^{n} V_{Zi}}; \quad e_{RZ} = \frac{\sum_{j=1}^{m} V_{Xj} x_j}{\sum_{j=1}^{m} V_{Zj}} \tag{18.2}$$

where V_{Zj} and V_{Xj} are the element strength in the Z and X directions, respectively.

Peak response displacements at the opposite sides of an asymmetric building do not occur simultaneously, and they do not correspond to peak torsional response. Then, it is not possible to define exact analytical methods for simple design. Beyer et al. (2008) and Castillo (2004) give the maximum response displacements of structures with stiffness eccentricities with or without strength eccentricity in Z direction for a wall building (Fig. 18.2)

$$\Delta_{Zi} = \Delta_{z,\text{CM}} \pm \theta\,(x_i - e_{VX}) \tag{18.3}$$

taking care with signs. The maximum displacement is then the sum of C_M displacement ($\Delta_{z,\text{CM}}$) plus the torsional component.

It can be noted that in Eq. 18.3 the strength eccentricity is used. The twist angle θ to be considered in the equation is found considering the total building strength in the direction considered, V_{BZ}, and the effective rotational stiffness $J_{R,\text{eff}}$ as

$$\theta = \frac{V_{BZ} \cdot e_{RX}}{J_{R,\text{eff}}} \tag{18.4}$$

where

$$J_{R,\text{eff}} = \frac{1}{\mu_{\text{sis}}} \cdot \sum_{i=1}^{n} k_{\text{el},Zi}(x_i - e_{RX})^2 + \sum_{j=1}^{m} k_{\text{el},Xj}(z_j - e_{RZ})^2 \tag{18.5}$$

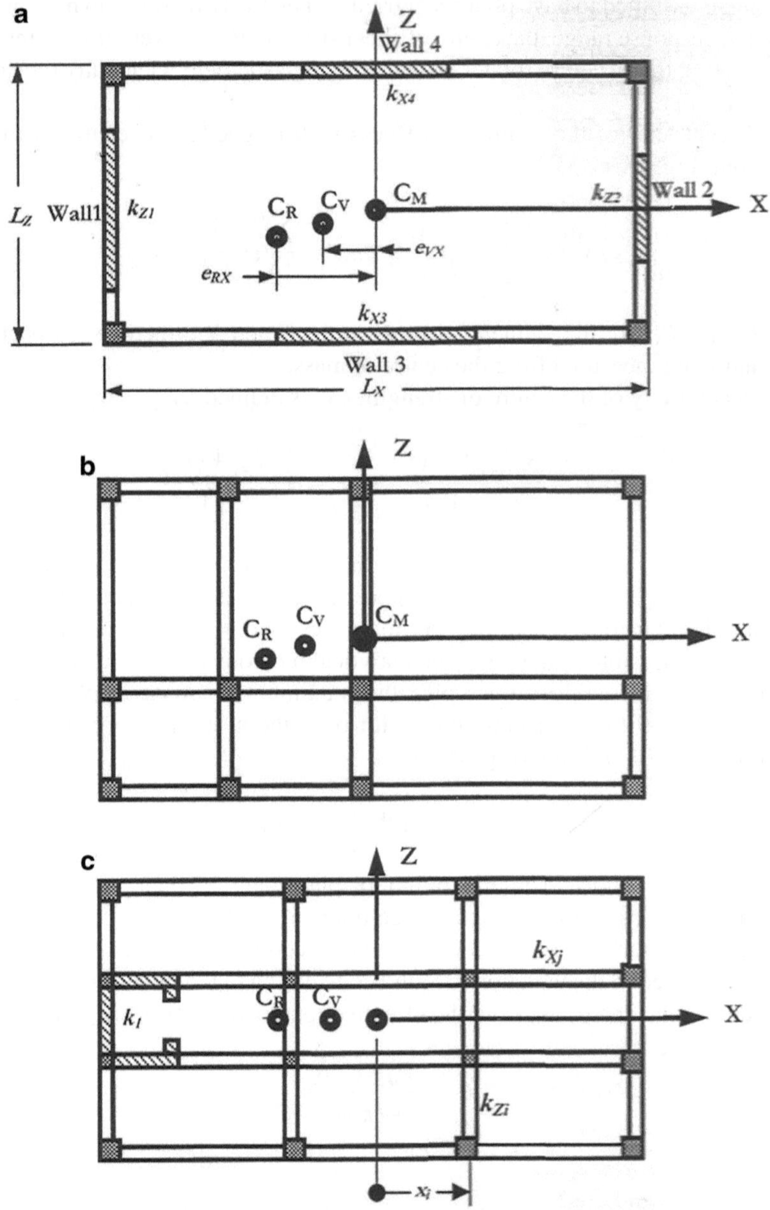

Fig. 18.1 Structures with plan eccentricities (Priestley et al. 2007). (**a**) Structural wall building. (**b**) Eccentric frame building. (**c**) Frame with eccentric service core

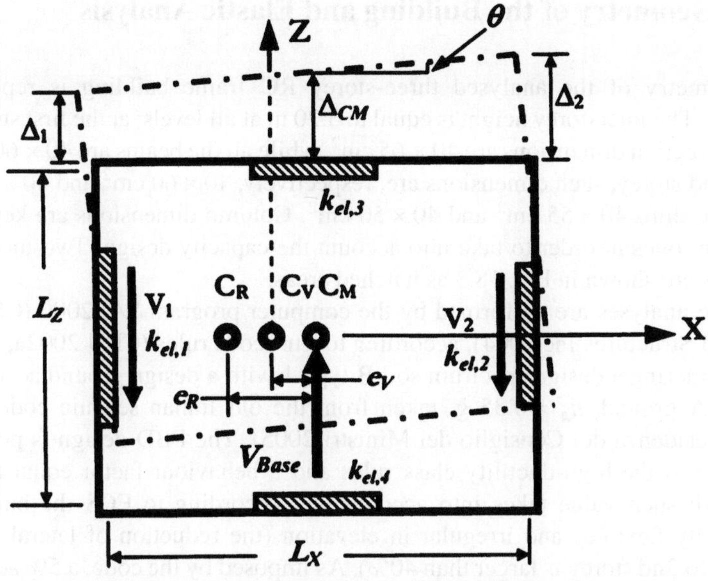

Fig. 18.2 Torsional response of an asymmetric wall building (Priestley et al. 2007)

In Eq. 18.5, the stiffness eccentricity is used, and the elastic stiffness of elements responding in the Z direction is reduced by the design system ductility, μ_{sys}. Since the transverse (X direction) elements are expected to remain elastic, or nearly elastic, their elastic stiffness is not reduced.

For frame buildings, the most common design situation will be that design displacements are governed by code drift limits. The code drift will apply to the element with the greatest displacement, including torsional effects, meaning that the design displacement at the building centre of mass, used in the SDOF design, will need to be reduced proportionally to torsional displacements. Then, the design displacement for C_M is given by

$$\Delta_{ZM} = \Delta_{i,\text{cr}} - \theta \left(x_{i,\text{cr}} - e_{VX} \right) \tag{18.6}$$

where $\Delta_{i,\text{cr}}$ is the drift-controlled displacement of the critical element.

In general, then, it will be necessary to use an iterative approach to determine the design displacement when torsional effects are significant, since θ depends on J_R, e_R and e_V, which in turn depend on the relative strengths assigned to the lateral force-resisting elements in both orthogonal directions and on the system ductility.

18.3 Geometry of the Building and Elastic Analysis

The geometry of the analysed three-storey RC frame building is reported in Fig. 18.3. The interstorey height is equal to 3.20 m at all levels; at the first storey, the columns section dimensions are 40×65 cm^2, while all the beams are 40×60 cm^2; at the second storey, such dimensions are, respectively, 40×60 cm^2 and 40×55 cm^2, and at the third 40×55 cm^2 and 40×50 cm^2. Column dimensions are kept larger than beam ones in order to take into account the capacity design. Two-metre-wide balconies are shown in Fig. 18.3 as hatched areas.

Elastic analyses are performed by the computer program SAP2000 (CSI Computer and Structures Inc 2004), according to Eurocode rules (CEN 2002a, b, 2004) and considering a design spectrum soil B type 1 with a design ground acceleration on type A ground, a_g =0.35 g, taken from the old Italian seismic code OPCM 3431 (Presidenza del Consiglio dei Ministri 2005). The FBD design is performed according to the high-ductility class rules and a behaviour factor equal to 2.4 is computed; such value takes into account that according to EC8 the building is "torsionally flexible" and irregular in elevation (the reduction of lateral stiffness from 1st to 2nd storey is larger than 40%). As imposed by the code, a 5% accidental eccentricity is considered; consequently, 4 models are analysed with the centre of mass placed in four different positions, indicated in the following with respect to the initial position: "dx" at (0.45,0), "sx" at (−0.45,0), "sup" at (0,1.05) and "inf" at (0,−1.05).

Concrete characteristic cylinder strength equal to $f_{ck} = 30$ N/mm^2 and steel characteristic yielding strength equal to $f_{yk} = 430$ N/mm^2 are adopted.

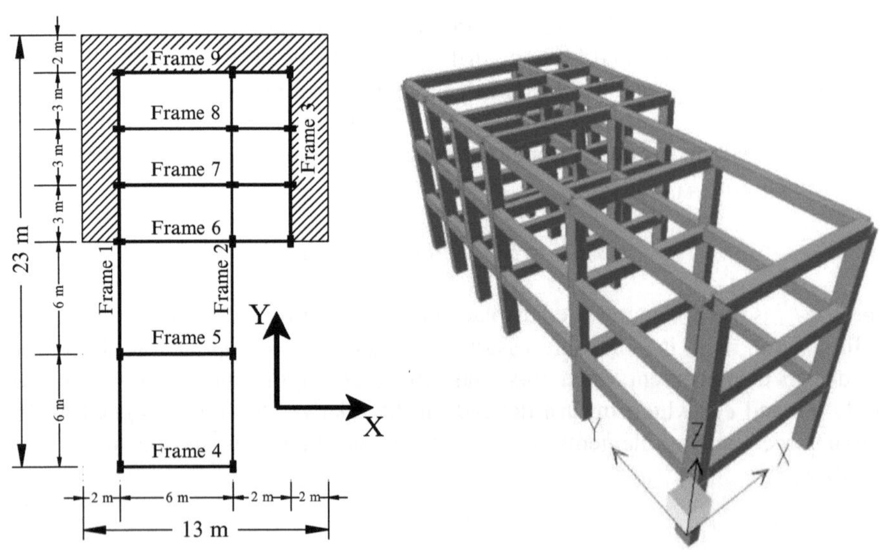

Fig. 18.3 Geometry of the analysed building

18.4 Specifications on Direct Displacement-Based Design Applied to a RC Irregular In-Plan 3D Multistorey Structure

For irregular in-plan buildings, the design displacement passes through the evaluation of top C_M one that depends by displacement of resisting critical (which exhibits the greatest displacements) plane frame by Eq. 18.3.

In the initial design phase, it is necessary to make assumptions on stiffnesses and structural strengths of members, even though they are still unknown. In order to reduce torsional effects, at the beginning, the strength eccentricity is assumed equal to zero; consequently, θ is assumed equal to zero. The aim of the design is generally to reduce the strength eccentricity in order to minimise the inelastic twist. The design displacement is derived under this assumption; it is the same along the two orthogonal directions. The 3D frame is computed considering it is made of different frames.

Starting by these assumptions, the design displacement Δ_d at the effective height can be computed. The yield displacement of the system is calculated doing a weighted average, with respect to the shear strength at the first floor, of the yielding displacements of the plane frames. Consequently, the ratio V_i/V_{Tot} between i frame shear strength and the building total shear strength is to be assumed: in this case, at the first iteration, a uniform distribution of the strength is assigned, assuming e_V and θ are equal to zero (the centre of mass is computed). The yielding displacement of the system is

$$\Delta_{y,\mathrm{sys}} = \sum_{i=1}^{n} \frac{V_i}{V_{\mathrm{Tot}}} \cdot \Delta_{y,i} \tag{18.7}$$

where $\Delta_{y,i}$ is the displacement of the i frame and $\Delta_{y,\mathrm{sys}}$ is the system displacement. The $\Delta_{y,i}$ is computed as in the case of 2D frames; even though the plan frames present irregular spans, overturning moments of single spans are assumed equal.

By the design and yielding displacements, the ductility, the equivalent viscous damping and, finally, the base shear, to be distributed along the building height, are computed for each direction.

The yielding moments at the column base are computed assuming the strength eccentricity equal to zero, as

$$M_0 = 0.6 \cdot H_1 \cdot \frac{V_i}{V_{\mathrm{Tot}}} \cdot V_{\mathrm{base}} \tag{18.8}$$

where H_1 is the interstorey height at the first level.

The following step is to perform the analysis in order to compute the strength and the stiffness of the frames; a secant stiffness is calculated as ratio between storey shear and displacement.

The base shear is distributed in the two directions with separated analyses, using SAP 2000 program (CSI Computer and Structures Inc 2004). The analysis under seismic forces is performed without taking into account seismic gravitational loads. The modelling of the building and of member stiffness in the two directions is different: plastic hinges at the frame base are only considered in the horizontal load direction and the members have secant stiffness only along this direction; in the other direction, members are elastic. In this last case, the lateral stiffness, which is needed in order to compute the torsional twist, is evaluated assuming a force distribution according to EC8 previsions.

The top floor centre of mass displacement, $\Delta_{n,\mathrm{sys}}$, is evaluated by Eq. 18.6, where $\Delta_{i,\mathrm{cr}}$ is the top floor displacement of the frame at the greatest distance from C_V. The design displacement at the effective height is calculated by

$$\Delta_{d,\mathrm{sys}} = \Delta_{n,\mathrm{sys}} \cdot \frac{\Delta_{\mathrm{He,frame}}}{\Delta_{n,\mathrm{frame}}} \tag{18.9}$$

where $\Delta_{\mathrm{He,frame}}$ and $\Delta_{n,\mathrm{frame}}$ are the displacement at the effective height and at the top, respectively, of the most stressed frame.

For each frame, the overturning moment is computed at each bay in order to evaluate the yielding displacement and, by Eq. 18.7, the system yielding displacement. Consequently, the ductility, the damping and, then, the SDOF base shear can be computed.

The procedure can be iterated updating, for each iteration, the strength and then the stiffness of the members. The procedure ends when the base shears of two consecutive iterations are almost equal.

EC2 provisions concerning the detailing are applied; while the detailing provided by EC8 is not considered, because this is related to the force method and it has to assure the required ductility according to q factor assumption. Then its application is inappropriate for DDBD.

18.5 Comparison Between the Reinforcement Obtained by FBD and DDBD

In this section, a preliminary comparison between the two design methods, FBD and DDBD, is carried out. The comparison is performed in terms of mean longitudinal reinforcement ratio for each floor, for beams (Fig. 18.4) and columns (Fig. 18.5) at end zones.

The two different design methods provide very different reinforcement ratios: the DDBD allows a reinforcement saving of about 70% for beams and 50% for columns.

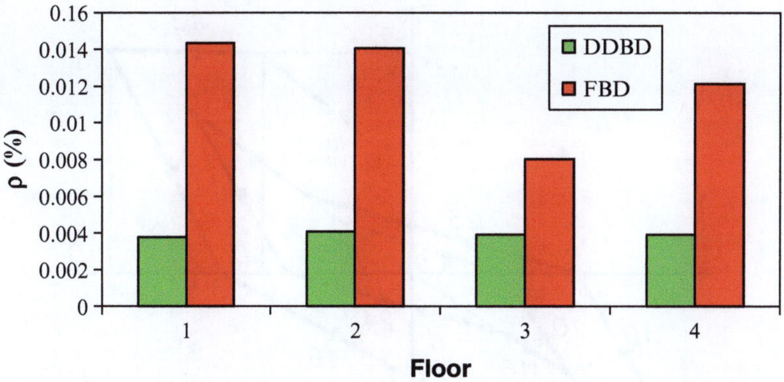

Fig. 18.4 Percentage of beam reinforcement according to FBD and DDBD

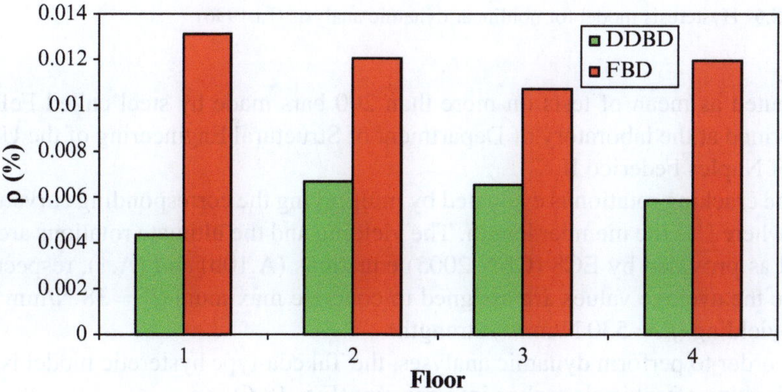

Fig. 18.5 Percentage of column reinforcement according to FBD and DDBD

18.6 Nonlinear Analyses

Nonlinear analyses are performed by means of the computer program CANNY99 (Li 1996). Nonlinearity regards flexural rotations, while all the other deformations are assumed linear. Both beams and columns are characterised by lumped plasticity models; in the latter case, for each section, two independent nonlinear springs are assigned, one for each orthogonal direction. No axial force-bending moment interaction is considered at plastic hinge.

Bending moment springs are characterised by a trilinear skeleton curve, defined by cracking and yielding moment and corresponding rotations; the post-yielding stiffness is assumed equal to zero. An elastic perfectly plastic steel stress-strain diagram is considered, characterised by a yielding strength equal to 530 MPa,

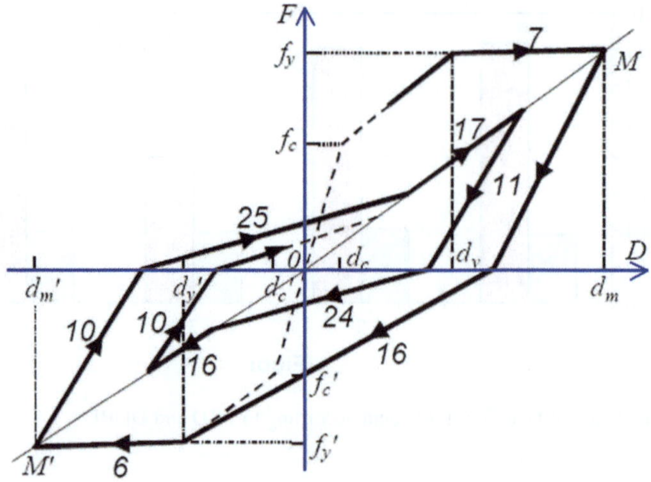

Fig. 18.6 Hysteresis model for nonlinear dynamic analysis (Li 1996)

computed as mean of tests on more than 200 bars made by steel called FeB44K, performed at the laboratory of Department of Structural Engineering of the University of Naples Federico II.

The cracking rotation is evaluated by multiplying the corresponding curvature by L/6, where L is the member length. The yielding and the ultimate rotations are evaluated as provided by EC8 (CEN 2005) equations (A.10b) and (A.1), respectively, where the average values are assigned to concrete maximum ($f_c = 38$ N/mm^2) and steel yielding ($f_y = 530$ N/mm^2) strength.

In order to perform dynamic analyses, the Takeda-type hysteretic model is used; the pinching effect is also taken into account (Fig. 18.6).

Both the horizontal components of a set of 7 earthquakes, i.e. 14 natural records, are used for nonlinear dynamic analyses whose results are shown herein.

The selected natural accelerograms are (Maddaloni et al. 2012):

- Northern and central Iran, Iran, 16/09/1978 (cod.000187)
- Montenegro, Yugoslavia, 15/04/1979 (cod.000196)
- Montenegro, Yugoslavia, 15/04/1979 (cod.000199)
- Montenegro (aftershock), Yugoslavia, 24/05/1979 (cod.000230)
- Campano Lucano, Italy, 23/11/1980 (cod.000291)
- South Iceland, Iceland, 17/06/2000 (cod.0006263)
- South Iceland (aftershock), Iceland, 21/06/2000 (cod.0006334)

They satisfy the EC8 provisions; the mean of zero period spectral response acceleration values (calculated from individual time histories) should not be smaller than the value of $a_g \cdot S$ concerning the considered site; in the range of periods between $0.2T_1$ and $2T_1$, where T_1 is the fundamental period of the structure, in the direction where the accelerogram is applied, no value of the mean 5% damping elastic spectrum, calculated from all time histories, should be less than 90% of the corresponding value of the 5% damping elastic response spectrum; if the response is obtained from at least 7 nonlinear time histories analyses, the average of response quantities should be used as the design value of the action effect E_d in relevant verifications. Only one record (cod. 000187) is scaled by a low scale factor equal to 1.08, in order to match in the prescribed range the code spectrum used for the design of the analysed building, assigning $q = 1$.

According to the direction of registration of the accelerograms, a direction, *X* or *Y*, is given to the two components of each earthquake; the analyses are performed applying, for each earthquake, the *X* component along the longitudinal direction of the building and the *Y* component along the orthogonal direction.

Top centre of mass displacements and demand/capacity ratio in terms of total chord rotation at beam and column ends are shown. For each of the four models obtained moving the centre of mass, the average of the seven maximum results obtained applying the seven earthquakes is considered; the maximum effect on the four models is always considered.

18.7 Result and Comparisons

The results of nonlinear dynamic analyses carried out on models designed according to FBD and DDBD are compared in terms of displacements and rotational ductility demand, and such demand is compared to capacity at the ultimate limit state, which is equal to 3/4 Θ_u.

Figure 18.7 shows the frame top displacements absolute and normalised with respect to CM displacement, along *X* and *Y* directions.

It can be observed that NLDA on "DDBD structure" gives absolute displacements larger than NLDA on "FBD structure". But the torsional behaviour of the structure designed according to DDBD is better: in *X* direction, the displacement of the frame 4 (critical frame) is about 15% lower. The effect of torsional twist on *Y* direction maximum displacements is not significant.

Figure 18.8 shows interstorey drifts of the two structures evaluated at CM along *X* and *Y* directions. The limit drift, assumed equal to 2.5%, is indicated by a red line.

Even though the DDBD provides a more deformable structure than FBD, the limit drift is not attained.

In Fig. 18.9, demand/capacity ratios in terms of total chord rotations at beam (black numbers) and column (red numbers) end sections of frames 1 and 3 in *X* direction are presented.

NLDA on both FDB and DDBD structures provides all ratios lower than one; consequently, it is confirmed that the building designed according to either EC8 provisions by elastic modal response spectrum analysis at the ultimate limit state or DDBD is verified at the same limit state by nonlinear dynamic analysis performed according to EC8. Furthermore, also in terms of ductility demand, as already noted

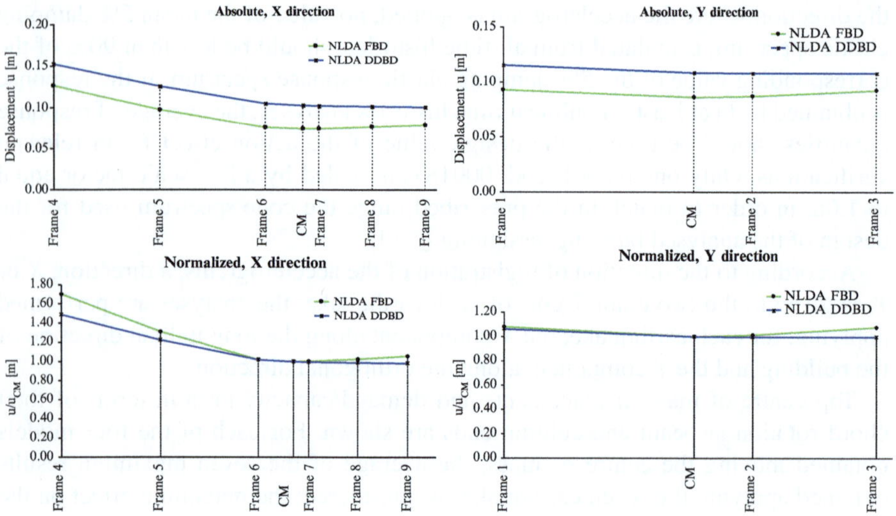

Fig. 18.7 Absolute and normalised frame top displacements along *X* and *Y* directions

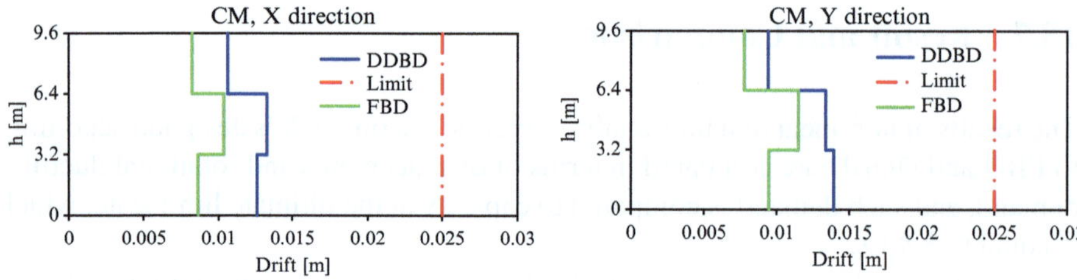

Fig. 18.8 Drifts at CM along *X* and *Y* directions

in terms of displacements, NLDA FBD provides lower values with respect to NLDA DDBD; the demand/capacity ratios in terms of maximum rotational ductility at the beam and column base end on "DDBD structure" are about twice the corresponding ratios obtained on "FBD structure".

18.8 Conclusions

In this chapter, the application of the direct displacement-based design (DDBD) to the multistorey irregular in-plan frame structures is discussed. A case study is presented in order to extend and validate the method. The design of a torsionally flexible system is carried out according to DDBD and its seismic performances are compared, by nonlinear dynamic analyses, to those of the same building designed according to FBD rules of EC8.

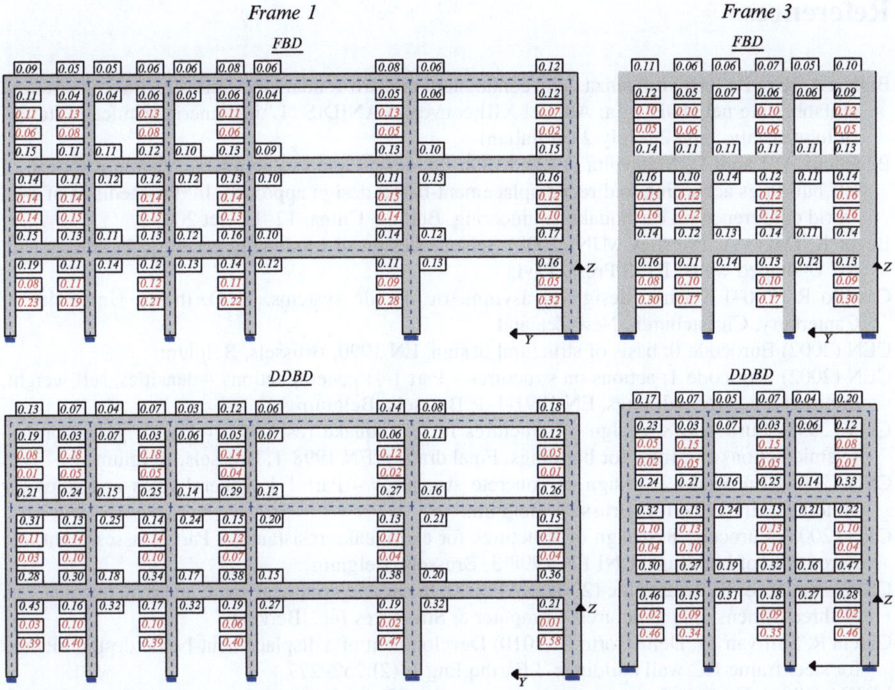

Fig. 18.9 Frame 1 and 3: demand/capacity ratio in terms of maximum rotational ductility at the element ends

The main conclusions of the study are:

- The two structures present very different reinforcement ratios: the design by DDBD leads to a saving of about 70% of beam reinforcement and of 50% in the case of columns.
- Even though its reinforcement is largely reduced, the building designed by DDBD shows a better nonlinear behaviour under strong earthquakes: the torsional twist is lower.
- Even though its reinforcement is largely reduced, nonlinear dynamic analyses confirm that also the building designed according to DDBD satisfies the ultimate limit state.

Acknowledgements This research has been partially funded by Italian Department of Civil Protection in the frame of the national project DPC-ReLUIS 2010–2013. The authors thank Eng. Angelo Farnetano for the precious collaboration in the execution of the numerical analyses.

References

Belleri A, Riva P (2009) Influenza delle connessioni pilastro-fondazione e pilastro-trave di strutture prefabbricate nel DDBD. In: Atti del XIII convegno ANIDIS "L' ingegneria sismica in Italia", Bologna, Italy, June 28–July 2 (in Italian)

Benedetti A, Landi L, Malavolta D (2008) On the design and evaluation of seismic response of RC buildings according to direct displacement-based design approach. In: Proceedings of 14th world conference on earthquake engineering, Beijing, China, 12–17 Oct 2008

Beyer K, Dazio A, Priestley MJN (2008) Seismic design of torsionally eccentric buildings with RC U-shaped walls. IUSS Press, Pavia

Castillo R (2004) Seismic design of asymmetric ductile systems. Ph.D. thesis, University of Canterbury, Christchurch, New Zealand

CEN (2002) Eurocode 0: basis of structural design, EN 1990, Brussels, Belgium

CEN (2002) Eurocode 1: actions on structures – Part 1-1: general actions – densities, self weight, imposed loads for buildings, EN 1991-1-1, Brussels, Belgium

CEN (2003) Eurocode 8: design of structures for earthquake resistance – Part 1: general rules, seismic actions and rules for buildings, Final draft, prEN 1998-1, Brussels, Belgium

CEN (2004) Eurocode 2: design of concrete structures – Part 1-1: general rules and rules for buildings, EN 1992-1-1, Brussels, Belgium

CEN (2005) Eurocode 8: design of structures for earthquake resistance – Part 3: assessment and retrofitting of buildings, UNI EN 1998-3, Brussels, Belgium

CSI Computer & Structures Inc (2004) SAP2000: linear and nonlinear static and dynamic analysis of three-dimensional structures. Computer & Structures Inc., Berkeley

Garcia R, Sullivan TJ, Della Corte G (2010) Development of a displacement-based design method for steel frame-RC wall buildings. J Earthq Eng 14(2):252–277

Li KN (1996) Canny 99: a computer program for 3D nonlinear dynamic analysis of building structures – technical manual and users' manual. Canny Consultants Pte Ltd., Singapore

Maddaloni G, Magliulo G, Manfredi G (2012) Effect of the seismic input on non-linear response of r/c building structures. Adv Struct Eng 15(10):1861–1877

Maley TJ, Sullivan TJ, Della Corte G (2010) Development of a displacement-based design method for steel dual systems with buckling-restrained braces and moment-resisting frames. J Earthq Eng 14(S1):106–140

Paulay T (2001) Some design principles relevant to torsional phenomena in ductile buildings. J Earthq Eng 5(3):273–308

Pettinga JD, Priestley MJN, Pampanin S, Christopoulos C (2007) The role of inelastic torsion in the determination of residual deformations. J Earthq Eng 11(1):133–157

Presidenza del Consiglio dei Ministri (2005) Ordinanza n. 3431 del 3 maggio 2005: Ulteriori modifiche ed integrazioni all'Ordinanza del Presidente del Consiglio dei Ministri n. 3274 del 20 marzo 2003, Italia (in Italian)

Priestley MJN (1993) Myths and fallacies in earthquake engineering – conflicts between design and reality. Bull NZ Natl Soc Earthq Eng 26(3):329–341

Priestley MJN (1998) Displacement-based approaches to rational limit states design of new structures. In: Proceedings of European conference of earthquake engineering, Paris, France

Priestley MJN, Calvi GM, Kowalsky MJ (2007) Displacement-based seismic design of structures. IUSS Press, Pavia

Rizzato M, Tullini N, Laudiero F (2009) Analisi di pareti in c.a. mediante modelli a fibre: confronti tra l'approccio agli spostamenti ed alle forze. In: Atti del XIII Convegno ANIDIS "L' ingegneria sismica in Italia", Bologna, Italy, 28 June–2 July (in Italian)

Part V
Seismic Control and Monitoring of Irregular Structures

Chapter 19
A Simple Methodology for the Seismic Passive Control of Irregular 3D Frames Using Friction Dampers

Yael Daniel, Oren Lavan, and Robert Levy

Abstract This chapter presents a simple performance-based design methodology of an analysis/redesign type for sizing and allocating friction dampers within existing 3D irregular framed structures in order to control seismic performance. The methodology allows the determination of brace stiffness of all added damping devices, under a constant predetermined slip displacement, using analysis tools only. The methodology allows the design of an irregular framed structure of any type and is nondependent of the type of irregularity. Using this methodology, fully stressed design solutions are obtained in an example of a linear 3D setback frame undergoing an ensemble of ground motions.

19.1 Introduction

Seismic design of structures is a very intriguing and important issue in the field of structural engineering. Severe earthquakes are not a very common event and are random in nature, making them hard to predict and design for, and once they occur, life-threatening situations and major structural destruction are likely to happen. When seismic structural protection had begun, life safety was the leading guide for design. However, in modern design, the emphasis has shifted to performance-based design, as modern seismic design philosophies acknowledge the economical

Y. Daniel • O. Lavan (✉)
Faculty of Civil and Environmental Engineering, Technion – Israel Institute of Technology, Haifa 32000, Israel
e-mail: yaeldan@tx.technion.ac.il; lavan@tx.technion.ac.il

R. Levy
Department of Structural Engineering, Faculty of Engineering Sciences, Ben-Gurion University of the Negev, P.O.Box 653, Beer-Sheva 84105, Israel
e-mail: levyrob@bgu.ac.il

O. Lavan and M. De Stefano (eds.), *Seismic Behaviour and Design of Irregular and Complex Civil Structures*, Geotechnical, Geological and Earthquake Engineering 24, DOI 10.1007/978-94-007-5377-8_19,

benefits of such designs. Structural performance is highly related to damage states within the structure. The most commonly used method of damage analysis in structures is the use of damage indices (DIs). A rich description of DIs may be found in the state-of-the-art review of Williams and Sexsmith (1995). Some response measures often associated with damage are drifts and energy dissipation, and structural performance can be efficiently achieved by controlling these parameters.

When considering different irregularities of structures (either vertical irregularities, such as setback structures or horizontal asymmetric irregularities) undergoing ground motions, special attention has to be given to the behavior of the structure due to these irregularities, such as large torsional effects. Many works regarding different aspects of these phenomena can be found in the extensive reviews of Rutenberg and De Stefano 1997; De Stefano and Pintucchi 2008 as well as references within.

One way to control the responses and enhance seismic performance within an existing structure is by adding external control systems. Clearly, the design of structures with added control systems should be cost-effective as well as practical and intuitive for the use of practicing engineers. Passive friction dampers that dissipate energy through Coulomb friction forces created between two solids during slippage are considered in this chapter. The hysteretic model is a stick-slip one, with the slip displacement being smaller than the yielding displacement of the brace, to ensure the device slips. The hysteretic curves of these dampers are usually rectangular, show large dissipation abilities, and exhibit nearly no fade over a large number of cycles (Symans et al. 2008; Soong and Dargush 1997).

Several methodologies for the allocation or of hysteretic dampers exist. Among those are nonoptimal design methodologies which predetermine the damping distribution (Pall and Marsh 1982; Filiatrault and Cherry 1989, 1990; Cherry and Filiatrault 1993; Levy et al. 2001, 2005; Choi and Kim 2006; Lee et al. 2008). Methodologies that consider full nonlinear behavior of the structure for the design of friction dampers are based on genetic algorithm (GA), which is a very general and robust, yet very slow converging methodology. Several works using GA for implementation of friction and yielding dampers are available, including (Moreschi and Singh 2003; Wongprasert and Symans 2004; Dargush and Sant 2005; Ok et al. 2008; Farhat et al. 2009; Lavan and Dargush 2009). The drawback of using GAs is that the process is computationally expensive.

Most of the works mentioned above apply to plane structures or symmetric structures and do not necessarily cater to the special demands of 3D irregular structures. The first established attempt to study the control of seismic response of plan-asymmetric structures by using friction dampers was in Pekau and Guimond (1991), followed by a series of papers concerning the same concept (Pekau and Mastrangelo 1992; Martin and Pekau 1996; Pekau et al. 2000). The objective was to study the ratio of the sum of stiffness and slip load of the two dampers to the sum of stiffness and yielding force of the original structure under different eccentricities of the structures. In a series of papers (de la Llera et al. 2005; Vial et al. 2006), the empirical center of balance (ECB) concept was introduced for the case of plan-symmetric structures. This concept, based on which weak torsional balance, imposes a decoupling of translation and rotations within the plan.

This chapter proposes to use an analysis/redesign methodology for the optimal allocation and sizing of friction dampers within existing irregular 3D structures. This methodology has been applied to a plane symmetric frame (Daniel et al. 2010). Whereas no formal proof of optimality is provided, it is conjectured here that the solution is optimal based on prior work by the authors on allocation and sizing of viscous dampers where indeed such algorithms converge to an optimal solution.

19.2 Problem Formulation

19.2.1 Equations of Motion

The differential equations of motion for a non-yielding MDOF structure with added hysteretic damping devices under seismic loading are formulated as

$$\mathbf{M}\ddot{\mathbf{x}}(t) + \mathbf{C}\dot{\mathbf{x}}(t) + \mathbf{K}\mathbf{x}(t) + \mathbf{f}_{\mathbf{s0}}\left(\mathbf{x}(t), \dot{\mathbf{x}}(t)\right) = -\mathbf{M}\mathbf{e}a_g(t)$$
$$\mathbf{x}(0) = \mathbf{0}, \dot{\mathbf{x}}(0) = \mathbf{0} \tag{19.1}$$

where **M**, **C**, and **K** are the mass, inherent damping, and stiffness matrices of the bare structure in inter-story drift DOFs, respectively; $a_g(t)$ is the ground motion's acceleration vector, **e** is the excitation direction matrix with values of zero and one; $\ddot{\mathbf{x}}(t)$, $\dot{\mathbf{x}}(t)$, and $\mathbf{x}(t)$ are the relative inter-story acceleration, velocity, and displacement vectors between the DOFs and the ground; and $\mathbf{f}_{\mathbf{s0}}\left(\mathbf{x}(t), \dot{\mathbf{x}}(t)\right)$ is the restoring force vector of the nonlinear damping devices. A similar formulation may be used in case the structure, too, is expected to yield, where the expression for $\mathbf{K}\mathbf{x}(t)$ is replaced by another restoring force $\mathbf{f}_{\mathbf{s}}\left(\mathbf{x}(t), \dot{\mathbf{x}}(t)\right)$, which represents the nonlinear restoring forces of the frame.

A bilinear hysteretic representation of the yielding devices is suggested (e.g., Sivaselvan and Reinhorn 2000), for which the restoring force can be divided into two contributions: linear and nonlinear as the following

$$\mathbf{f}_{\mathbf{s0}}\left(\mathbf{x}(t), \dot{\mathbf{x}}(t)\right) = \mathbf{K}_{\alpha 0}\mathbf{x}(t) + \mathbf{f}_{\mathbf{h0}}\left(\mathbf{x}(t), \dot{\mathbf{x}}(t)\right) \tag{19.2}$$

where $\mathbf{K}_{\alpha 0}$ represents the stiffness matrix of the dampers (in drift DOFs) after yielding, and $\mathbf{f}_{\mathbf{h0}}\left(\mathbf{x}(t), \dot{\mathbf{x}}(t)\right)$ is the hysteretic force vector of the yielding dampers in local drift coordinates with zero secondary slope. The hysteretic rule for the nonlinear contribution of the restoring force can be represented in the following differential form

$$\dot{\mathbf{f}}_{\mathbf{h0}}(t) = f\left(\dot{\mathbf{x}}(t), \mathbf{f}_{\mathbf{h0}}(t), \mathbf{k}_0\right), \mathbf{f}_{\mathbf{h0}}(0) = 0 \tag{19.3}$$

where $\mathbf{k}_0$ is the vector of added brace stiffness. In conclusion, the following equations of motion (in drift coordinates) of a non-yielding structure with friction dampers are formulated:

$$\begin{array}{ll} \mathbf{M}\ddot{\mathbf{x}}(t) + \mathbf{C}\dot{\mathbf{x}}(t) + \mathbf{K}\mathbf{x}(t) + \mathbf{f}_{\mathbf{h}0}\left(\mathbf{x}(t), \dot{\mathbf{x}}(t)\right) = -\mathbf{M}\mathbf{e}a_g(t); & \mathbf{x}(0) = \mathbf{0}, \dot{\mathbf{x}}(0) = \mathbf{0} \\ \dot{\mathbf{f}}_{\mathbf{h}0}(t) = f\left(\dot{\mathbf{x}}(t), \mathbf{f}_{\mathbf{h}0}(t), \mathbf{k}_0\right); & \mathbf{f}_{\mathbf{h}0}(0) = 0 \end{array} \tag{19.4}$$

19.2.2 Performance Measures

The performance of the structure is evaluated on the basis of damage. This is done using any existing DI or any other response found to portray structural damage due to seismic action. It seems that in the case of structures which stay within the elastic zone following a seismic incident, damage is strongly related to the amount of inter-story drifts developed. Otherwise, it seems more appropriate to use an accumulated-natured performance measure, relating to both drifts and hysteretic energy. Added friction dampers help reduce these performance measures, and the measure of cost and effectiveness of these devices is by the amount of added stiffness. The amount of added stiffness represents the cost of the retrofit, in addition to the fact that minimum added stiffness to the structure results in smaller added forces to the structure and smaller accelerations.

19.2.3 Formal Optimization Problem

The problem at hand is formulated as an optimization problem for which the objective function minimizes the total amount of added friction dampers under the constraints of maximal added damping and maximal performance measures according to the desirable performance. In this study, the slip displacement is predetermined, and the friction damper's parameter is chosen as the brace stiffness. The optimization problem can be formulated in the following scheme

$$\begin{gathered} \min_{\mathbf{k}_0} J = \sum K_0 = \mathbf{k}_0{}^T \cdot \mathbf{1} \\ s.t.\ pi \leq 1.0 \\ \mathbf{0} \leq \mathbf{k}_0 \leq \mathbf{k}_{0,\max} \end{gathered} \tag{19.5}$$

where $\mathbf{k}_0$ and $\mathbf{k}_{0,\max}$ are the vectors of added brace stiffness and maximal allowable added stiffness, and pi is a normalized inter-story performance index of interest. The performance index should be directly related to a desired measure of performance, such as a DI. A formulation of the nonlinear equations of motion is needed for the evaluation of the performance index. Equation (19.4) is used for the case of a linear structure with friction dampers.

19.2.4 Fully Stressed Design (FSD) Characteristics

Designs that are based on fully stressed characteristics go back to the classical design of trusses under static loads, whereby the weight is minimized for a given allowable stress. For that problem, it had been widely accepted that the optimal design yields a statically determinate fully stressed design, with members out of the design having strains smaller than the allowable. This result appeared in the literature as early as 1900 (Cilley 1900) as: "*A statically determined framework of included figure is the most economic form of a framework of given indeterminate figure for the support of a given loading*." It was later shown (Levy 1985) that this design is a Karush-Kuhn-Tucker point and, therefore, an optimal design. Later on, it was shown that some dynamic optimal designs also possess "fully stressed" characteristics. Levy and Lavan (2006) considered the minimization of total added damping to framed structures subjected to ground accelerations while constraining various inter-story responses. Their optimal solutions attained by formal optimization indicated that: "*At the optimum, damping is assigned to stories for which the local performance index has reached the allowable value. Stories with no assigned damping attain a local performance index which is lower or equal to the allowable*." That is, the optimal solutions attained "fully stressed" characteristics.

Based on past experience of the authors in similar problems, it is conjectured here that the optimal solution to friction damper allocation and sizing problem in framed structures (the solution of Eq. (19.5)) possesses FSD characteristics, that is: *At the optimum, friction dampers are assigned to locations for which the inter-story performance index of interest has reached the allowable value under a given ensemble of ground motions.*

19.2.5 Analysis/Redesign Algorithm

Solutions to optimization problems, which possess fully stressed characteristics, are efficiently achieved iteratively using a two-step algorithm in each iteration cycle. In the first step, an analysis is performed for a given preliminary design, whereas in the second step, the design is changed using a recurrence relationship that targets fully stressedness. The recurrence relation can be generally written as

$$x_k^{(i+1)} = x_k^{(i)} \cdot \left(\frac{pi_k^{(i)}}{pi_{\text{allowable}}} \right)^P \tag{19.6}$$

where x_k is the value of the design variable k, pi_k is the performance measure of interest for variable k, $pi_{allowable}$ is the allowable performance measure, i is the iteration number, and P is the convergence parameter. In the optimal design of trusses, for example, the practitioner would assume initial values for the cross sections and run an analysis. Then, based on the attained performance measures and their allowable values, the cross section of each member could be redesigned using Eq. (19.6). The process is repeated until convergence. The advantages of the analysis/redesign algorithm include its simplicity, the need to use analysis tools only, and the fairly small computational effort that lies in the small number of iterations required for convergence. Such an analysis/redesign procedure will be utilized here to attain fully stressed designs where the stiffness of friction dampers is to be determined. The design methodology is described in detail in the next section.

19.3 Design Methodology

The following is a step-by-step procedure for the optimal allocation and sizing of friction dampers. This methodology is based on envelope peripheral inter-story drifts, for which the drifts at each iteration cycle are evaluated for all ground motion within the ensemble. This method may be a bit wasteful when considering the amount of needed analyses. Methods for preselecting an active ground motion and sequentially adding additional ground motions into an active set as the iterative process continues are available and may be used to reduce the number of analyses. However, it is needed to find a preselection method suitable for the problem of nonlinear hysteretic dampers.

Step 1 Perform time-history analysis on the bare frame for all ground motions within the chosen ground motion ensemble and check violation of the allowable performance measures.

Step 2 If any of the performance measures exceed the allowable, add friction dampers at the peripheral frames with a predetermined slip displacement and an arbitrary initial stiffness.

Step 3 Perform time-history analysis on the braced frame for all ground motions within the ensemble and evaluate the peripheral inter-story drifts.

Step 4 Using Eq. (19.6), redesign each brace's stiffness according to the envelope of inter-story peripheral drifts, which is the maximum drift of all ground motions in the specific peripheral location.

Step 5 Repeat steps 3–4 until convergence of all brace stiffness is reached.

19.4 Example

The following 8-story 3-bay setback reinforced-concrete (RC) frame structure (Fig. 19.1) introduced by Tso and Yao (1994) is retrofitted using friction dampers under the LA 10 in 50 ground motion ensembles in the "y" direction. A uniform distributed mass of 0.75 t/m^2 is taken. The column dimensions are 0.5 m by 0.5 m for frames 1 and 2 and 0.7 m by 0.7 m for frames 3 and 4. The beams are 0.4 m wide and 0.6 m tall. 5% Rayleigh damping for the first and second modes is used. Drift is the performance measure of choice in this example as the structure was assumed to behave linearly; thus, no energy-based measure was needed. The maximal allowable drift was 0.035 m (1% of story height).

The design variables are the locations and stiffnesses of the individual friction dampers. The dampers are to potentially be located in the peripheral frames, where they are most effective (as these are the locations within the plan where drifts will be maximal). Dampers will be assigned to peripheral frames 1 (lower 4 floors, middle bay), 3 (upper 4 floors, middle bay), 4 (middle bay), 5 (right-hand bay),

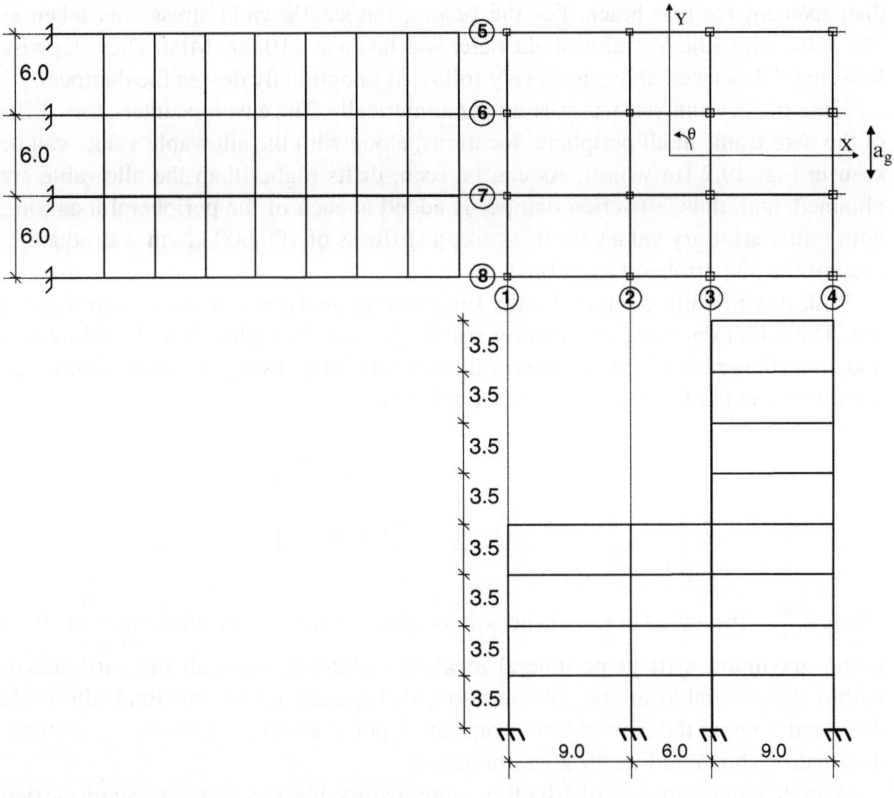

Fig. 19.1 Eight-story setback structure

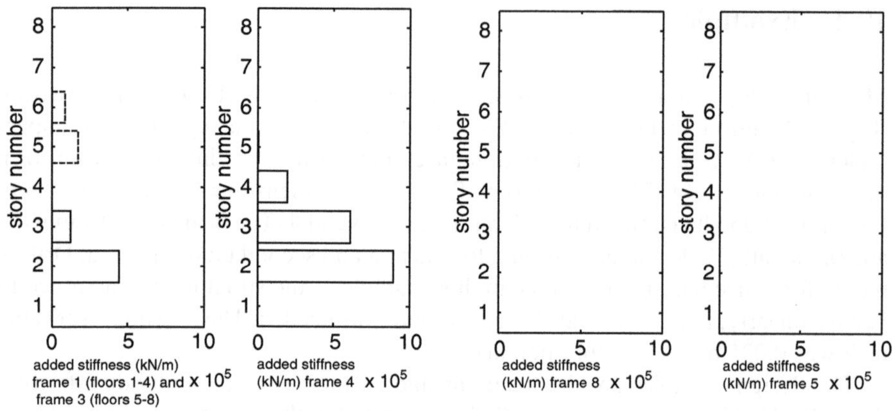

Fig. 19.2 Final stiffness of the friction damper's brace

and 8 (right-hand bay). The yield displacement for each damper was computed such that the damper slips at a displacement which equals to 90% of the yield displacement for that brace. For the braces, the steel's yield stress was taken as 235 MPa, while the modulus of elasticity was taken as 210,000 MPa. The "stepwise flowchart" described above is closely followed to optimally design the dampers.

Time-history analysis was carried out numerically. The envelope inter-story drifts of the bare frame at all peripheral locations, along with the allowable value, can be seen in Fig. 19.2 (in white). As can be seen, drifts higher than the allowable are obtained, and, thus, a friction damper is added at each of the peripheral locations, with initial arbitrary values (in this case, a stiffness of 100,000 kN/m was added in each of the 32 peripheral locations).

With this initially damped frame, time-history analysis was once again carried out. The initial damping distribution still leads to drifts higher than the allowable, and, thus, the brace stiffness was reevaluated iteratively, using the analysis/redesign algorithm and the following recurrence relation

$$k_{0,i}^{(n+1)} = k_{0,i}^{(n)} \cdot \left(\frac{\max\limits_{\text{all EQs}} \left| \text{drift}_i^{(n)} \right|}{\text{drift}_{\text{allowable}}} \right)^p \tag{19.7}$$

where $k_{0,i}^{(n)}$ is the added brace stiffness at location i at the nth iteration, $\max\limits_{\text{all EQs}} \left|\text{drift}_i^{(n)}\right|$ is the maximum drift of peripheral location i obtained from all the earthquakes within the ensemble at the nth iteration, $\text{drift}_{\text{allowable}}$ is the maximal allowable drift, and p being the convergence constant. Upon convergence, the brace stiffness distribution shown in Fig. 19.2 was obtained.

With this final damping distribution, time-history analysis was once again carried out for all ground motions within the chosen ensemble, and the envelope inter-story

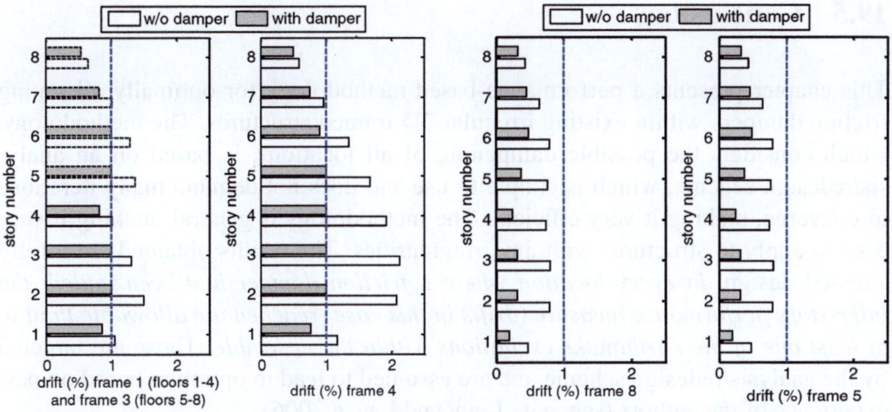

Fig. 19.3 Peripheral envelope drifts and allowable values (*dashed*) of the bare and final damped structure under the LA 10 in 50 ground motion ensembles

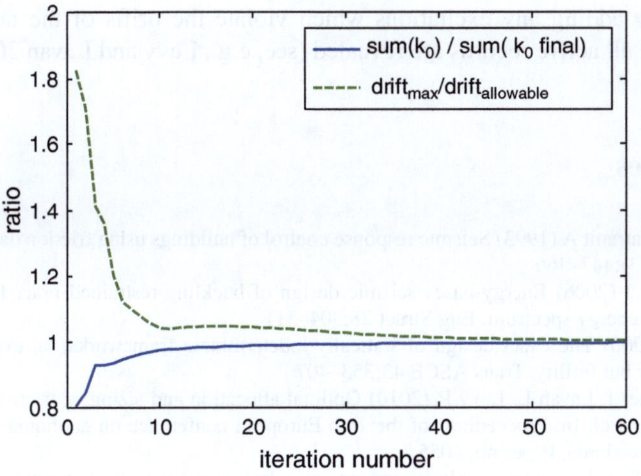

Fig. 19.4 Convergence of sum of stiffness and maximal inter-story drift

drifts obtained at all peripheral locations can be seen in Fig. 19.3 (in gray). As can be seen from Figs. 19.2 and 19.3, dampers were only added to locations where the inter-story drifts equal the allowable, making the obtained solution a FSD.

Figure 19.4 presents the convergence of the design variables (brace stiffnesses) and the performance measure (inter-story drifts). As can be seen in Fig. 19.4, although the initial guess was very far from optimum, convergence is practically reached within less than 50 iterations. This was obtained for a convergence power value of $p = 1$ for iterations one to ten and $p = 0.5$ for all iterations after that.

19.5 Conclusions

This chapter presents a performance-based methodology for optimally allocating friction dampers within existing irregular 3D framed structures. The methodology, which considers the possible dampening of all locations, is based on an analysis/redesign scheme, which is simple to use and does not demand many iterations to converge, making it very efficient. The methodology is general, making it very easy to apply to structures with any irregularities. The results obtained are a fully stressed design, *in every location where a friction damper had been added*; *the inter-story performance measure* (*drifts in this case*) *reached the allowable limit in at least one of the earthquake excitations within the ensemble*. Those are targeted by the analysis/redesign scheme and are assumed to lead to optimum based on past experience of the authors (see, e.g., Levy and Lavan 2006).

The methodology presented herein uses the envelope of drifts obtained from all ground motions within the ensemble, which may impose many unnecessary analyses. If needed to make the methodology even less computationally heavy, it is suggested to use a method of preselecting an active ground motion and then sequentially adding any excitations which violate the drifts of the newly braced frame, until all active records are revealed (see, e.g., Levy and Lavan 2006).

References

Cherry S, Filiatrault A (1993) Seismic response control of buildings using friction dampers. Earthq Spectra 9(3):447–466

Choi H, Kim J (2006) Energy-bases seismic design of buckling-restrained braced frames using hysteretic energy spectrum. Eng Struct 28:304–311

Cilley FH (1900) The exact design of statically indeterminate frameworks, an exposition of its possibility but futility. Trans ASCE 43:353–407

Daniel Y, Nsier I, Lavan L, Levy R (2010) Optimal allocation and sizing of friction dampers for seismic control. In: Proceedings of the 14th European conference on earthquake engineering, Ohrid, Macedonia, Paper no. 1055

Dargush GF, Sant RS (2005) Evolutionary aseismic design and retrofit of structures with passive energy dissipation. Earthq Eng Struct Dyn 34:1601–1626

de la Llera JC, Almazan JL, Vial IJ (2005) Torsional balance of plan-asymmetric structures with frictional dampers: analytical results. Earthq Eng Struct Dyn 34:1089–1108

De Stefano M, Pintucchi B (2008) A review of research on seismic behaviour of irregular building structures since 2002. Bull Earthq Eng 6:285–308

Farhat F, Nakamura S, Takahashi K (2009) Application of genetic algorithm to optimization of buckling restrained braces for seismic upgrading of existing structures. Comput Struct 87: 110–119

Filiatrault A, Cherry S (1989) Parameters influencing the design of friction damped structures. Can J Civ Eng 16:753–766

Filiatrault A, Cherry S (1990) Seismic design spectra for friction-damped structures. J Struct Eng 116(5):1334–1355

Lavan O, Dargush GF (2009) Multi-objective evolutionary seismic design with passive energy dissipation systems. J Earthq Eng 13(6):758–790

Lee SH, Park JH, Lee SK, Min KW (2008) Allocation and slip load of friction dampers for a seismically excited building structure based on storey shear force distribution. Eng Struct 30:930–940

Levy R (1985) On the optimal design of trusses under one loading condition. Q Appl Math 43(2):129–134

Levy R, Lavan O (2006) Fully stressed design of passive controllers in framed structures for seismic loadings. J Struct Multidisciplin Optim 32:485–498

Levy R, Marianchik E, Rutenberg A, Segal F (2001) A simple approach to the seismic design of friction damped braced medium-rise frames. Eng Struct 23:250–259

Levy R, Lavan O, Rutenberg A (2005) Seismic design of friction-damped braced frames based on historical records. Earthq Spectra 21(3):761–778

Martin L, Pekau OA (1996) Plan-wise distribution of slip load in friction damped eccentric structures. In: Proceedings of the 11th world conference on earthquake engineering, Acapulco, Mexico, Paper no. 554

Moreschi LM, Singh MP (2003) Design of yielding metallic and friction dampers for optimal seismic performance. Earthq Eng Struct Dyn 32:1291–1311

Ok SY, Song J, Park KS (2008) Optimal design of hysteretic dampers connecting adjacent structures using multi-objective genetic algorithm and stochastic linearization method. Eng Struct 30:1240–1249

Pall AS, Marsh C (1982) Response of friction damped braced frames. J Struct Div ASCE 108:1313–1323

Pekau OA, Guimond R (1991) Controlling seismic response of eccentric structures by friction dampers. Earthq Eng Struct Dyn 20:505–521

Pekau OA, Mastrangelo E (1992) Seismic performance of friction damped asymmetric structures. In: Proceedings of the 10th world conference on earthquake engineering, Balkema, Rotterdam, pp 4129–4132

Pekau OA, Dasgupta B, Bedair H (2000) Improved deployment of friction dampers in asymmetric multi-story buildings. In: Proceedings of the 12th world conference on earthquake engineering, Auckland, New Zealand, Paper no. 1144

Rutenberg A, De Stefano M (1997) On the seismic performance of yielding asymmetric multistory buildings: a review and case study. In: Fajfar P, Krawinkler H (eds) Seismic design methodologies for the next generation of codes. Balkema, Rotterdam, pp 299–310

Sivaselvan MV, Reinhorn AM (2000) Hysteretic models for deteriorating inelastic structures. J Eng Mech 126(6):633–640

Soong TT, Dargush GF (1997) Passive energy dissipation systems in structural engineering. Wiley, Chichester

Symans MD, Charney FA, Whittaker AS, Constantinou MC, Kircher CA, Johnson MW, McNamara RJ (2008) Energy dissipation systems for seismic applications: current practice and recent development. J Struct Eng 134(1):3–11

Tso WK, Yao S (1994) Seismic load distribution in buildings with eccentric setback. Can J Civ Eng 21:50–62

Vial IJ, de la Llera JC, Almazan JL, Ceballos V (2006) Torsional balance of plan-asymmetric structures with frictional dampers: experimental results. Earthq Eng Struct Dyn 35:1875–1898

Williams MS, Sexsmith RG (1995) Seismic damage indices for concrete structures: a state-of-the-art review. Earthq Spectra 11(2):319–349

Wongprasert N, Symans MD (2004) Application of a genetic algorithm for optimal damper distribution within the nonlinear seismic benchmark building. J Eng Mech 130(4):401–406

Chapter 20
An Energy-Based Design Method for Buildings with Supplemental Damping and Nonlinear Behavior

Raul Barrón and A. Gustavo Ayala

Abstract This chapter presents an energy-based procedure for the design of supplemental damping. The average energy dissipated pear cycle at each interstory of the framing system is used as the response parameter to be reduced. The response is obtained from a stochastic approach using the equivalent linearization method and the transfer-matrix method. As illustrative example, a building with asymmetries in-plan and in-elevation subjected to bidirectional ground excitation is considered. In this example, it is observed that the proposed energy-based procedure produces better designs in terms of the optimal placement and the amount of damping required to reduce the structural response. It is concluded that the placement of a small amount of dampers, optimally located in the building, can diminish its structural response in an efficient and economical manner leading to structures with diminished hysteretic energy dissipation and consequently reduced structural damage.

20.1 Introduction

Past years have witnessed an accelerated growth in the development and application of energy dissipation devices, such as viscoelastic, viscous fluid, friction, and metallic dampers, to reduce the effects of intense earthquake and wind demands

R. Barrón
Facultad de Ingeniería, Universidad Autónoma de Zacatecas, Av. López V. No 801,
Mexico City, Zacatecas 98000, Mexico
e-mail: raul.barron1@gmail.com

A.G. Ayala (✉)
Instituto de Ingeniería, Universidad Nacional Autónoma de México,
Mexico City D.F 04510, Mexico
e-mail: gayalam@iingen.unam.mx

O. Lavan and M. De Stefano (eds.), *Seismic Behaviour and Design of Irregular and Complex Civil Structures*, Geotechnical, Geological and Earthquake Engineering 24, DOI 10.1007/978-94-007-5377-8_20,

upon structures. This chapter presents a procedure for designing the supplemental damping along with a procedure for the stochastic analysis of hysteretic structures. The dampers are designed based on energy considerations allowing the supplemental damping to take a great portion of the hysteretic energy. With this approach, more damping is provided to the interstories where the inelastic effects are more notorious. This leads to a more efficient use of the energy dissipation devices to control the performance of a structure up to the desired limits.

Asymmetries in-plan and in-elevation are considered with a bidirectional ground excitation characterized by its cross-spectral density. The analysis procedure is based on the stochastic equivalent linearization method as was proposed in Caughey (1959, 1963) and on the transfer-matrix technique (Clough and Penzien 2003).

It is important to mention that the stochastic linearization method is used as an approximation to the problem as, up till now, there is not an exact analytic method to analyze hysteretic systems within a random vibrations context.

A numerical example is given to illustrate the use of the design procedure proposed in this chapter. The example is a 6-story steel building with in-plan and in-elevation asymmetries. The supplemental damping considered is of the viscous-fluid type.

20.2 Procedure of Analysis

The model used to analyze the buildings assumes that their structure is formed by independent plane frames interconnected by the floor diaphragms considered rigid in their own plane. Each frame is idealized by a stick model with a single resisting element per story. The masses of the building are lumped at the mass center of the floors. Under these assumptions, the building's response is described by three degrees of freedom per level, i.e., the lateral displacements in the X and Y directions and the rotation θ with respect to a vertical axis passing through its mass center (see Fig. 20.1). Thus, the displacement vector for the n floor is $\mathbf{u}_n = \{u_{Xn}\ u_{Yn}\ \theta_n\}^T$. The displacements are absolute; i.e., they are referenced to the ground level.

To allow considering the asymmetry in height, the center of mass of each floor is not aligned along a vertical axis. As shown later in this chapter, in the transfer-matrix method, this requires translating the state variables of the lower level of two consecutive levels to the point aligned with the mass center of the upper level as shown by the broken line axis in Fig. 20.1.

The equation of motion of the $n+1$ floor relates the inertial forces with the shear forces of the adjacent upper and lower interstory as shown in the following equation

$$[M_{n+1}]\{\ddot{\mathbf{u}}_{n+1}\} = \{\mathbf{q}_{n+1}\} - \{\mathbf{q}'_n\}, \tag{20.1}$$

where M_{n+1} is a 3 by 3 diagonal matrix with the mass and the moment of inertia of the $n+1$ floor, $\mathbf{q}'_n = \{q_{Xn}\ q_{Yn}\ T_n\}^T$ is the resultant shear-force vector containing

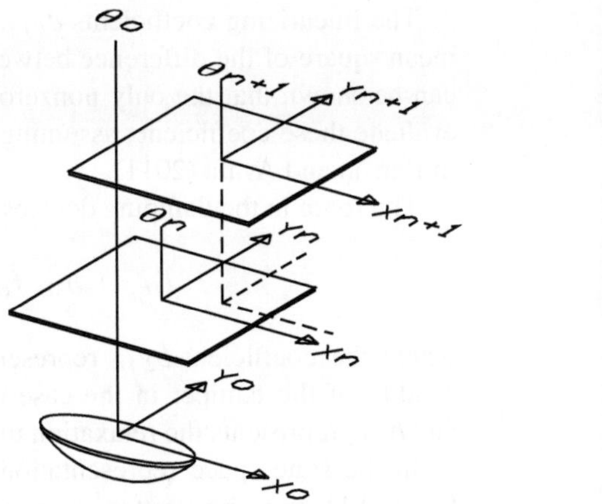

Fig. 20.1 Degrees of freedom of two consecutive levels and the ground level

the X and Y components and the torsional moment T_n. The prime indicates that their corresponding values have been modified when translated to a position aligned with the vertical axis of the level above.

The individual shear forces of the resisting elements have three components: a linear term, $\alpha_{jn}\, k_{jn}\, \hat{u}_{jn}$, a hysteretic term, $(1 - \alpha_{jn})k_{jn}z_{jn}$, and a damping term $f_{D\,jn}$ from the damping device, where $\hat{u}_{jn} = u_{jn} - u_{jn-1}$ and $\hat{v}_{jn} = v_{jn} - v_{jn-1}$ are, respectively, the element relative displacement and velocity, k_{jn} is the initial stiffness, and α_{jn} is the post-yielding stiffness ratio, $0 \leq \alpha_{jn} \leq 1$. The shear-force vector $\mathbf{q}'_n$ can be represented by the following equation

$$\mathbf{q}'_n = [K_n]\{\mathbf{u}_n - \mathbf{u}_{n-1}\} + [C_n]\{\mathbf{v}_n - \mathbf{v}_{n-1}\} + [P_{Zn}]\{kz_n\} + [P_{Dn}]\{f_{Dn}\}, \quad (20.2)$$

where $[K_n]$ is the stiffness matrix obtained by adding the projected individual stiffnesses k_{jn}, $[C_n]$ is the damping matrix representing the inherent damping, $\{kz_n\}$ is the vector containing the products $(1 - \alpha_{jn})k_{jn}z_{jn}$ on each row, and $[P_{Dn}]$ and $[P_{Zn}]$ are, respectively, the projection matrix of the dampers and the hysteretic variables containing on each row the direction cosines of each frame and the perpendicular distance from its axis to the mass center of the level $n+1$, i.e., the jth row is $[\cos\theta_j \;\; \sin\theta_j \;\; d_j]$.

The variables z_{jn}, which define the assumed hysteretic behavior, are modeled by a first-order nonlinear differential equation. In this research, the Bouc-Wen model without degradation is used (Bouc 1967; Wen 1976). Taking the coefficients β and γ equal to 0.5 and an odd exponent n equal to 3 reduces the original equation to

$$\dot{z}_{jn} = \frac{\hat{v}_{jn} - 0.5(|\hat{v}_{jn}|z^3{}_{jn} + \hat{v}_{jn}|z^3{}_{jn}|)}{u^3_{y\,jn}}, \quad (20.3)$$

where $u_{y\,jn}$ is the yield displacement. A linear approximation of Eq. (20.2) is

$$\dot{z}_{jn} = a_{0jn} + a_{1jn}\hat{u}_{jn} + a_{2jn}\hat{v}_{jn} + a_{3jn}z_{jn}. \quad (20.4)$$

The linearizing coefficients $a_{i\,jn}$, $i = 0,1,2,3$ are determined by minimizing the mean square of the difference between the original and its linear approximation. It can be shown that the only nonzero coefficients are a_{2j} and a_{3j}. The formulas to evaluate these coefficients assuming a truncated gaussian distribution are obtained in Barrón and Ayala (2011).

The force in the damping devices is modeled with the following linear equation

$$f_{D_{jn}} + b_{3jn}\dot{f}_{D_{jn}} = b_{1jn}\hat{u}_{jn} + b_{2jn}\hat{v}_{jn}, \tag{20.5}$$

where the coefficient $b_{2\,jn}$ represents the viscous damping, $b_{1\,jn}$ represents the rigidity of the damper in the case of viscoelastic dampers, or yielding dampers, and $b_{3\,jn}$ represents the relaxation time of the damper.

In the state space representation, the equation of motion for the level $n+1$, Eq. (20.1), may be written as a system of nine first-order ordinary differential equations in the variables $\{\mathbf{u}_{n+1}\ \mathbf{v}_{n+1}\ \mathbf{q}_{n+1}\}^T$ and $\{\mathbf{u}'_n\ \mathbf{v}'_n\ \mathbf{q}'_n\}^T$ as shown next

$$\begin{bmatrix} D_t & -I & 0 \\ 0 & -M_{n+1}D_t & I \\ K_n & C_n & 0 \end{bmatrix} \begin{Bmatrix} \mathbf{u}_{n+1} \\ \mathbf{v}_{n+1} \\ \mathbf{q}_{n+1} \end{Bmatrix} = \begin{bmatrix} 0 & 0 & 0 \\ 0 & 0 & I \\ K_n & C_n & I \end{bmatrix} \begin{Bmatrix} \mathbf{u}'_n \\ \mathbf{v}'_n \\ \mathbf{q}'_n \end{Bmatrix} - \begin{bmatrix} 0 \\ 0 \\ P_{Zn} \end{bmatrix} \{kz_n\} - \begin{bmatrix} 0 \\ 0 \\ P_{Dn} \end{bmatrix} \{f_{Dn}\}, \tag{20.6}$$

where I is the 3×3 identity matrix and D_t is a diagonal matrix with first-order differential operators on its diagonal to take the first derivative of the variables with respect to time.

Equations (20.4), (20.5) and (20.6) form a system of simultaneous ordinary differential equations. The solution of these equations is carried out in the frequency domain using the transfer-matrix method (Clough and Penzien 2003), which requires evaluating their Fourier transform to transform them on a system of linear algebraic equations for each frequency.

Let $\mathbf{U}_n(\omega) = \{\mathbf{U}_n\ \mathbf{V}_n\ \mathbf{Q}_n\}^T$ be the Fourier transform of the state vector $\mathbf{u}_n(t) = \{\mathbf{u}_n\ \mathbf{v}_n\ \mathbf{q}_n\}^T$; from the transformed equations of motion, the state vector $\mathbf{U}_{n+1}$ can be solved and expressed in terms of the state vector of the adjacent lower floor $\mathbf{U}_n$ through the matrix $[T_{n+1}]$ known as the transfer matrix from level n to level $n+1$. The analysis of the building structure is carried out level by level by means of a sequence of transfer-matrix multiplications beginning with the transfer matrix of the state variables at the ground level $\mathbf{U}_0$, which allows the establishment of a relationship between the state variables of each level with the state variables of the ground as shown in Eq. (20.6):

$$\{\mathbf{U}_{n+1}\} = [T_{n+1}T_n \ldots T_1]\{\mathbf{U}_0\}. \tag{20.7}$$

It is important to mention that for the first level, the ground displacements and velocities are obtained by integrating the ground accelerations. In the frequency domain, this is performed with the following equations: $\mathbf{U}_o = (i\omega)^2\{u_{\ddot{X}_o}\ u_{\ddot{Y}_o}\ \ddot{\theta}_o\}^T$

and $\mathbf{V}_o = (i\omega)\{u_{\ddot{X}o}\ u_{\ddot{Y}o}\ \ddot{\theta}_o\}^T$. The boundary conditions at the roof are zero shear forces. Introducing these conditions in the system of equations for the last level allows the evaluation of the base shear force Q_0. For this sake, it is assumed that the ground acceleration components are Dirac delta function so that their Fourier transform is unity; therefore, the state variables $\mathbf{u}_n(t)$ represent the *unit-impulse response functions,* and their Fourier transform $\mathbf{U}_n(\omega)$ are the corresponding *transfer functions*.

Having calculated the transfer function of the ground-state variables, $\mathbf{U}_0$, the transfer function $\mathbf{U}_n$ of each floor can readily obtained by means of Eq. (20.7), and its covariance matrix can be evaluated with the following equation

$$[\mathbf{\Gamma}_{u_n u_n}] = \int_{-\infty}^{\infty} \mathbf{U}_n[\mathbf{S}_{\ddot{X}\ddot{Y}}]\mathbf{U}_n^* d\omega, \tag{20.8}$$

where $\mathbf{U}_n^* =$ conjugate $(\mathbf{U}_n)$ and $[\mathbf{S}_{\ddot{X}\ddot{Y}}]$ is the matrix that contains the auto- and cross-spectral densities of the ground accelerations.

20.3 Design of the Dampers

The procedure for designing the dampers requires evaluating the mean amplitude of the vibration cycles $D_{\max}$. For this sake, the displacement probability density function of local maxima as proposed in Cartwright and Longuet-Higgins (1956) is used. The dissipated energy at a given interstory of a certain frame consist of three terms: the hysteretic energy E_{H}, the energy dissipated by the inherent camping E_{ID}, and the energy dissipated by the damper itself, E_{D}. In terms of the expected values for a random response, or for a harmonic vibration, it can be shown that the dissipated energy per cycle can be evaluated with the following equations (Chopra 2011):

Hysteretic dissipated energy:

$$E_{\mathrm{H}} = \text{area under the force-displacement curve of amplitude } D_{\max}.$$

Energy dissipated by the inherent damping in the structure:

$$E_{\mathrm{ID}} = \frac{2\pi\zeta K \bar{D}_m^2 \omega}{\omega_n}.$$

Energy dissipated in a viscous type damper:

$$E_{\mathrm{D}} = \pi c\omega \bar{D}_m^2,$$

where ω is the central frequency in the power spectral density of the ground acceleration, ω_n and ζ are, respectively, the natural frequency of the structure and the damping ratio of the structure without dampers, c is the damper constant, and K is the interstory stiffness.

The mean local maxima amplitude D_{max} can be reduced up to the desired limit D_0 by increasing the constant c of the damper. As the energy introduced by the earthquake depends on the amplitude of the response, and this is what is aimed to be reduced, the design method requires an iterative procedure which keeps reducing the mean amplitude D_{max} until the objective value is reached, i.e., $D_{\text{max}} = D_0$. The steps of the iterative process are the following:

1. Calculate the target energy E_0 in a cycle of vibration of reduced amplitude D_0, which is made of the hysteretic energy E_{H} plus the energy dissipated by the inherent damping in the structure.

$$E_0 = \frac{E_{\text{H}}^{D_0} + 2\pi\zeta k D_0 \omega}{\omega_n}$$

2. Calculate the total energy E_{T} in a cycle of amplitude D_{max} dissipated by the inherent damping, the hysteresis, and the damper itself.

$$E_{\text{T}} = E_{\text{ID}}^{\bar{D}_{\text{max}}} + E_{\text{H}}^{\bar{D}_{\text{max}}} + E_{\text{D}}^{\bar{D}_{\text{max}}}$$

3. With the difference of energies $E_{\text{T}} - E_0$, calculate the new damper coefficient c.

$$c = \frac{E_{\text{T}} - E_0}{\pi\, \omega\, u_0^2}$$

4. Repeat steps 2–3 until convergence on the value of c is reached for each damper.

20.4 Numerical Example

A 6-story steel building with in-plan and in-elevation asymmetries is used to illustrate the design procedure for the supplemental damping; see Fig. 20.2.

Table 20.1 presents the floor masses and their respective center of mass and the sections of beams and columns.

The inherent damping of the structure was modeled as a Rayleigh type using as parameter those obtained with the frequencies of the first and third modes using for them 3% damping ratios. The base of the building was assumed rigid, and P-Δ effects were neglected. The building was designed to have a ductile behavior using the strong columns-weak beams criterion. This approach guaranties that the plastic hinges are formed at beams. Plastic hinges in columns were only allowed at their bases in the ground floor.

The analysis procedure proposed in this investigation models the planar frames as shear-beam-like structures with a hysteretic element at each interstory to represent its stiffness and energy dissipation capacity. The interstory stiffnesses were

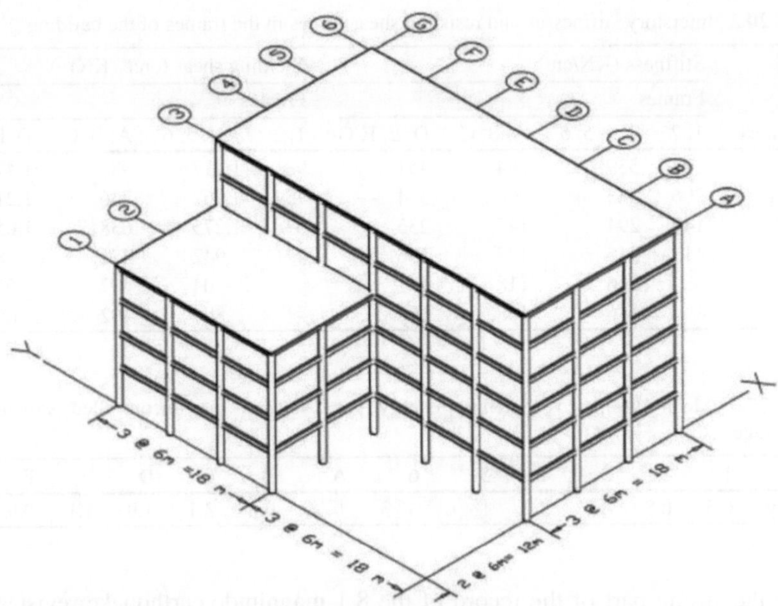

Fig. 20.2 Isometric view and in-plan dimensions of the building structure

Table 20.1 Floor masses, their location, and the section of beams and columns

Level	Mass (kg) × 10^3	Mom. of inertia (kg m^2) × 10^3	Xcm (m)	Ycm (m)	Beams W section	Columns circular section diameter and thickness
1	691	252,461	17.25	20.25	16 × 40	16 × 1/2
2	691	252,461	17.25	20.25	16 × 40	16 × 1/2
3	691	252,461	17.25	20.25	16 × 31	16 × 3/8
4	648	209,544	18.0	19.8	16 × 26	16 × 3/8
5	518	69,984	21.0	18.0	16 × 26	16 × 3/8
6	389	52,488	21.0	18.0	16 × 26	16 × 3/8

calculated from a static linear elastic analysis with lateral forces proportional to the first mode. The yielding shear forces were calculated considering that the plastic moments of the beams are distributed by half between the upper and lower columns to which they are connected. The shear force was obtained by dividing the sum of the moments at the ends of a column by the interstory height. For the first interstory, it was considered that the top moments are 75% of their corresponding plastic moment. Table 20.2 shows the stiffnesses and yielding shear forces as calculated.

The ground acceleration is assumed as a vector random process with components in X and Y direction. This vector process is assumed stationary gaussian with zero-mean. The unilateral auto- and cross-power spectral densities were obtained

Table 20.2 Interstory stiffnesses and resisting shear forces in the frames of the building

	Stiffness (KN/cm)				Yielding shear force (KN)			
	Frames				Frames			
Inter-story	1, 2	3, 4, 5, 6	A, B, C	D, E, F, G	1, 2	3, 4, 5, 6	A, B, C	D, E, F, G
1	294	539	294	451	981	1,717	981	1,472
2	206	383	206	314	734	1,462	736	1,216
3	147	294	147	235	638	1,275	638	1,059
4	118	245	127	186	471	942	471	785
5	–	216	118	137	–	804	402	667
6	–	186	108	108	–	804	402	667

Table 20.3 Mean dissipated hysteretic energy per cycle in the uncontrolled structure in percentage

Frame	1	2	3	4	5	6	A	B	C	D	E	F	G
Energy	0.5	0.8	4.3	8.9	18.9	37.3	14.2	6.8	3.1	1.9	1.1	0.9	1.3

using the strong part of the record of the 8.1 magnitude earthquake registered at the building of Minister of Communications and Transportations (SCT) in Mexico City during the 1985 Michoacan earthquake. The strong portion of the record was evaluated based on the Arias intensity; its duration is 45 s.

20.4.1 Presentation and Discussion of the Results

In what follows, a comparison of the performances of the buildings with and without dampers is presented. The building was designed to have a ductile behavior that implies large displacements in the building's frames; therefore, the hysteretic energy dissipated by the uncontrolled structure (with no dampers) is large. The mean dissipated hysteretic energy per cycle by the structure is 130 kJ. The distribution of this energy in the frames, in percent, is given in Table 20.3.

From this table, it can be observed that the frames 5, 6, and A dissipate the larger amount of energy. It appears that the best locations to place the dampers are those frames. However, if the dampers are placed only on those frames, the structure may become unbalanced because the dampers are not distributed evenly around the building. With the procedure herein proposed, one can assume that the dampers are placed at each interstory of each frame. The procedure by itself will assign the required amount of damping according to the dissipated energy at each interstory; i.e., it will add more damping to the interstories with the larger amount of hysteretic energy. Table 20.4 presents the value of the assigned damping to each interstory. From this table, it can be observed that dampers are placed in the outermost frames where they are more effective. It is also shown the mean displacement of local maxima, $D_{\max}$, for the uncontrolled and controlled structures. The target

Table 20.4 Location and coefficients of the added dampers and mean displacement of local maxima in the uncontrolled and controlled structures

	Damping coefficient (KN-s/cm)			Initial displacement (cm) Uncontrolled structure			Final displacement (cm) Controlled structure		
Interstory	6	A	G	6	A	G	6	A	G
6	0	0	0	0.9	0.9	0.7	0.6	0.7	0.7
5	0	0	0	3.0	2.2	1.3	1.6	1.4	1.1
4	60	0	0	3.3	3.1	1.7	2	2.0	1.6
3	84	44	0	3.4	3.3	1.9	2	2	1.9
2	146	0	11	3.5	3.1	1.9	2	1.9	2
1	0	0	0	2.9	2.6	1.7	1.9	1.8	1.5

Table 20.5 A comparison of the dissipated energies in the uncontrolled and the controlled structures

Dissipated energy (kilojoules/cycle)		
Structure	Uncontrolled	Controlled
Inherent damping	120	70
Added damping	0	130
Hysteresis	130	20
Total	250	220

displacement D_0 was set to 2 cm. From this table, it can be seen that, although the initial displacements are larger than 2 cm in several interstories, the design procedure assigns the adequate amount of damping to reduce all the displacements to be less or equal to 2 cm.

A comparison of the various energies dissipated in the uncontrolled and the controlled structures is presented in Table 20.5. It can be seen that the hysteretic energy is reduced 85% in the controlled structure. Also, it can be seen that the supplemental damping takes a larger amount of the dissipated energy.

Figure 20.4a, b shows the hysteresis loops of the interstory 2 of frame 6 in the uncontrolled and controlled structures. It can be observed that the maximum displacement, which in the uncontrolled structure is close to 8 cm, is reduced to 4.5 cm in the controlled structure. The maximum displacement in the controlled structure is a little bit larger than the yield displacement of 4.3 cm; this makes the structure to behave almost linearly as can be seen on its hysteretic plot in Fig. 20.4b.

For the shortness of this chapter, additional results are brevity mentioned as follows: (1) In the frame 6, at roof level, the mean maximum displacement was reduced from 34 to 23 cm (see Fig. 20.3); this represents a 32% reduction. (2) Regarding to the floor rotations, which are not shown in this chapter, they were reduced 35%. (3) Finally, there was a large reduction in the total mean dissipated hysteretic energy per cycle; this was reduced 85%.

Fig. 20.3 Displacement profile of frame 6 for the controlled and uncontrolled structures

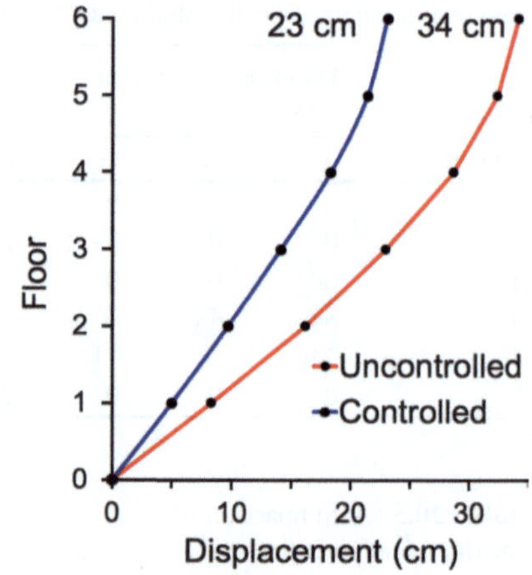

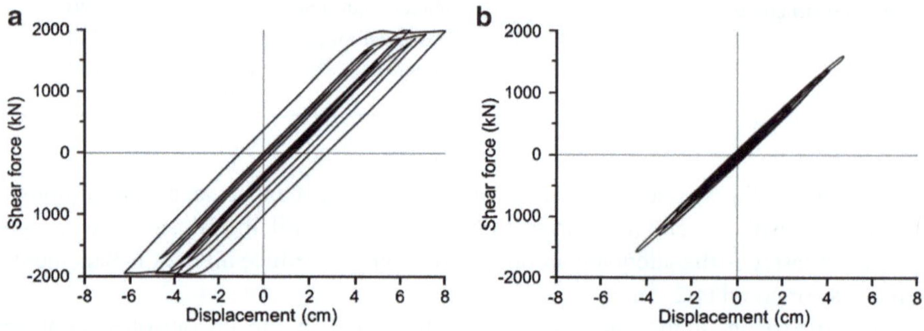

Fig. 20.4 (a) Hysteresis loops of the uncontrolled structure (b) Hysteresis loop of the controlled structure

20.5 Conclusions

In this chapter, a method to evaluate the probabilistic seismic response of an asymmetric building with nonlinear behavior and a procedure to design the supplemental damping aimed to reduce this response are presented. To demonstrate the application of these two, an illustrative numerical example of a steel building with in-plan and in-elevation asymmetries is used. From the analysis of the results of this example, the following conclusions may be derived.

- The strategically positioning of the damper allows a more efficient use of them, which results in a considerably reduction of the structural response at a minimum cost. For example, the rotations of the levels were reduced up to 35%, the reduction of displacements was of the order of 32%, and the energy dissipated by hysteresis was reduced up to 85%.

- It is confirmed that the criterion of transferring the largest part of the energy dissipated by the structure to the supplemental damping is a good procedure to dimension the dampers; with this criterion, more damping is proportioned to the interstories with the larger energy dissipation.

References

Barrón R, Ayala AG (2011) An energy-design method for buildings with supplemental damping and nonlinear behavior. In: Proceedings of the 6th EWICS, Haifa, Israel

Bouc R (1967) Forced vibration of mechanical systems with hysteresis. In: Proceedings of the 4th conference on non-linear oscillations, Prague

Cartwright DE, Longuet-Higgins MS (1956) The statistical distribution of the maxima of a random function. Proc R Soc A 237:212–232

Caughey TK (1959) Response of a nonlinear string to random load. J Appl Mech 26:341–344

Caughey TK (1963) Equivalent linearization techniques. J Acoustic Soc Am 35:1706–1711

Chopra AK (2011) Dynamics of structures: theory and applications to earthquake engineering, 4th edn. Prentice Hall, Englewood Cliffs

Clough RW, Penzien J (2003) Dynamics of structures, 3rd edn. Computers & Structures Inc, Berkeley

Wen YK (1976) Method of random vibration of hysteretic systems. J Eng Mech ASCE 102(EM2):249–263

Chapter 21
Mitigation of the Seismic Response of Structures with Vertical Stiffness and Strength Irregularity Using Supplemental Dampers

Jack Wen Wei Guo and Constantin Christopoulos

Abstract As part of the Canadian Seismic Research Network's (CSRN) newly developed performance spectra-based design methodology for buildings with supplemental damping, this chapter proposes an index for quantifying the degree of vertical stiffness/strength irregularity based on elastic modal properties. Using this index, a damper distribution technique is developed to correct irregularities while adding damping to control the seismic response to meet a given set of performance targets. Nonlinear time-history analyses of simplified 4-, 9- and 15-storey irregular shear buildings with hysteretic and linear viscous and viscoelastic dampers are used to verify and to establish the limits of application of the proposed distribution technique. It is found that the procedure offers simplicity and sufficient accuracy for design purposes if the structure meets the criteria identified in this study. Retrofit examples using the proposed methodology are also presented for a 9-storey irregular steel moment frame in Canada.

21.1 Introduction

Supplemental damping technologies have demonstrated tremendous potential for the seismic upgrade of structures due to their versatility and cost-effectiveness compared to traditional stiffening and strengthening techniques. In order for practising engineers to utilize supplemental damping effectively in seismic upgrades of low- to mid-rise frame structures, a simplified approximate design method based on performance spectra has recently been proposed (Guo and Christopoulos 2011). Central to this approach is the identification of a target SDOF with specific

J.W.W. Guo • C. Christopoulos (✉)
University of Toronto, 35 St. George St, Toronto ON M5S 1A4, Canada
e-mail: wen.guo@utoronto.ca; c.christopoulos@utoronto.ca

O. Lavan and M. De Stefano (eds.), *Seismic Behaviour and Design of Irregular and Complex Civil Structures*, Geotechnical, Geological and Earthquake Engineering 24, DOI 10.1007/978-94-007-5377-8_21,

damping parameters on the performance spectrum (P-Spectrum) and the subsequent transformation of this SDOF into an MDOF system with supplemental dampers to achieve a given set of performance targets. A major issue associated with such a transformation is the existence of vertical irregularities in the original structure (herein called the base frame), which can significantly alter the global response of the MDOF structure from the design target.

Vertical mass, stiffness and strength irregularities have been studied by a number of researchers in the past. It has been previously established by Valmundsson and Nau (1997), Tremblay et al. (2005) and Michalis et al. (2006) that mass irregularity is least detrimental in causing undesirable structural response when compared to stiffness and strength irregularities. Furthermore, Valmundsson and Nau (1997) and Michalis et al. (2006) found that stiffness and strength can increase local and global displacements significantly. Using shear structures, Al-Ali and Krawinkler (1998) confirmed the drift amplifications due to stiffness and strength irregularities and found that irregularities near the base and mid-height of the structure tend to generate higher displacement demands than the same irregularity located at the top of the structure. Studies on real vertically irregular frame structures (Magliulo et al. (2002), Das and Nau (2003) and Athanassiadou (2007)) showed that real frames show smaller drift amplification than shear structures since storey mechanisms (i.e. column sway) do not form frequently and redistribution of storey shear occurs. Studies based on structural models with weak-beam-strong-columns by Chintanapakdee and Chopra (2003) as well as shear wall structures by Miranda and Reyes (2002) reached similar conclusions regarding the decrease in drift amplification due to redistribution of forces.

While there is good evidence suggesting that the effects of irregularities in stiffness and strength can be effectively mitigated by forcing a beam-hinge mechanism through proper capacity design, this assumption is often unconservative when dealing with existing structures constructed prior to modern seismic codes. This study primarily concerns with the global displacements and forces of the upgraded irregular structure designed using the newly proposed performance spectra-based method. An index based on elastic properties is proposed to quantify vertical irregularity. Based on this index, damper distributions are modified in order to mitigate seismic responses to the predetermined performance targets.

21.2 Modelling Irregularity

In order to capture the column sway mechanism in older structures, simplified shear building models with constant storey mass are used in this study to represent the idealized MDOF structure. Furthermore, following the assumption that stiffness and strength are in direct proportion (Priestley et al. 2007), a complete description of stiffness and strength of the shear MDOF can be determined from knowing the fundamental period, mode shape and the shear strength V_f, expressed as the % of

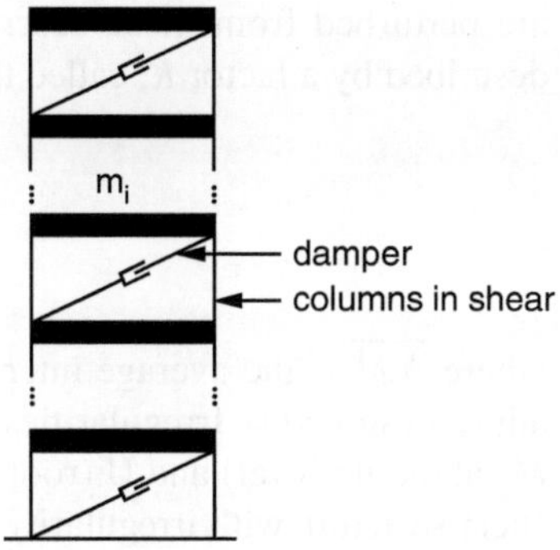

Fig. 21.1 Idealized shear structure with dampers

elastic base shear computed by a traditional spectral analysis. When $V_f = 100\%$, the structure remains linear elastic for the given ground excitation. Figure 21.1 illustrates the shear structure considered in this study.

In many of the previous studies, a structure is considered irregular if it violates the requirements on the distribution of mass, stiffness and strength in the applicable building codes (i.e. UBC in Valmundsson and Nau (1997), Das and Nau (2003), EC8 in Athanassiadou (2007)). The binary classification of regular/irregular in codes acts to restrict the usage of certain analysis methods in the iterative design process. In order to find the limit of application for the proposed direct supplemental damper distribution method, a binary classification is not sufficient and different degrees of irregularity need to be quantified. In this study, a perfectly regular structure is defined as a structure that has a linear or straight fundamental mode shape. Hence, given the base frame period T_f and strength V_f expressed as a fraction of the elastic seismic base shear, the storey lateral stiffness and strength are given by

$$K_{f,i} = \left(\frac{2\pi}{T_f}\right)^2 \frac{\sum_{j=i}^{n} m_j \phi_j^1}{\Delta\phi_i^1} \tag{21.1}$$

$$V_{f,1} = V_f \Gamma_f^1 S_\mathrm{a}\left(T_f\right) g \sum m_i \phi_i^1;\ V_{f,i} = V_{f,1}\left(\frac{K_{f,i}}{K_{f,1}}\right) \tag{21.2}$$

where m_i, ϕ_i^1 and $\Delta\phi_i^1$ are the storey seismic mass, first mode ith modal ordinate and the ith modal inter-storey drift, respectively. $S_a\left(T_f\right)$, Γ_f^1 and g are the spectral acceleration, modal participation factor and acceleration due to gravity. For a regular n-storey structure with constant height, the first mode shape is defined as

$$\phi_i^1 = \frac{i}{n};\quad \Delta\phi_i^1 = \frac{1}{n} \tag{21.3}$$

In this study, irregular stiffness and strength distributions are obtained by Eqs. (21.1) and (21.2) using mode shapes whose modal inter-storey drifts, $\Delta\phi_i^1$'s, are perturbed from those described in (21.3). Specifically, a given irregularity is described by a factor K, called the drift amplification factor defined as follows

$$K = \max\left(\frac{\Delta\phi_i^1}{\overline{\Delta\phi_i^1}}\right) \tag{21.4}$$

where $\overline{\Delta\phi_i^1}$ is the average inter-storey drift excluding storey i and its immediately adjacent storey(s). Irregularities are also categorized into types L (base of structure), M (at the jth level) and U (roof of structure) by their locations. $\Delta\phi_j^1$ in an n-storey shear structure with irregularity at ith storey have the following properties:

1. $\Delta\phi_j^1 = \text{const.}\ (1 \le j \le n; j \ne i, i \pm 1)$
2. $\Delta\phi_j^1 = K\Delta\phi_i^1\ (j = 1)$
3. $\Delta\phi_j^1 = \frac{(K+1)\Delta\phi_i^1}{2}\ (1 \le j = i \pm 1 \le n)$

From weak-beam-strong-column structural models, it was shown that an irregularity at storey i tends to influence adjacent storeys (Chintanapakdee and Chopra 2003). Equation (21.3) above describes a linear variation of drift between the irregularity and its adjacent floors to the rest of the structure. Due to the dependency of stiffness and strength, K provides a quantitative measure of stiffness and strength irregularity for the proposed damper distribution procedure. By manipulating Eqs. (21.1), (21.3) and (21.4), the stiffness reduction e_k necessary to generate a drift concentration K at the location of irregularity for an n-storey shear structure can be calculated as follows

$$e_k = \frac{2\left(\left(\frac{n-1}{2K}\right)(3K+n-1)+1\right)}{n(n+1)} - 1 \text{ (Type L)}$$
$$e_k = \frac{(n-j-1)(4K+j+n-2)+7K+4j-5}{K(n(n+1)-j(j-1))} - 1 \text{ (Type M)}$$
$$e_k = \frac{2n+3K-3}{2Kn} - 1 \text{ (Type U)} \tag{21.5}$$

The curves generated by Eq. (21.5) are shown in Fig. 21.2 for different values of n. The irregularity is assumed to be at the mid-height for type M.

Based on the shear building model, for the same levels of drift amplification, type U irregularity can tolerate more reduction in stiffness than both types L and M. Furthermore, as the number of storeys increases, the influence of a single irregularity tends to decrease. The drift amplification is relatively stable with respect to the stiffness reduction when the reduction is less than roughly 15%, which corresponds to $K \le 1.3$. For this reason, $K = 1.3$ can be considered as a limit of regularity beyond which modification of damper design would be required to account for irregularities.

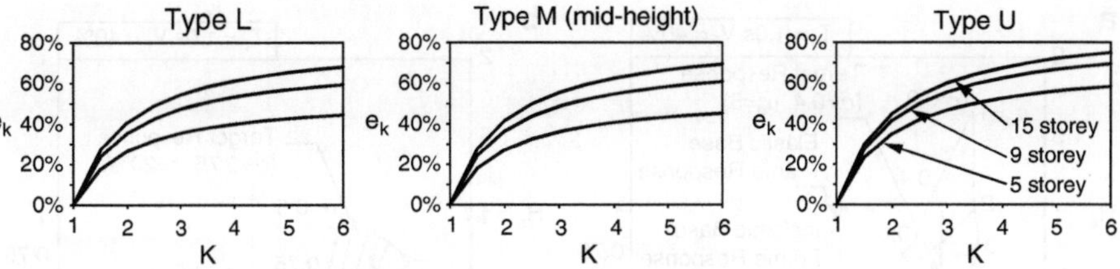

Fig. 21.2 Stiffness reductions at irregularity vs. elastic drift amplification

21.3 Proposed Distribution Procedures

A brief description of the performance spectra-based design method is first outlined. Performance spectra (P-Spectra) are graphical tools showing the response of an inelastic SDOF system with a certain supplemental damping characteristic. P-Spectra for hysteretic damper, viscous damper and viscoelastic damper can be generated automatically using a computer script or derived using an equivalent linearization formulation described in Guo and Christopoulos (2011). Each P-Spectrum corresponds to a base frame with period T_f and strength V_f, expressed as a % of the elastic base shear $mS_a(T_f)$, where m and $S_a(T_f)$ are the mass and spectral acceleration, respectively. Figure 21.3 shows sample P-Spectra generated from nonlinear time-history analysis of systems equipped with (a) hysteretic and (b) linear viscous/viscoelastic dampers.

In Fig. 21.3, the response quantities R_d and R_a are defined as the damped system displacement and base shear (proportional to acceleration) normalized to the spectral displacement and base shear of the elastic base frame with period T_f. R_s is the fraction of R_d that is retained as residual displacement. Once the target displacement, base shear and residual displacement are specified, the required damper properties can be read directly from the P-Spectra. Hysteretic dampers are defined by the damper ductility μ_d and stiffness ratio α, which represents the fraction of the total initial stiffness contribution of the damped structure by the base frame. Viscoelastic dampers are defined by α and the damping factor ξ, which equals the damping ratio when the system period is T_f. Once these parameters are identified, a transformation is applied to compute the required damper properties at each storey of the MDOF structure in order to achieve the performance targeted on the P-Spectrum. For all dampers (except viscous fluid dampers), a non-zero supplemental stiffness $K_{d,i}$ is assigned based on

$$K_{d,i} = \left(\frac{2\pi}{T_f}\right)^2 \frac{\sum_{j=i}^{n} m_j d_j^1}{\Delta d_i^1} - K_{f,i} \geq 0 \tag{21.6}$$

$$T_i = T_f \sqrt{\alpha} \tag{21.7}$$

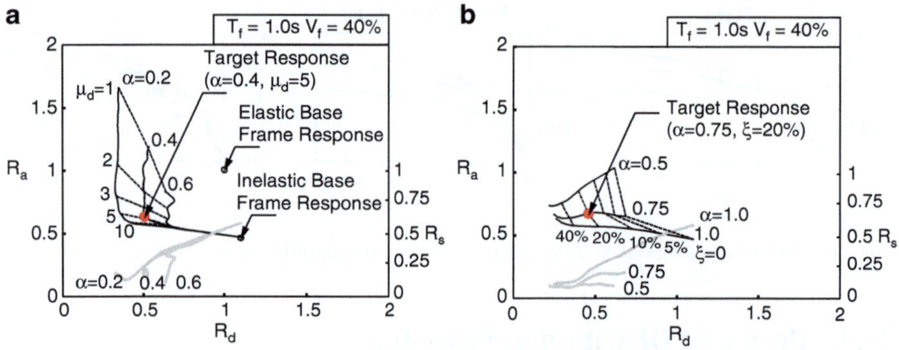

Fig. 21.3 P-Spectra for (**a**) hysteretic dampers and (**b**) viscous/viscoelastic dampers

where m_i, d_i^1 and $K_{f,i}$ are the ith storey mass, design mode shape (designer choice) and lateral stiffness contribution from the base frame and $\Delta d_i^1 = d_i^1 - d_{i-1}^1$. Equation (21.6) computes the difference between the required initial stiffness of the braced structure (given by initial period T_i and design mode shape d_i^1) and the available stiffness of the base frame. Assuming that irregularities are absent, the activation load $V_{d,i}$ of the hysteretic dampers are given by

$$V_{d,i} = \Gamma_1 S_d\left(T_f\right) R_d \Delta d_i^1 \left(\frac{K_{d,i}}{\mu_d}\right) \geq 0 \tag{21.8}$$

where Γ_1 and $S_d(T_f)$ are the first modal participation factor and the spectral displacement of the system with period T_f. The term $\Gamma_1 S_d\left(T_f\right) R_d \Delta d_i^1$ is the target inter-storey drift Δ_{di} and $K_{d,i}/\mu_d$ is the secant lateral stiffness of the damper.

For viscous and viscoelastic dampers, the required viscous constants c_i's are functions of the target viscous damping parameter ξ found on the P-Spectra:

$$c_i = \frac{2\xi M_1\left(\frac{2\pi}{T_f}\right) K_{d,i}}{\sum K_{d,i}\left(\Delta d_i^1\right)^2} \tag{21.9}$$

In (21.9), M_1 is the first mode modal mass. For viscous dampers, $K_{d,i} = 0$ and the design mode shape is made equal to the first mode shape of the base frame. The viscous constants are obtained from (21.9) by replacing $K_{d,i}$ by a fictitious stiffness $\overline{K_{d,i}}$ computed using (21.6) using an arbitrary $T_i < T_f$. $\overline{K_{d,i}}$'s are not real stiffness but are numerical tools for obtaining stiffness proportional damping as discussed in Christopoulos and Filiatrault (2006).

Equation (21.6) accounts for the difference in required and available stiffness for regular and irregular stiffness distributions. Furthermore, the activation load in hysteretic dampers in Eq. (21.8) can be altered to correct the strength irregularity using a more general expression

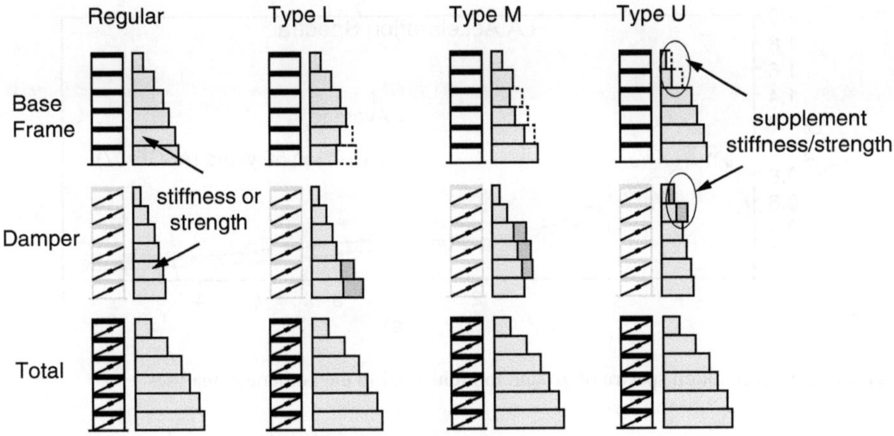

Fig. 21.4 Conceptual illustration of supplemental damper design

$$V_{dIR,i} = V_{d,i} + V_{\mathrm{reg},f,i} - V_{f,i} \geq 0 \tag{21.10}$$

where $V_{d,i}$ is given by (21.8), $V_{\mathrm{reg},f,i}$ is given by (21.2) using mode shape of (21.3) and $V_{f,i}$ is the actual strength in the base frame. For shear structures, it is given by (21.2) using actual mode shapes of the irregular base frame computed from an eigenvalue analysis. Equation (21.10) assigns activation loads of hysteretic dampers to ensure that the total storey strength in an irregular structure is equal to that of the regular structure, which is $V_{d,i} + V_{\mathrm{reg}f,i}$. The procedure of supplementing missing stiffness and strength in an irregular structure using dampers is illustrated in Fig. 21.4.

When Eqs. (21.6) and (21.10) give negative values of stiffness and strength, the structure should be considered highly irregular and either requires additional retrofit (possibly involving softening/weakening) to correct its irregularity prior to applying supplemental dampers or a different set of design targets should be used. Finally, modifying the damper distribution for irregularity using (21.6) and (21.10) can impair the energy dissipation properties of the system. Hence, the first mode secant damping of the damped MDOF computed as the dissipated energy over the elastic strain energy at the target displacement should not be less than 80% of the damping in the target SDOF. The expressions to evaluate these parameters are given in Guo and Christopoulos (2011).

21.4 Irregularity Limits

In order to establish limits of K such that Eqs. (21.6), (21.9) and (21.10) produce designs that satisfy the performance targets, extensive numerical analyses of shear

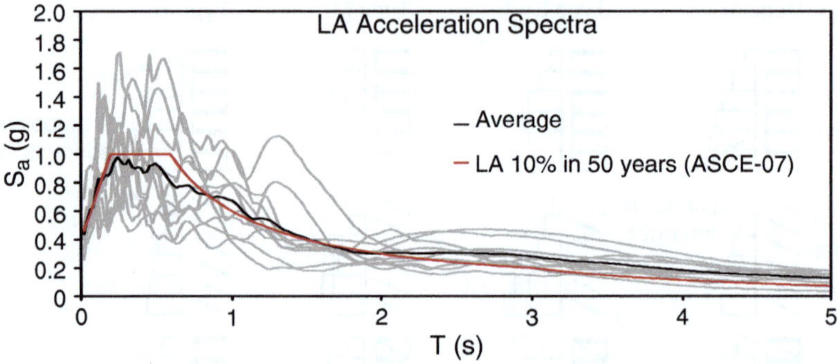

Fig. 21.5 Acceleration spectra of ground motions used in the numerical analyses

structures are conducted. For these analyses, a suite of ten recorded ground motions scaled to the LA design spectra, shown in Fig. 21.5, are used.

Using (21.6), (21.9) and (21.10), 9 different damper designs (3 α's, 3 μ_d's for hysteretic damper and 3 ξ's for viscoelastic damper) for 4- to 15-storey shear structures with T_f ranging from 0.5 to 2.0 s and V_f from 10 to 40% are analysed. Each analysis is repeated for irregularities at the base (type L), mid-height (type M) and roof (type U) with K set to 2.0, 4.0 and 6.0. The roof displacement and base shear are normalized to the elastic responses and are compared to the target $R_{d,}R_v$ (equal to R_a less the viscous contribution) and R_a on the P-Spectra.

To conserve space, only the responses of type L irregularities are shown for hysteretic and viscoelastic dampers in Fig. 21.6. As indicated by Fig. 21.6, the responses match the predictions reasonably well when K is less than 4.0 for hysteretic dampers and less than 2.0 for viscoelastic dampers. While some hysteretic designs with $K = 6$ achieved good performance, the vast majority of these designs performed poorly and are not shown because they failed to meet the inequality in (21.6) and (21.10). Hysteretic dampers can tolerate more stiffness and strength irregularity because they supplement both stiffness and strength. On the other hand, viscoelastic dampers only supplement stiffness through Eq. (21.6) and have no cap on the damper forces. It is possible that an increase in supplemental stiffness to compensate the initial stiffness deficiency can result in a higher storey strength than what is required to achieve regularity. This "overcompensation" in turn causes the irregularity to shift to locations where supplemental stiffness was decreased to fix the initial stiffness irregularity. Based on the numerical analysis, overcompensation is much more likely to occur for base frames with strength $V_f < 20\%$ because the total storey strength capacity is highly sensitive to added capacity of the viscoelastic spring. Hence, the criteria that $K \leq 2.0$ and $V_f \geq 20\%$ is imposed to ensure satisfactory performance for structures with viscoelastic dampers. While it was found that systems with viscous fluid dampers can occasionally give good performance, the trend is not consistent. For some ground motions, particularly records that induce a few very large inelastic cycles, stiffness and strengths are more

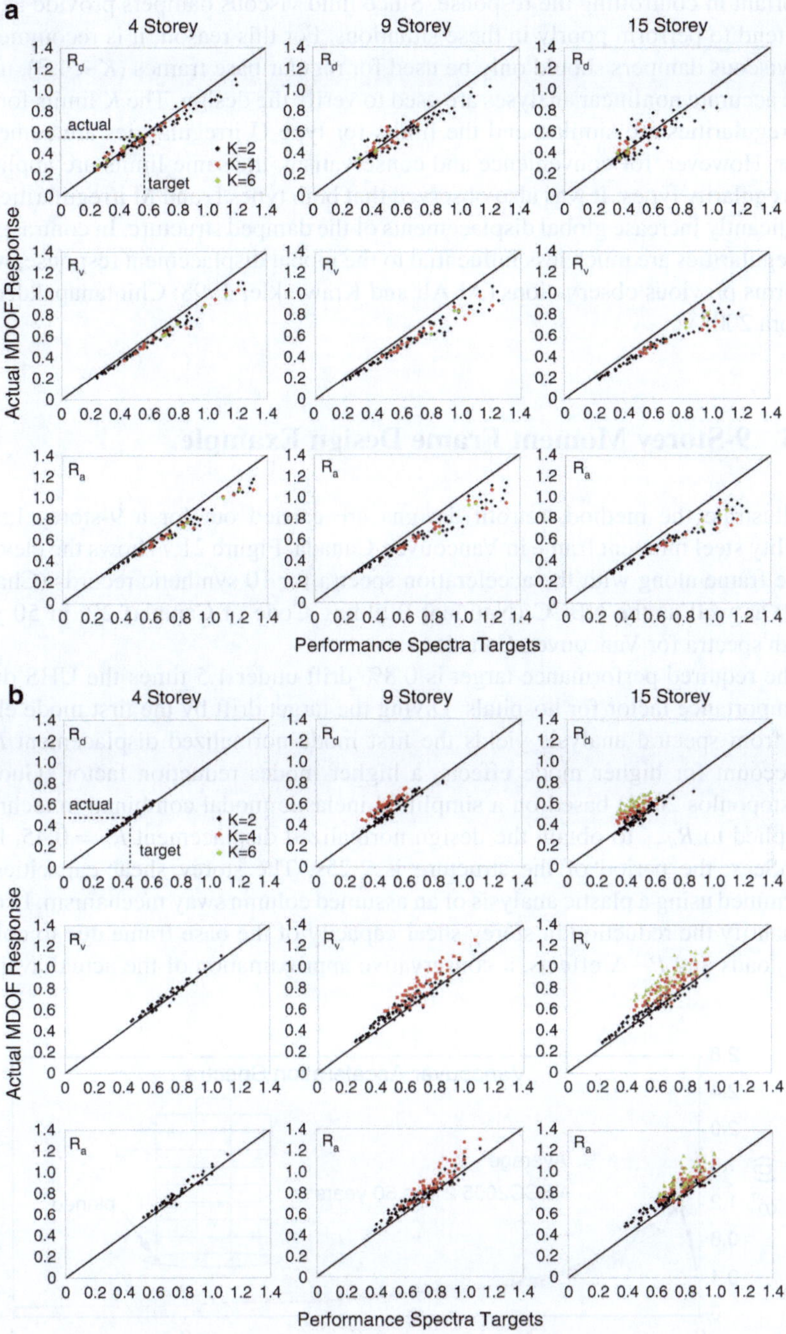

Fig. 21.6 Responses of type L MDOF with (**a**) hysteretic and (**b**) viscoelastic dampers

important in controlling the response. Since fluid viscous dampers provide neither, they tend to perform poorly in these situations. For this reason, it is recommended that viscous dampers should only be used for regular base frames ($K < 1.3$), unless more accurate nonlinear analyses are used to verify the design. The K limits for type M irregularities are similar, and the limits for type U irregularities are somewhat larger. However, for convenience and conservatism, the same limits are applied to all irregularity types. It was also observed that both types L and M irregularities can significantly increase global displacements of the damped structure. In contrast, type U irregularities are much less influential to the global displacement response, which confirms previous observations (Al-Ali and Krawinkler 1998; Chintanapakdee and Chopra 2003).

21.5 9-Storey Moment Frame Design Example

To illustrate the method, retrofit designs are carried out for a 9-storey limited ductility steel moment frame in Vancouver, Canada. Figure 21.7 shows the elevation of the frame along with the acceleration spectra for 10 synthetic records (Charette 2009) as well as the NBCC (National Building Code of Canada) 2% in 50 years design spectra for Vancouver, Canada.

The required performance target is 0.8% drift under 1.5 times the UHS due to the importance factor for hospitals. Diving the target drift by the first mode elastic drift from spectral analysis yields the first mode normalized displacement R_{d-1}. To account for higher mode effects, a higher modes reduction factor (Guo and Christopoulos 2011) based on a simplified inelastic modal combination technique is applied to R_{d-1} to obtain the design normalized displacement $R_d = 0.45$. From OpenSees, the period of the structure is 2.25s. The storey shear capacities are determined using a plastic analysis of an assumed column sway mechanism. In order to quantify the reduction in storey shear capacity of the base frame due to column axial loads and $P-\Delta$ effects, a conservative approximation of the actual available

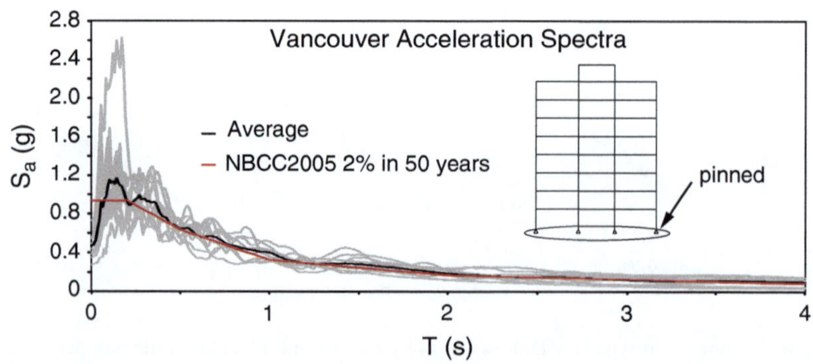

Fig. 21.7 Acceleration spectra of ground motions and elevation of 9-storey frame

Table 21.1 Summary of properties of elastic frame E1 and retrofit R1

	Base frame (E1—pinned)				R1	
i	$d_i^{(21.3)}$	$\Delta_{di}(\mathrm{mm})^{(21.8)}$	$K_f\left(\frac{\mathrm{kN}}{\mathrm{mm}}\right)^{(21.1)}$	$V_f\ (\mathrm{kN})^{(21.11)}$	$K_d\left(\frac{\mathrm{kN}}{\mathrm{mm}}\right)^{(21.6)}$	$V_d\ (\mathrm{kN})^{(21.10)}$
9	1.00	22.2	3.9	130	0	0
8	0.90	24.5	21.7	450	0	0
7	0.79	24.5	21.0	371	39.6	321
6	0.68	24.5	31.0	706	53.6	240
5	0.57	24.5	31.4	602	73.4	656
4	0.46	24.5	35.0	917	86.0	551
3	0.35	24.5	33.3	792	100.0	891
2	0.24	24.5	29.2	1,604	112.5	209
1	0.12	27.6	13.1	511	116.9	1,545

$^{(x)}$ denotes the equation number for which the calculations are based on

lateral strength is made by reducing the shear strength of the plastic analysis $V_{f,i}$ by the following

$$V_{f,i}{}^{*} = V_{f,i}\left(1 - \frac{\sum P_{\mathrm{c}}}{\sum A_{\mathrm{c}} f_{\mathrm{y}}}\right) - \frac{\sum_{j=i}^{n}\left(\left(\sum P_{\mathrm{c}}\right)\theta_{\mathrm{t}}\left(H_j - H_{i-1}\right)\right)}{h_i} \tag{21.11}$$

where P_{c}, A_{c} and f_{y} are the column axial force, column sectional area and yield strength. θ_t is the target design drift of 0.8%, H_i and h_i are floor height measured from the ground and storey height, respectively. From Eqs. (21.11) and (21.4), $V_f = 28\%$ and $K = 4.33$, which is close enough to the allowable irregularity limit for hysteretic dampers. However, viscous/viscoelastic dampers are not recommended. Alternatively, fixing the base columns results in a new base frame with $T_f = 1.78$ s, $V_f = 58\%$ and $K = 1.68$. Using the P-Spectra, a hysteretic damper solution R1 ($\alpha = 0.2$; $\mu_d = 2.0$) for the pinned frame as well as a hysteretic damper solution R2 ($\alpha = 0.2$; $\mu_d = 10$) and a viscoelastic damper solution R3 ($\alpha = 0.75$; $\xi = 22.7\%$) for the fixed base frame are selected. The design calculations are summarized in Tables 21.1 and 21.2.

Figure 21.8 shows the average drifts of R1 compared to the pinned elastic base frame E1 and the average drifts of R2 and R3 compared to the fixed elastic base frame E2. The inelastic responses of the undamped frames are not shown because the frames collapsed for almost all of the records that are considered.

Significant reduction in response can be seen in all retrofits compared to the undamped elastic structures E1 and E2. The first mode targets in Fig. 21.8 are based on the R_d used for design, which have been reduced to account for higher modes so that the actual storey drifts will meet the performance target of 0.8%. Figure 21.8 shows that the drifts in storey 7 of R1 and storey 2 in R2 are somewhat over the target of 0.8%. This is due to inaccuracies in the estimate of the frame strength and higher mode contributions in the inelastic structure during the design process. Further fine-tuning is required to improve the responses. By dividing the roof drift and base shear

Table 21.2 Summary of properties of elastic frame E2 and retrofits R2 and R3

	Base frame (E2—fixed)				R2		R3	
i	$d_i^{(21.3)}$	Δ_{di}(mm) $^{(21.8)}$	$K_f\left(\frac{kN}{mm}\right)^{(21.1)}$	V_f (kN)$^{(21.11)}$	$K_d\left(\frac{kN}{mm}\right)^{(21.6)}$	V_d (kN)$^{(21.10)}$	$K_d\left(\frac{kN}{mm}\right)^{(21.6)}$	$c_d\left(\frac{kNs}{mm}\right)^{21.9}$
9	1.00	19.7	4.1	130	0	0	0	0
8	0.90	21.8	22.0	450	0	0	0	0
7	0.79	21.8	21.1	371	75.6	320	4.7	2.1
6	0.68	21.8	31.2	706	104.0	256	4.8	2.2
5	0.57	21.8	31.6	602	135.9	605	13.0	5.9
4	0.46	21.8	35.5	917	157.8	478	16.0	7.3
3	0.35	21.8	35.8	792	177.2	753	21.0	9.5
2	0.24	21.8	40.8	1,604	185.6	35	19.6	8.8
1	0.12	24.5	48.8	1,284	158.8	375	6.5	3.0

$^{(x)}$ denotes the equation number for which the calculations are based on

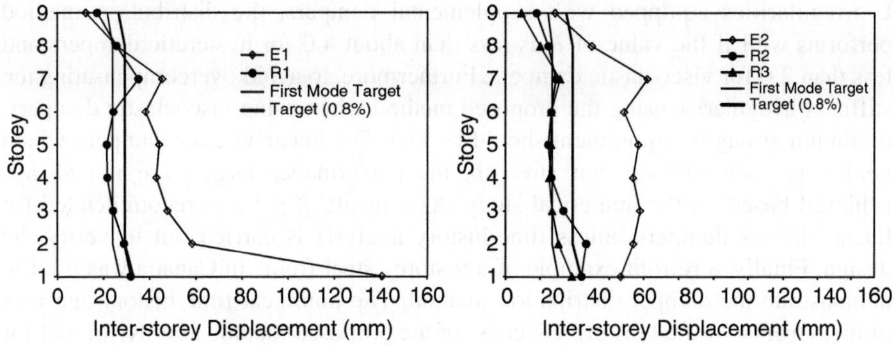

Fig. 21.8 Average inter-storey drifts of R1, R2 and R3

Table 21.3 Summary of global responses compared to P-Spectra targets

	R_d*		R_V/R_a**	
Retrofits	Target	Actual	Target	Actual
R1	0.46	0.48	1.21	0.98
R2	0.45	0.46	0.63	0.67
R3	0.47	0.43	0.71	0.74

* R_d calculated at roof, **R_V = base shear for R1 and R2, R_a = base shear for R3

forces of the retrofits R1, R2 and R3 by the roof drift and base shear of the elastic base frames E1 and E2, the global normalized responses are computed. Table 21.3 compares the global responses obtained from nonlinear time-history analysis with the P-Spectra targets.

The displacement targets are generally well achieved, while the base shear targets are somewhat less accurate. The 20% overestimate of the actual normalized shear in R1 is due to the very different higher mode contributions to shear in the pinned elastic frame and the more regular retrofitted structure. This example illustrates that the retrofits resulting from the proposed direct design method met or performed close to the performance targets and can be a convenient and powerful tool for practical designs of structures with supplemental dampers.

21.6 Conclusions

A method compatible with performance spectra-based seismic design that distributes supplemental dampers to a multistorey structure with existing vertical stiffness and strength irregularities to achieve performance targets is proposed. An index called the drift amplification factor $K \geq 1$ has been proposed to quantify the degree of irregularity in stiffness and strength assuming that these quantities are proportional. Based on parametric analyses of systems with types L, M and

U irregularities equipped with supplemental dampers, the distribution method performs well if the value of K is less than about 4.0 for hysteretic dampers and less than 2.0 for viscoelastic dampers. Furthermore, to avoid overcompensating the stiffness irregularity using the proposed method when using viscoelastic dampers, minimum strength requirement should be met. For linear viscous dampers which neither provide stiffness nor strength, the performance targets are not always achieved based on the numerical study. As a result, $K \leq 1.3$ is recommended for linear viscous dampers unless time-history analysis is carried out to verify the design. Finally, a retrofit example of a 9-storey steel frame in Canada was used to demonstrate the damper distribution method. The nonlinear time-history analyses of the retrofits verified the effectiveness of the proposed method as a useful tool for the sizing and distribution of supplemental dampers.

References

Al-Ali AK, Krawinkler H (1998) Effects of vertical Irregularities on seismic behavior of building structures. Report 130, Stanford University

Athanassiadou CJ (2007) Seismic performance of R/C plane frames irregular in elevation. Eng Struct 30:1250–1261

Charette KG (2009) Effets des Mouvements Sismiques sur les Structures en Acier de la Categorie des Constructions Conventionnelles. Dissertation, Ecole Polytechnique de Montreal (in French)

Chintanapakdee C, Chopra AK (2003) Seismic response of vertically irregular frames: response history and modal pushover analyses. J Struct Eng 130:1177–1185

Christopoulos C, Filiatrault A (2006) Principles of passive supplemental damping and seismic isolation. Viscous and viscoelastic dampers. IUSS Press, Pavia

Das S, Nau JM (2003) Seismic design aspects of vertically irregular reinforced concrete buildings. Earthq Spectra 19:455–477

Guo JWW, Christopoulos C (2011) Performance spectra-based design method for structures with supplemental damping. Research report, University of Toronto

Magliulo G, Ramasco R, Realfonzo R (2002) A critical review of seismic code provisions for vertically irregular frames. In: Proceeding of the 3rd European workshop on the seismic behaviour of irregular and complex structures, CD ROM, Florence

Michalis F, Dimitrios V, Manolis P (2006) Evaluation of the influence of vertical irregularities on the seismic performance of nine-storey steel frame. Earthq Eng Struct Dyn 35:1489–1509

Miranda E, Reyes CJ (2002) Approximate lateral drift demands in multistory buildings with nonuniform stiffness. J Struct Eng 128:840–849

Priestley MJN, Calvi GM, Kowalsky MJ (2007) Displacement-based seismic design of structures. Analysis tools for direct displacement-based design. IUSS Press, Pavia

Tremblay R, Merzouq S, Izvernari C, Alexieva K (2005) Application of the equivalent static force procedure for the seismic design of multi-storey buildings with vertical mass irregularity. Can J Civ Eng 32:561–568

Valmundsson EV, Nau JM (1997) Seismic response of building frames with vertical structural irregularities. J Struct Eng 123:30–41

Chapter 22
Allocation and Sizing of Multiple Tuned Mass Dampers for Seismic Control of Irregular Structures

Yael Daniel and Oren Lavan

Abstract This chapter presents a methodology for the optimal design of multiple tuned mass dampers (TMDs) in 3D irregular buildings. The objective function minimizes the total mass of all added TMDs while constraints are added to limit the total accelerations experienced at the edges of the floors in the direction parallel to each edge. The formulation of the design methodology relies on optimality criteria conjectured herein; hence, a two-stage iterative analysis/redesign procedure, that is based on analysis tools only, is resulted. The methodology applies to all types of irregularity, which allows the application of the methodology in a practical design process.

22.1 Introduction

Seismic protection of structures is an important issue in structural design due to its threatening consequences. Often, it is required that the design of the structure provide even more than life safety, promising a certain level of serviceability following a severe earthquake, while allowing for a defined level of damage, i.e., performance-based design. In performance-based design, it is often desired to limit important responses such as inter-story drifts, total accelerations, residual drifts, and hysteretic energy.

There is ample literature on the reduction of structural responses to earthquakes through passive control. Several passive damping devices are available, including viscous, viscoelastic, metallic, and friction dampers (see, e.g., Soong and Dargush 1997; Christopoulos and Filiatrault 2006; Takewaki 2009). For wind vibration

---

Y. Daniel (✉) • O. Lavan
Faculty of Civil and Environmental Engineering, Technion – Israel Institute of Technology, Haifa 32000, Israel
e-mail: yaeldan@tx.technion.ac.il; lavan@tx.technion.ac.il

O. Lavan and M. De Stefano (eds.), *Seismic Behaviour and Design of Irregular and Complex Civil Structures*, Geotechnical, Geological and Earthquake Engineering 24, DOI 10.1007/978-94-007-5377-8_22,

control of tall buildings, tuned mass dampers (TMDs) are often effectively used (e.g., McNamara 1977). Details about TMDs and their applications may be found in the fine works of Den-Hartog (1940), Warburton (1982), and Soong and Dargush (1997), only to name a few. As wind response of buildings is kept within the linear range and is usually dominated by a single mode, TMDs indeed provide a very efficient solution. Seismic action, on the other hand, may cause yielding of the structure, which can jeopardize the action of TMDs due to their detuning. In addition, in seismic vibrations, no single distinct frequency dominates the behavior, but rather many frequencies, including the ones of higher modes. Those two obstacles have led many researchers to be hesitant in using TMDs for seismic structural applications (e.g., Kaynia et al. 1981; Sladek and Klingner 1983). Nonetheless, provided those obstacles are overcome, TMDs could provide a very promising alternative for multi-hazard mitigation for both winds and earthquakes.

The two aforementioned obstacles can often be overcome quite easily. In some cases where TMDs are used, inter-story drifts may already be reduced below yield drifts, and thus the structure remains elastic. A particular example to that could be found in cases where the static portion of wind loading dominates the lateral load design which results in a stiff and strong structure. Here, of course, the reduction of other structural responses (e.g., total accelerations) to both winds and earthquakes may also be desired. In addition, even if yielding does occur, causing the effective structure's stiffness to shift and thus the TMD to detune, the use of a semi-active TMD (SATMD) has been proposed (e.g., Nagarajaiah and Sonmez 2007; Roffel et al. 2010). Another approach that is based on passive control could split the TMD to several TMDs, each tuned to a slightly different frequency within a bandwidth close to the natural frequency of the main system, thus reducing the detuning effect and allowing design robustness (e.g., Xu and Igusa 1992), and in the case of 3D structures, (Jangid and Datta 1997; Li and Qu 2006). Of course, this approach is suitable only for cases where yielding is limited.

As for the second obstacle, regarding the multimodal seismic response, several solutions were proposed. One solution is the use of an active TMD (ATMD) which uses a control law to alter the frequency of the device at each moment (e.g., Abdel-Rohman 1984). This solution, however, requires a large external power supply to be activated, which may be costly and may force a reliability issue during an actual earthquake. Another possible solution, that is adopted herein, is the use of multiple TMDs (MTMDs), each tuned to a different frequency. These MTMDs could be distributed along the structure and located at locations which will optimize the control of the structure. The idea of using MTMDs tuned to various natural frequencies of the structure is not new. Clark (1988) indicated that a single TMD cannot significantly reduce the motion created due to seismic excitations, while MTMDs can substantially reduce motion. Moon (2010) shows a practical application of vertically distributed MTMDs in tall buildings for reducing wind-induced vibrations and offers a method of distributing them by mode shape.

In the literature, there are not many methodologies available for the design of MTMDs of various frequencies and locations in seismic application. In their pioneering work, Chen and Wu (2001) use a frequency-based transfer function as a response measure of the multimodal vibration problem of structures and allocate multimodal MTMDs using a sequential search technique. Luo et al. (2009) minimized a dynamic magnification factor of the first mode obtained based on the transfer function of the structure's response in frequency domain. Lin et al. (2010) proposed a two-stage frequency domain-based optimal design of MTMDs taking into consideration both the structural response and the TMD stroke. Fu and Johnson (2011) suggest using passive MTMDs with a vertical distribution of mass, where each story is assigned with one TMD of which its parameters are optimized as to minimize the sum of inter-story drifts. As for methodologies for designing 3D asymmetric structures, several methodologies using genetic algorithms exist (Singh et al. 2002; Ahlawat and Ramaswamy 2003; Desu et al. 2007). Other optimization methods for similar problems were taken in (Lin et al. 1999, 2011). Some of these methodologies only allow for dampers to dampen a single mode. While the above methodologies present a huge step forward, there is still no methodology which allows the possible dampening of all modes and leads to a desired performance in small computational efforts while using analysis tools only.

This chapter presents a simple performance-based design methodology for the allocation and sizing problem of multimodal MTMDs in structures undergoing seismic excitations. In many cases where TMDs are considered for wind mitigation, the static portion of wind loading dominates the lateral load design to result a stiff and strong structure. Hence, with the addition of TMDs, inter-story drifts may not be the main response of concern, and the reduction of total accelerations may become of major importance. Hence, the objective function minimizes the total mass of all added TMDs, while limiting the total accelerations experienced at the edges of the floors in the direction parallel to each edge. The methodology is based on a simple iterative analysis/redesign procedure where, first, an analysis is performed for a given design and then redesign of the TMDs is performed according to recurrence relations. The redesign first determines the mass of all dampers at a given location based on local acceleration measures. It is then distributed between dampers tuned to various frequencies. The proposed performance-based methodology is simple, relies on analyses tools only, generally applies to any irregular 3D problem, and possesses fast convergence.

22.2 Problem Formulation

22.2.1 Equations of Motion

Following Soong and Dargush (1997), the equations of motion of a MDOF system can be represented in state-space notation as

$$\dot{\hat{\mathbf{x}}}(t) = \mathbf{A} \cdot \hat{\mathbf{x}}(t) + \mathbf{B} \cdot a_g(t); \quad \mathbf{y}(t) = \mathbf{CC} \cdot \hat{\mathbf{x}}(t) \tag{22.1}$$

where $\hat{\mathbf{x}}^{\mathrm{T}} = \left[\mathbf{x}^{\mathrm{T}}\ \dot{\mathbf{x}}^{\mathrm{T}}\right]$; $\mathbf{x} \in R^{N}$, $\hat{\mathbf{x}} \in R^{2N}$ is the state variable vector, $\mathbf{x}(t)$ is the displacement vector between the DOFs and the ground, $a_g(t)$ is the ground motion's acceleration, a dot represents the derivative with respect to time, and $\mathbf{y}(t)$ is the output vector of the system, whose entries are responses of interest. Those responses are a linear combination of the state variables and the input forces (in our case $\mathbf{y}(t) = \mathbf{x}(t)$ gives an output of displacements, absolute acceleration will be accounted for later). The matrices **A**, **B**, and **CC** are defined as following

$$\mathbf{A} = \begin{bmatrix} \mathbf{0}_{N\times N} & \mathbf{I}_{N\times N} \\ -\mathbf{M}^{-1}\mathbf{K} & -\mathbf{M}^{-1}\mathbf{C} \end{bmatrix} \quad \mathbf{B} = \begin{bmatrix} \mathbf{0}_{N\times 1} \\ -\mathbf{e} \end{bmatrix} \quad \mathbf{CC} = \begin{bmatrix} \mathbf{I}_{N\times N} & \mathbf{0}_{N\times N} \end{bmatrix} \tag{22.2}$$

where **M**, **C**, and **K** are the mass, inherent damping, and stiffness matrices of the structure according to the chosen N degrees of freedom (DOFs), respectively, **e** is the excitation direction vector, **I** is the identity matrix, and **0** is a zero matrix of appropriate dimensions as noted. It should be noted that for the sake of presentation, Eq. (22.1) and the following methodology are presented using a single input (component of the ground motion).

22.2.2 Performance Measures

As previously mentioned, there are many cases where inter-story drifts and structural damage levels under a severe ground motion obtain acceptable values. In these situations, total accelerations are to be limited.

Added TMDs help control the responses of the structure, and the measure of cost of this controlling system is by the amount of added mass. As more mass is added to the structure, the retrofit is said to be more expensive and thus less cost-effective.

22.2.3 Problem Formulation

This work makes use of a stochastic description of the ground motion. The problem is formulated as an optimization problem for which the objective function minimizes the total amount of added masses in the TMDs under constraints of maximal performance measures. The design variables are the mass of each TMD, located in parallel to the edges of all floors. The root-mean-square (RMS) total accelerations at all peripheral locations of all floors are taken as the performance measures, as they are the largest accelerations expected within story limits. Those locations are shown in Fig. 22.1 as $(\mathbf{x}_{\mathrm{pyl}})_n$, $(\mathbf{x}_{\mathrm{pyr}})_n$, $(\mathbf{x}_{\mathrm{pxt}})_n$, and $(\mathbf{x}_{\mathrm{pxb}})_n$ and are the peripheral coordinates in the "y," "y," "x," and "x" directions at the left, right, top, and bottom edges of floor n, respectively. The remaining variables in Fig. 22.1 will be explained subsequently. That is, the constraints are on the total accelerations at the edges of all floors in the directions parallel to each edge. The optimization problem is thus formulated as

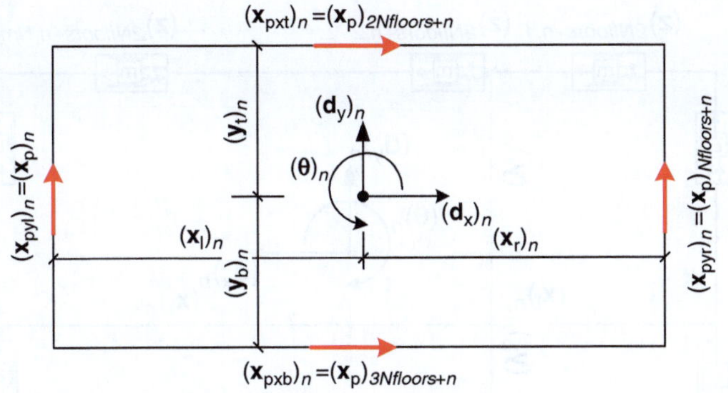

Fig. 22.1 Definition of dynamic DOFs and peripheral coordinates of the *n*th floor

$$\min_{(\mathbf{m}_{\mathrm{TMD}})_{l,f}} J = \sum_{l}^{\substack{\text{all}\\ \text{locations}}} \sum_{f}^{\substack{\text{all}\\ \text{frequencies}}} (\mathbf{m}_{\mathrm{TMD}})_{l,f}$$

$$\text{s.t.} \quad \frac{\mathrm{RMS}\left(\left(\ddot{\mathbf{x}}_{\mathrm{p}}^{\mathrm{t}}\right)_l\right)}{a_{\mathrm{all}}^{\mathrm{RMS}}} \leq 1.0 \quad \forall l = 1, 2, \ldots, N_{\mathrm{locations}} \tag{22.3}$$

where $(\mathbf{m}_{\mathrm{TMD}})_{l,f}$ is the mass of the TMD located at peripheral location l tuned to frequency f, $N_{\mathrm{locations}}$ is the number of locations to be dampened, $a_{\mathrm{all}}^{\mathrm{RMS}}$ is the allowable RMS total acceleration, and $\mathrm{RMS}\left(\left(\ddot{\mathbf{x}}_{\mathrm{p}}^{\mathrm{t}}\right)_l\right)$ is the root mean square of the total acceleration at location l (the lth term of $\mathrm{RMS}\left(\ddot{\mathbf{x}}_{\mathrm{p}}^{\mathrm{t}}\right)$). Such reference to a component of a vector or a matrix, i.e., $(\cdot)_l$, will be used throughout the chapter.

22.3 Proposed Solution Scheme

22.3.1 Fully Stressed Design

Designs that are based on fully stressed characteristics go back to the classical design of trusses under static loads, whereby the weight is minimized for a given allowable stress. For that problem, it had been widely accepted that the optimal design yields a *statically determinate fully stressed design, with members out of the design having strains smaller than the allowable*. This result appeared in the literature as early as 1900 (Cilley 1900). This has been proven in several occasions, using various approaches (Cilley 1900; Mitchell 1904; Levy 1985).

Recently, it was shown that some dynamic optimal designs also possess "fully stressed" characteristics. Lavan and Levy (2005); Levy and Lavan (2006) considered

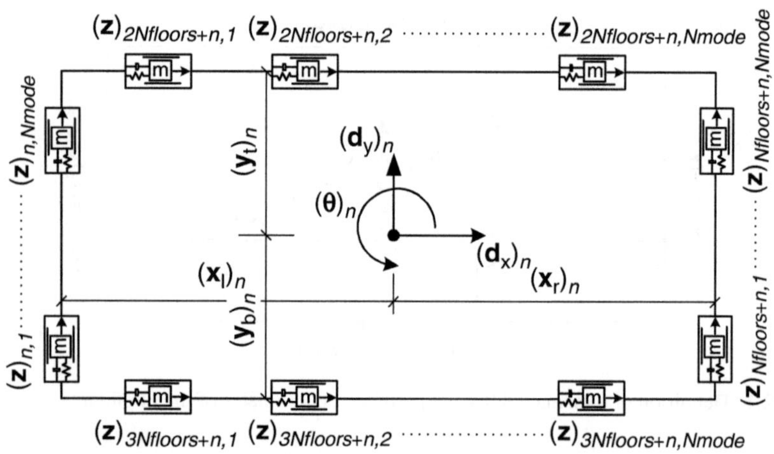

Fig. 22.2 Locations of TMDs at the floor n and their associated z DOFs

the minimization of total added viscous damping to frame structures subjected to ground accelerations while constraining various inter-story responses. Their optimal solutions attained by formal optimization indicated that "At the optimum, damping is assigned to stories for which the local performance index has reached the allowable value. Stories with no assigned damping attain a local performance index which is lower or equal to the allowable." That is, the optimal solutions attained "fully stressed" characteristics.

Based on past experience of the authors in similar problems, it is conjectured here that the optimal solution to MTMD allocation and sizing in structures under a stochastic ground acceleration input (solution of Eq. (22.3)) possesses fully stressed design (FSD) characteristics, i.e., "At the optimum, TMDs are assigned to peripheral locations for which the RMS total acceleration has reached the allowable value under the considered input acceleration PSD. In addition, at each location to which TMDs are added, TMDs of a given frequency are assigned only to frequencies for which the output spectral density is maximal."

Potential locations for TMDs are located at the edges of the floors, as their lines of action are in direction parallel to the edges (Fig. 22.2). Those are actually the same locations where total accelerations are to be limited.

Stage one of the conjecture imposes that for all peripheral locations with masses within the design, the total acceleration equals the allowable one, while all peripheral locations with zero masses (outside the design) have an acceleration equal to or less than the allowable. This is illustrated on the left-hand side of Fig. 22.3, which presents the concept on a selected peripheral frame. The second stage of the conjecture imposes that for all dampers at a peripheral location where the acceleration equals the allowable one, and are within the design, the output spectral densities are maximal (with respect to ω) and equal. As for masses outside

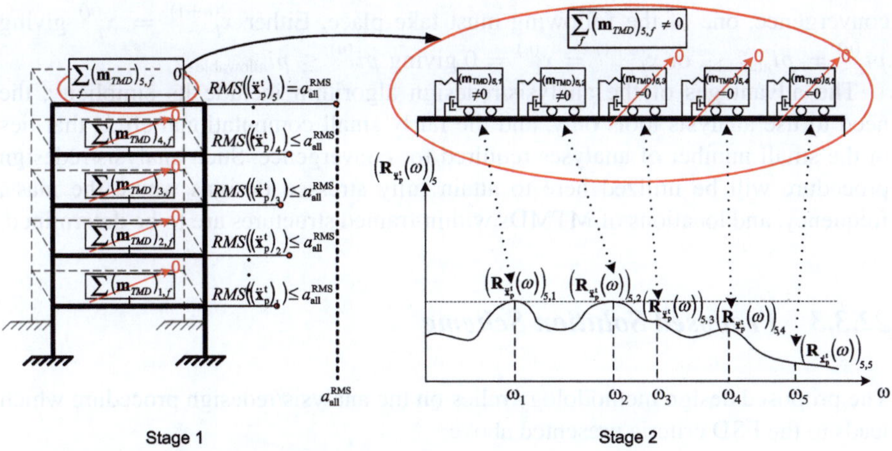

Fig. 22.3 Illustrations of conjecture

of the design at this DOF, the output spectral density is less than maximal. This is illustrated on the right-hand side of Fig. 22.3.

The above conjecture suggests that the tuning frequency of each TMD is searched for among all possible frequencies. However, for practical reasons, it is reasonable to assume that these frequencies are in the vicinity of the bare structure's frequencies, and thus the tuning of TMDs could be in relation to the bare structure's natural frequency.

22.3.2 Analysis/Redesign Algorithm

Solutions to optimization problems, which possess fully stressed characteristics, are efficiently achieved iteratively using a two-step algorithm in each iteration cycle. In the first step, an analysis is performed for a given preliminary design, whereas in the second step, the design is changed using a recurrence relationship that targets fully stressedness. The recurrence relation can be generally written as

$$x_l^{(n+1)} = x_l^{(n)} \cdot \left(\frac{pi_l^{(n)}}{pi_{\text{allowable}}} \right)^P \tag{22.4}$$

where x_l is the value of the design variable associated with the location l, pi_l is the performance measure of interest for the location l, $pi_{\text{allowable}}$ is the allowable value for the performance measure, n is the iteration number, and P is a convergence parameter. Fully stressedness is obtained from using Eq. (22.4) since upon convergence, one of the following must take place. Either $x_l^{(n+1)} = x_l^{(n)}$ giving $pi_l^{(n)} = pi_{\text{allowable}}$ or $x_l^{(n+1)} = x_l^{(n)} = 0$ giving $pi_l^{(n)} \leq pi_{\text{allowable}}$.

The advantages of the analysis/redesign algorithm include its simplicity, the need to use analysis tools only, and the fairly small computational effort that lies in the small number of analyses required for convergence. Such analysis/redesign procedure will be utilized here to attain fully stressed designs where the mass, frequency, and locations of MTMDs within framed structures are to be determined.

22.3.3 Proposed Solution Scheme

The proposed design methodology relies on the analysis/redesign procedure which leads to the FSD criteria presented above:

Step 1: An allowable RMS acceleration is chosen. The mass, damping, and stiffness matrices of the structure are assembled.

Step 2: Solution of the eigenvalue problem determines the structure's natural frequencies and mode shapes.

Step 3: A power spectral density (PSD) for the input acceleration is chosen. Examples of such input spectra are stationary white noise, which gives a constant PSD, and the Kanai-Tajimi (1957) PSD. The PSD is fitted to represent real ground motions. This is done by fitting its parameters to a frequency-based spectrum representing the decomposition of earthquakes into frequency components (e.g., a FFT spectrum). For each DOF, the transfer function of total acceleration of the bare frame is evaluated using Eq. (22.5). This transfer function represents the ratio between the sinusoidal output amplitude and the sinusoidal input amplitude with frequency ω. For total accelerations, it can be shown that the appropriate transfer vector, $\mathbf{H}_{\ddot{\mathbf{x}}^t}(j\omega)$, is

$$\mathbf{H}_{\ddot{\mathbf{x}}^t}(j\omega) = -\mathbf{M}^{-1} \cdot (j\omega\mathbf{C} + \mathbf{K}) \cdot \mathbf{H}_{\mathbf{x}}(j\omega) \quad (22.5)$$

where $j = \sqrt{-1}$ and $\mathbf{H}_{\mathbf{x}}(j\omega)$ is the displacement transfer vector given by (Kwakernaak and Sivan 1991)

$$\mathbf{H}_{\mathbf{x}}(j\omega) = \mathbf{CC} \cdot (j\omega\mathbf{I} - \mathbf{A})^{-1} \cdot \mathbf{B} \quad (22.6)$$

This transfer function is transformed to peripheral coordinates using

$$\mathbf{H}_{\ddot{\mathbf{x}}^t_p}(j\omega) = \mathbf{T} \cdot \mathbf{H}_{\ddot{\mathbf{x}}^t}(j\omega) \quad (22.7)$$

where $\mathbf{H}_{\ddot{\mathbf{x}}^t_p}(j\omega)$ is the structure's transfer function of total accelerations in peripheral coordinates and $\mathbf{T}$ is a transformation matrix from the original DOFs to peripheral coordinates. The output spectral densities of the peripheral accelerations, $\left(\mathbf{R}_{\ddot{\mathbf{x}}^t_p}(\omega)\right)_l$, are evaluated using

$$\left(\mathbf{R}_{\ddot{\mathbf{x}}_\mathrm{p}^\mathrm{t}}(\omega)\right)_l = \left|\left(\mathbf{H}_{\ddot{\mathbf{x}}_\mathrm{p}^\mathrm{t}}(j\omega)\right)_l\right|^2 \cdot S(\omega) \tag{22.8}$$

where $S(\omega)$ is the input PSD, $\left|\left(\mathbf{H}_{\ddot{\mathbf{x}}_\mathrm{p}^\mathrm{t}}(j\omega)\right)_l\right|^2 = \left(\mathbf{H}_{\ddot{\mathbf{x}}_\mathrm{p}^\mathrm{t}}(j\omega)\right)_l \cdot \left(\mathbf{H}^*_{\ddot{\mathbf{x}}_\mathrm{p}^\mathrm{t}}(j\omega)\right)_l$, $\left(\mathbf{H}_{\ddot{\mathbf{x}}^\mathrm{t}}(j\omega)\right)_l$ is the lth term of $\mathbf{H}_{\ddot{\mathbf{x}}^\mathrm{t}}(j\omega)$, and $\left(\mathbf{H}^*_{\ddot{\mathbf{x}}^\mathrm{t}}(j\omega)\right)_l$ is its complex conjugate.

The area under the output spectral density curve equals the mean-square response (Newland 1993), and, thus, the root mean square (RMS) of total accelerations at peripheral coordinate l, $\mathrm{RMS}\left(\ddot{\mathbf{x}}_\mathrm{p}^\mathrm{t}\right)_l$,taking into consideration the contribution of all frequencies to the total response, is derived using

$$\mathrm{RMS}\left(\ddot{\mathbf{x}}_\mathrm{p}^\mathrm{t}\right)_l = \sqrt{2 \cdot \int_0^\infty \left(\mathbf{R}_{\ddot{\mathbf{x}}_\mathrm{p}^\mathrm{t}}(\omega)\right)_l \mathrm{d}\omega} \tag{22.9}$$

Step 4: If for any peripheral coordinate, l, the RMS acceleration obtained is larger than the allowable RMS acceleration, MTMDs are added to suppress the acceleration produced. Each TMD of mass $(\mathbf{m}_{\mathrm{TMD}})_{l,f}$ is assigned with a DOF for its displacement relative to the ground, $(\mathbf{z})_{l,f}$. Here, the subscript l stands for its location while the subscript f stands for its frequency. The location, l, is corresponding to the peripheral coordinate $(\mathbf{x}_\mathrm{p})_l$ the TMD is attached to. At each location, N_{mode} TMDs are added, to suppress N_{mode} original frequencies of the structure, where N_{mode} is the number of modes to potentially be controlled. Thus, generally, a total of $N_{\mathrm{mode}} \cdot N_{\mathrm{locations}}$ dampers are potentially added (Fig. 22.2).

The response of each mode could be evaluated based on a single-degree-of-freedom (SDOF) equivalent system. Hence, properties of the TMDs to dampen a certain mode could be set based on its SDOF representation. For the sake of simplicity in this chapter, Den-Hartog's (1940) properties were chosen. Nonetheless, more advanced criteria could easily be used with the proposed methodology. These Den-Hartog properties were derived for the optimal reduction of mass displacement of an SDOF system under external sinusoidal loading. They were later shown to also reduce the maximum total acceleration response of the mass of an SDOF system undergoing a harmonic base excitation (Warburton 1982). In the case of optimal Den-Hartog properties, the following initial properties are taken for the dampers:

1. The initial mass of all TMDs located at each peripheral coordinate is taken as certain predetermined percentage of the structure's mass (say 1%) and divided equally between the TMDs

$$(\mathbf{m}_{\mathrm{TMD}})_{l,f} = \left(\frac{0.01 \cdot M_{\mathrm{structure}}}{N_{\mathrm{mode}}}\right) \tag{22.10}$$

where l represents the damper's location, f represents the mode dampened, and $M_{\mathrm{structure}}$ is the structure's total mass. The mass ratio $(\mu_{\mathrm{TMD}})_f$ of all TMDs tuned to frequency f is calculated as the ratio between the effective TMD mass of all

TMDs tuned to frequency f and the fth modal mass of the structure. This mass ratio is defined as

$$(\mu_{\text{TMD}})_f = \frac{\left(\varphi_f^{\text{T}} \cdot \mathbf{T}^{\text{T}} \cdot D\left((\mathbf{m}_{\text{TMD}})_f\right) \cdot \mathbf{T} \cdot \varphi_f\right)}{\left(\varphi_f^{\text{T}} \cdot \mathbf{M}_{\textbf{original}} \cdot \varphi_f\right)} \tag{22.11}$$

where φ_f is the fth mode shape of the bare structure, $\left[\mathbf{M}_{\textbf{original}}\right]$ is the bare frame's mass matrix, and $D\left((\mathbf{m}_{\text{TMD}})_f\right)$ is a diagonal matrix with the terms $(\mathbf{m}_{\text{TMD}})_{1:N_{\text{locations}},f}$ sitting on the diagonal.

2. Each TMD's stiffness is determined according to the frequency of the mode which is dampened by the TMD and is tuned to

$$(\omega_{\text{TMD}})_f = \frac{(\omega_{\text{n}})_f}{\left(1 + (\mu_{\text{TMD}})_f\right)} \tag{22.12}$$

where $(\omega_{\text{n}})_f$ is the frequency f to be dampened. The compatible stiffness is

$$(\mathbf{k}_{\text{TMD}})_{l,f} = (\mathbf{m}_{\text{TMD}})_{l,f} \cdot \left((\omega_{\text{TMD}})_f\right)^2 \tag{22.13}$$

3. Each TMD's modal damping ratio is determined according to

$$(\xi_{\text{TMD}})_f = \sqrt{\frac{3 \cdot (\mu_{\text{TMD}})_f}{\left(8 \cdot \left(1 + (\mu_{\text{TMD}})_f\right)^3\right)}} \tag{22.14}$$

and the matching damping coefficient

$$(\mathbf{c}_{\text{TMD}})_{l,f} = 2 \cdot (\mathbf{m}_{\text{TMD}})_{l,f} \cdot (\xi_{\text{TMD}})_f \cdot (\omega_{\text{n}})_f \tag{22.15}$$

Step 5: The mass, damping, and stiffness matrices of the damped frame are formulated. Note the mass of TMDs is to be added to the mass of the structure perpendicular to their original DOF (i.e., if a certain damper is used to reduce vibration in the "y" direction and thus its DOF is in the "y" direction, the mass of that TMD is added to the mass of the structure in the "x" direction of the story where it is situated).

Step 6: The peripheral RMS accelerations of the damped frame at all coordinates are evaluated using frequency domain analysis based on Eqs. (22.5), (22.6), (22.7), (22.8), and (22.9), using the newly updated matrices (note that in Eq. (22.9), it is needed to take only the first N components of the extended vector $\mathbf{H}_{\ddot{\mathbf{x}}^{\text{t}}}(j\omega)$ as now DOFs of TMDs are included in this vector).

Step 7: The TMD's mass is redetermined using two stages; the total mass of all dampers located at a given location is determined, followed by the distribution of that mass between all TMDs at that location, having various tuning frequencies. This is done according to the recurrence relationships described below. Following the change in mass, the stiffness and modal damping ratios of each TMD are also updated while keeping the Den-Hartog principles intact, using Eqs. (22.11), (22.12), (22.13), (22.14), and (22.15). The two-stage analysis/redesign procedure is carried out iteratively in the following way

Stage 1:

$$\left(\mathbf{m}_{\text{TMD,total}}^{(n+1)}\right)_l = \sum_{f=1}^{\substack{\text{all}\\ \text{frequencies}}} \left(\mathbf{m}_{\text{TMD}}^{(n+1)}\right)_{l,f} = \sum_{f=1}^{\substack{\text{all}\\ \text{frequencies}}} \left(\mathbf{m}_{\text{TMD}}^{(n)}\right)_{l,f} \cdot \left(\frac{\text{RMS}\left(\left(\ddot{\mathbf{x}}_{\text{p}}^{\text{t}\,(n)}\right)_l\right)}{a_{\text{all}}^{\text{RMS}}}\right)^P \tag{22.16}$$

where $(\cdot)^{(n)}$ is the value at iteration n, $\left(\mathbf{m}_{\text{TMD,total}}^{(n+1)}\right)_l$ is the total mass of all dampers at location l, and P is a constant which influences the convergence and convergence rate. A large P will result in a faster but less stable convergence of the above equation.

Stage 2:

$$\left(\mathbf{m}_{\text{TMD}}^{(n+1)}\right)_{l,f} = \left(\mathbf{m}_{\text{TMD}}^{(n)}\right)_{l,f} \left(\frac{\sqrt{\left(\mathbf{R}_{\ddot{\mathbf{x}}_{\text{p}}^{\text{t}}}^{(n)}\left((\omega_{\text{n}})_f\right)\right)_l}}{\max\limits_f\left(\sqrt{\left(\mathbf{R}_{\ddot{\mathbf{x}}_{\text{p}}^{\text{t}}}^{(n)}\left((\omega_{\text{n}})_f\right)\right)_l}\right)}\right)^P \cdot \frac{\left(\mathbf{m}_{\text{TMD,total}}^{(n+1)}\right)_l}{\sum\limits_{f=1}^{\substack{\text{all}\\ \text{frequencies}}} \left(\mathbf{m}_{\text{TMD}}^{(n)}\right)_{l,f} \left(\frac{\sqrt{\left(\mathbf{R}_{\ddot{\mathbf{x}}_{\text{p}}^{\text{t}}}^{(n)}\left((\omega_{\text{n}})_f\right)\right)_l}}{\max\limits_f\left(\sqrt{\left(\mathbf{R}_{\ddot{\mathbf{x}}_{\text{p}}^{\text{t}}}^{(n)}\left((\omega_{\text{n}})_f\right)\right)_l}\right)}\right)^P} \tag{22.17}$$

where $\left(\mathbf{R}_{\ddot{\mathbf{x}}_{\text{p}}^{\text{t}}}^{(n)}\left((\omega_{\text{n}})_f\right)\right)_l$ is the component of $\mathbf{R}_{\ddot{\mathbf{x}}_{\text{p}}^{\text{t}}}^{(n)}(\omega)$ at the location l evaluated at $\omega = (\omega_{\text{n}})_f$.

Step 8: Repeat steps 5–7 until convergence of the mass is reached.

22.4 Example

The following 8-story setback RC frame structure (Fig. 22.4) introduced by Tso and Yao (1994) is retrofitted using MTMDs for ground motions exciting the structure in the "y" direction. A uniform distributed mass of 0.75 ton/m^2 is taken. The column dimensions are 0.5 m by 0.5 m for frames 1 and 2 and 0.7 m by 0.7 m for frames 3 and 4. The beams are 0.4 m wide and 0.6 m tall. Five percent Rayleigh damping for the first and second modes is used. A 45% reduction of the RMS total acceleration in the bare structure is desired. Hence, an allowable peripheral RMS acceleration of 55% of the maximal peripheral RMS acceleration of the bare structure is adopted. The response is analyzed under a Kanai-Tajimi PSD with parameters fitted to the average FFT values of the SE 10 in 50 ground motion ensemble (Somerville et al. 1997). The design variables are the locations and properties of the individual tuned mass dampers. The dampers are to potentially be located in the peripheral frames, where they are most effective, and as the excitation is in the "y" direction only, dampers will be assigned only to the peripheral frames 1 (lower 4 floors), 3 (upper 4 floors), and 4 to dampen frequencies of modes which involve "y" and "θ." The steps described above are closely followed to optimally design the MTMDs.

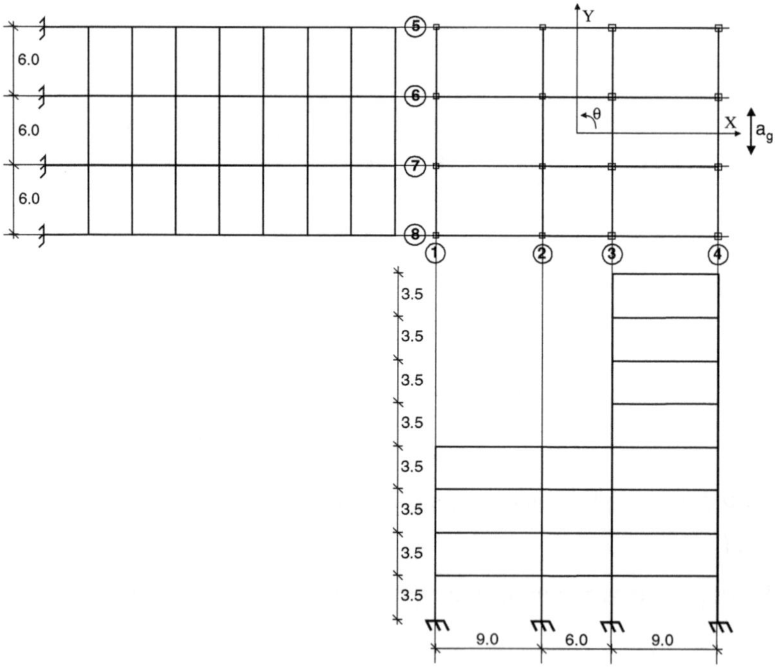

Fig. 22.4 Eight-story setback structure

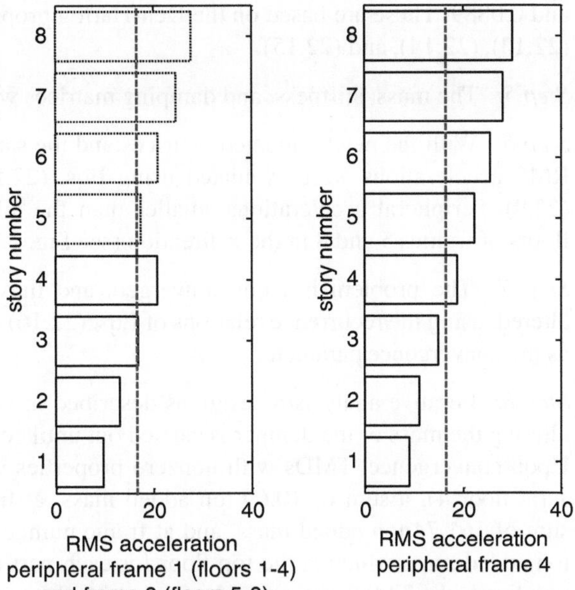

Fig. 22.5 Peripheral RMS accelerations of bare structure (*continuous* or *dotted*) and allowable values (*dotted*)

Step 1: The mass, inherent damping, and stiffness matrices of the frame were constructed.

Step 2: The natural frequencies of the structure were determined. The first few frequencies are (rad/s): 6.88 (x), 7.36 (y,θ), 10.37 (y,θ), 16.04 (x), 17.88 (y,θ), 22.61 (y,θ), 33.87 (x), 35.96 (y θ), and 43.48 (y,θ) where x,y and θ relate to the mode direction.

Step 3: The RMS accelerations of the undamped building at the peripheral frames in the y direction are presented in Fig. 22.5. Those were obtained using the Kanai-Tajimi PSD with parameters $\omega_g = 13$ rad/s, $\xi_g = 0.98$, and $S_0 = 1$. Those were determined by fitting the parameters ω_g and ξ_g to a spectrum of mean FFT values of the SE 10 in 50 ground motion ensemble scaled to $S_0 = 1.0$. The actual value of S_0 has no effect since the allowable RMS acceleration is determined by the percentage of reduction desired. The allowable RMS acceleration for all peripheral accelerations was earlier adopted as 55% of the maximum peripheral RMS acceleration of the bare frame, giving $a_{\mathrm{all}}^{\mathrm{RMS}} = 16.61$.

Step 4: 160 TMDs were added, as a first guess. Those are comprised of ten dampers each tuned to a different mode frequency (of modes related to "y" and "θ") at each of the 16 peripheral locations of frames 1, upper four floors of frame 3, and frame 4. The initial properties were a mass of 1.782 ton for each TMD; a frequency of (rad/s) 7.18, 10.20, 17.57, 22.15, 35.42, 42.48, 53.36, 65.92, 70.90, and 92.57 (TMDs tuned to dampen modes 2, 3, 5, 6, 8, 9, 11, 12, 13, and 15, respectively); and a damping ratio of 0.0931, 0.0778, 0.0802, 0.0860, 0.0739, 0.0909, 0.0811, 0.0851, 0.0728, and 0.0889. These are based on the Den-Hartog properties of Eqs. (22.11), (22.12), (22.13), (22.14), and (22.15).

Step 5: The mass, stiffness, and damping matrices were updated.

Step 6: With the newly updated matrices and the same PSD input, new peripheral RMS accelerations were evaluated using Eqs. (22.5), (22.6), (22.7), (22.8), and (22.9). Peripheral accelerations smaller than the allowable were attained for all floors of frames 5 and 8 in the x direction (see Fig. 22.4 for frame numbering).

Step 7: The problem has not converged, and thus the TMDs' properties were altered, using the recurrence relations of Eqs. (22.16) and (22.17), while using $P = 5$ as the convergence parameter.

Step 8: Iterative analysis/redesign as described in Eqs. (22.16) and (22.17) while altering the mass of the damper is carried out until convergence to allowable levels. Upon convergence, TMDs with nonzero properties were located at frame number 1 (at floor 4), a sum of 10.69 ton added mass; at frame number 3 (at floor 8), a sum of 161.74 ton added mass, and at frame number 4 (at floor 8), a sum of 0.08 ton added mass which is the top floor for each part of the setback frame. At floor number 4, the TMDs are set to dampen mode 3 ($m = 10.69$ ton, $k = 1079.5$ kN/m, $\xi = 0.1045$), while at floor number 8, the TMDs are set to dampen modes 2 at frame 3 ($m = 161.74$ ton, $k = 5859.35$ kN/m, $\xi = 0.2137$) and at frame 4 ($m = 0.08$ ton, $k = 2.98$ kN/m, $\xi = 0.2137$). All three assigned TMDs add up to 9.68% of the original structure's mass. For all practical reasons, the small TMD at frame number 4 (at floor 8) can be neglected.

Finally, an analysis of the retrofitted structure yields the peripheral RMS accelerations shown in Fig. 22.6. As can be seen, only locations that had reached the maximum allowable RMS total acceleration (Fig. 22.6) were assigned with added absorbers, making the solution obtained a FSD.

22.5 Conclusions

An analysis/redesign performance-based methodology for optimally allocating and sizing MTMDs in 3D irregular structures was presented. The proposed methodology considers the possible dampening of all modes of the structure, at all peripheral frames, thus eliminating the decision of what modes to dampen and where the TMDs should be allocated. The methodology is general and automatically takes into consideration the structural irregularities; thus, no special attention has to be given to these complexities. As shown, using MTMDs tuned to various frequencies can efficiently reduce total accelerations within the structure and bring them to a desired level, allowing for performance-based design. The advantages of this methodology are its simplicity of use and relying solely on analysis tools to solve

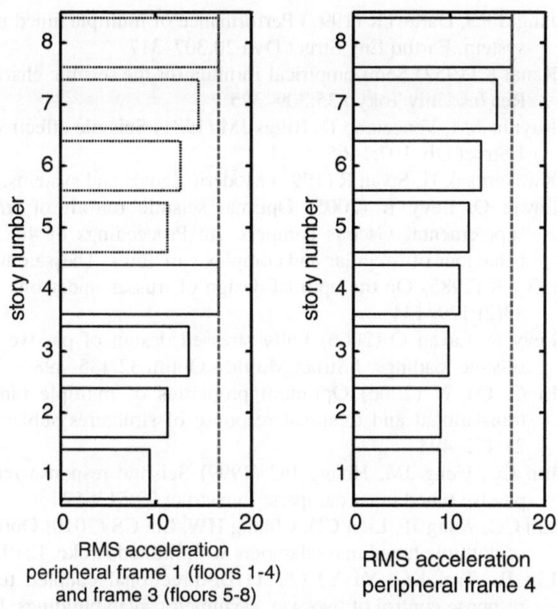

Fig. 22.6 Peripheral RMS accelerations of structure with final TMDs (*continuous* or *dotted*) and allowable values (*dashed*)

the allocation and sizing problem, with no assumptions or preselection of any design variable. These advantages make the proposed methodology attractive and efficient for practical use.

References

Abdel-Rohman M (1984) Optimal design of active TMD for buildings control. Build Environ 19(3):191–195

Ahlawat AS, Ramaswamy A (2003) Multiobjective optimal absorber system for torsionally coupled seismically excited structures. Eng Struct 25:941–950

Chen G, Wu J (2001) Optimal placement of multiple tune mass dampers for seismic structures. J Struct Eng 127(9):1054–1062

Christopoulos C, Filiatrault A (2006) Principles of supplemental damping and seismic isolation. IUSS Press, Milan

Cilley FH (1900) The exact design of statically indeterminate frameworks, an exposition of its possibility but futility. ASCE Trans 43:353–407

Clark AJ (1988) Multiple passive tuned mass dampers for reducing earthquake induced building motion. In: Proceedings of the 9th world conference on earthquake engineering, Tokyo-Kyoto, Japan, pp 779–784

Den-Hartog JP (1940) Mechanical vibrations, 2nd edn. McGraw-Hill Book Company Inc., New York

Desu NB, Dutta A, Deb SK (2007) Optimal assessment and location of tuned mass dampers for seismic control of a plan-asymmetrical buildings. Struct Eng Mech 26(4):459–477

Fu TS, Johnson EA (2011) Distributed mass damper system for integrating structural and environmental control in buildings. J Eng Mech. doi:10.1061/(ASCE)EM.1943-7889.0000211

Jangid RS, Datta TK (1997) Performance of multiple tuned mass dampers for torsionally coupled system. Earthq Eng Struct Dyn 20:307–317
Kanai K (1957) Semi-empirical formula for the seismic characteristics of the ground. Bull Earthq Res Ins Univ Tokyo 35:309–325
Kaynia AM, Veneziano D, Biggs JM (1981) Seismic effectiveness of tuned mass dampers. ASCE J Struct Div 107:1465–1484
Kwakernaak H, Sivan R (1991) Modern signals and systems. Prentice Hall, Englewood Cliffs
Lavan O, Levy R (2005) Optimal seismic retrofit of irregular 30 framed structures using supplemental viscous dampers. In: Proceedings of 4th European workshop on the seismic behaviour of irregular and complex structures, Thessaionica, Greece, paper no. 52, Aug 26–27
Levy R (1985) On the optimal design of trusses under one loading condition. Quart Appl Math 43(2):129–134
Levy R, Lavan O (2006) Fully stressed design of passive controllers in framed structures for seismic loadings. J Struct Multidis Optim 32:485–498
Li C, Qu W (2006) Optimum properties of multiple tuned mass dampers for reduction of translational and torsional response of structures subject to ground acceleration. Eng struct 28:472–494
Lin CC, Ueng JM, Huang EC (1999) Seismic response reduction of irregular buildings using passive tuned mass dampers. Eng struct 22:513–524
Lin CC, Wang JF, Lien CH, Chiang HW, Lin CS (2010) Optimum design and experimental study of multiple tuned mass dampers with limited stroke. Earth Eng Struct Dyn 39:1631–1651
Lin JL, Tsai KC, Yu YJ (2011) Bi-directional coupled tuned mass dampers for the seismic response control of two-way asymmetric-plan buildings. Earthq Eng Struct Dyn 40:675–690
Luo X, Ma R, Li G, Zhao D (2009) Parameter optimization of multi-mode vibration control system. In: International conference of measuring technology and mechatronics automation, IEEE Computer Society, pp 685–688
McNamara RJ (1977) Tuned mass dampers for buildings. ASCE J Struct Div 103:1785–1798
Mitchell AGM (1904) The limits of economy of Mateslal in framed structures. Philos Mag 6: 589–597
Moon KS (2010) Vertically distributed multiple tuned mass dampers in tall buildings: performance analysis and preliminary design. Struct Design Tall Spec Build 19:347–366
Nagarajaiah S, Sonmez E (2007) Structures with semiactive variable stiffness single/multiple tuned mass dampers. J Struct Eng 133(1):67–77
Newland DE (1993) An introduction to random vibrations, spectral & wavelet analysis. Prentice Hall, Harlow
Roffel AJ, Lourenco R, Narasimhan S (2010) Experimental studies on an adaptive tuned mass damper with real-time tuning capability. In: ASCE 19th analysis & computation specialty conference, 12–15 May 2010, pp 314–324
Singh MP, Singh S, Moreschi LM (2002) Tuned mass dampers for response control of torsional buildings. Earthq Eng Struct Dyn 31:749–769
Sladek JK, Klingner RE (1983) Effect of tuned-mass dampers on seismic response. ASCE J Struct Div 109:2004–2009
Somerville P, Smith N, Punyamurthula S, Sun J (1997) Development of ground motion time histories for phase 2 of the FEMA/SAC steel project. Report no. SAC/BD-97/04
Soong TT, Dargush GF (1997) *Passive* energy dissipation systems in structural engineering. Wiley, Chichester
Takewaki I (2009) Building control with passive dampers: optimal performance-based design for earthquakes. Wiley (Asia), Singapore
Tso WK, Yao S (1994) Seismic load distribution in buildings with eccentric setback. Can J Civil Eng 21:50–62
Warburton GB (1982) Optimum absorber parameters for various combinations of response and excitation parameters. Earthq Eng Struct Dyn 10:381–401
Xu K, Igusa T (1992) Dynamic characteristics of multiple substructures with closely spaced frequencies. Earthq Eng Struct Dyn 21:1059–1070

Chapter 23
The Possibility of a Continuous Monitoring of the Horizontal Buildings' Rotation by the Fibre-Optic Rotational Seismograph AFORS Type

Leszek R. Jaroszewicz, Zbigniew Krajewski, and Krzysztof P. Teisseyre

Abstract The design and investigation of the Autonomous Fibre-Optic Rotational Seismograph – AFORS as a system used for a continuous monitoring of the horizontal buildings rotation is presented in this chapter. Since AFORS utilizes the Sagnac effect for a direct measurement of rotational components it operates without any reference system designed to work during earthquakes. The presented system contains a special autonomous signal processing unit (ASPU) which optimizes operating of the rotation motions measurement, whereas a newly applied telemetric system based on the Internet (FORS-Telemetric Server) allows for a remote AFORS control. The laboratory investigation of the system named AFORS-2 indicated that with theoretical linear changes of sensitivity it keeps the accuracy no less than $4.8{\cdot}10^{-9}$ to $6.1{\cdot}10^{-8}$ rad/s in the frequency band from 0.83 to 106.15 Hz. In our opinion the above results are very optimistic for the future application of AFORS for the continuous monitoring of any rotational moving of the multi-storey buildings as well as the investigation of an in-plane irregularity. The first experimental results of the AFORS-2 operation in a real multi-storey building in Warsaw, Poland are also presented. On this basis the investigation of a new low cost system type AFORS is presented as the chapter conclusion.

L.R. Jaroszewicz (✉) • Z. Krajewski
Institute of Applied Physics, Military University of Technology, ul. gen. Sylwestra Kaliskiego 2, 00-908 Warsaw, Poland
e-mail: jarosz@wat.edu.pl; zkrajewski@wat.edu.pl

K.P. Teisseyre
Institute of Geophysics, Polish Academy of Sciences, ul. Księcia Janusza 64, 01-4552 Warsaw, Poland
e-mail: kt@igf.edu.pl

O. Lavan and M. De Stefano (eds.), *Seismic Behaviour and Design of Irregular and Complex Civil Structures*, Geotechnical, Geological and Earthquake Engineering 24, DOI 10.1007/978-94-007-5377-8_23,

23.1 Introduction

One of the most important problems of the seismic behaviour of irregular structures of buildings in-plane is the existence of difficulties with designing horizontal rotation of these structures. The existence of such rotations has a direct influence on torsional effects in structures as well as interstory drift. Since the application of new materials and technologies for building construction generates higher and more complicated engineering constructions, their safety needs a real rotations' monitoring as important measures in structural responses (Schreiber et al. 2009).

However, there is a limited number of sensors which can deliver direct data about rotational motion for such an analysis. The measurements of torsional response and interstory drift are reasonably easy on small scale models in a laboratory (Kao 1998), but are much more difficult in real structures. The first of them can be measured by using a pair of accelerometers and then dividing the differences in horizontal accelerations by the distance between them in a direction perpendicular to the measured motion. Then this has to be integrated twice with respect to time needed to give the torsional rotations (Schreiber et al. 2009). However, the inherent sensor drift and the small offset from zero in the absence of input signals are the important limitations of this technique. Similarly, for measuring interstory drifts, it is, in principle, possible to arrange a frame from the floor below to near the ceiling above to set up displacement transducers to measure the difference in displacements (McGinnis 2004). However, again far from the hardware complexity of this approach, it is also vulnerable to building deformations.

For the above reason the one of the most promised techniques seems to be sensors based on the Sagnac effect (Sagnac 1913). Its greatest strength is the fact that it does measure absolute rotations or oscillations, so that it does not require an external reference frame for its measurement. This means that it measures true rotations even during an earthquake where nothing remains static. The system design gives possibility for a proper choice of their accuracy as very wide mechanical oscillation frequencies by proper construction of its optical and electronic parts. Since it is an entirely optical device, it does not have the problems that characterize inertial mass transducers.

In this technique there is the wide scope of such sensors named FOG – fibre optic gyroscope, and some of them, commercially available have been already used (Schreiber et al. 2009; Franco-Anaya et al. 2008). However, their constructions are optimized for detection of angular changes rather than rotation speed. For the above reason in this chapter we conclude research and development according to the Autonomous Fibre-Optic Rotational Seismograph – AFORS, here AFORS-2, with optical parts based on the fibre-optic gyro construction whereas the special autonomous signal processing unit – ASPU optimizes its operation of the measurement of the rotation speed instead of the angular changes. The application of the new design telemetric system is based on Internet (FORS – Telemetric Server) and allows for the remote control system as it is shown in the presented examples of the system work. The presented results show that AFORS-2 with linear changes of the sensitivity protecting accuracy no less than $4.81 \cdot 10^{-9}$ to

$6.11 \cdot 10^{-8}$ rad/s in the frequency band from 0.83 to 106.15 Hz, is designed for a direct measurement of the rotational components of any structure in the plane of a sensor loop which may be horizontal plane of irregular structures of buildings. On this basis the first experimental results obtained in a real building structure as well as the theoretical investigation according to the new low cost system type AFORS for such investigation are presented.

23.2 AFORS General Specifications

Since a detailed description regarding the AFORS construction, calibration and laboratory investigation was presented before (Jaroszewicz et al. 2011), here we summarize only the above data with regard to the AFORS-2 example. The optical head of all constructed AFORS devices uses a fibre interferometer in a minimum optical gyro configuration (Jaroszewicz et al. 2006), as it is shown in the upper part of Fig. 23.1.

The application of the broadband low coherence superluminescent diode (SLD) gives possibility for a minimisation of polarization influence on the system operation by achieving light depolarization in a sensor loop (Krajewski et al. 2005). The use of the same arm of a coupler as an input/output way from a sensor loop gives

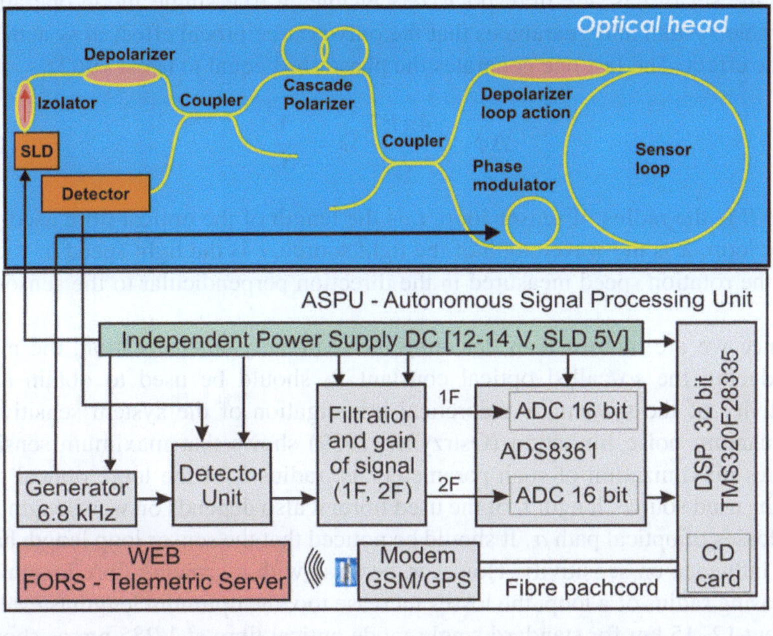

Fig. 23.1 General schema of the AFORS: *upper* – the optical head (generation of the Sagnac phase shift proportional to measured rotation rate Ω), *bottom* – Autonomous Signal Processing Unit (rotation calculation and recording)

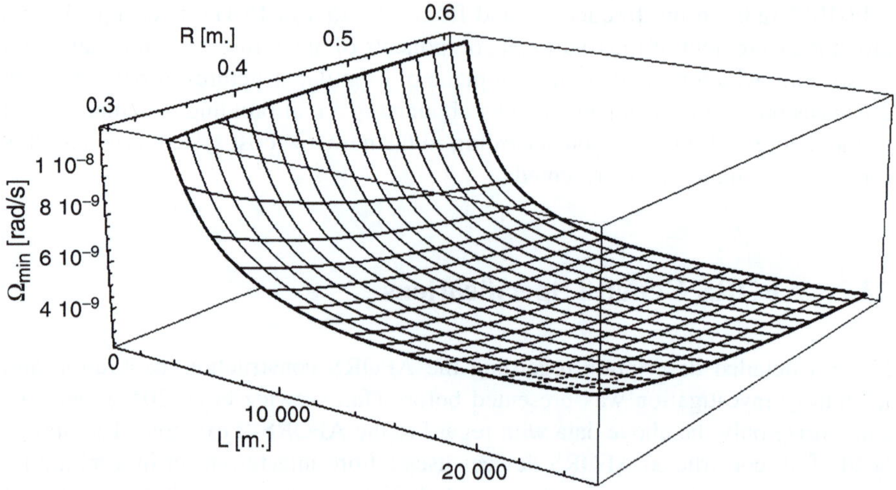

Fig. 23.2 AFORS resolution versus total optical length L and loop radius in 1 Hz detection band. Parameters for simulation: wavelength $\lambda = 1{,}285$ nm, fibre attenuation $\alpha = 0.45$ dB/km, optical path loss $\sigma = 14.5$ dB, optical power $P = 20$ mW

reciprocity condition for interfering beams, but an additional coupler is needed to separate a returned beam on detector, and a fibre optic isolator for SLD protection. Next, the set of cascade fibre polarizers enables a true single mode operation of the whole system and guarantees that the only nonreciprocal effect in system is the Sagnac effect. The last one generates the phase shift equal to (Post 1967):

$$\Delta\phi = \frac{4\pi RL}{\lambda c}\Omega = \frac{1}{S_o}, \tag{23.1}$$

where R is the radius of sensor loop, L is the length of the optical fibre used in the sensor loop, λ is the wavelength of the light source, c is the light speed in vacuum, Ω is the rotation speed measured in the direction perpendicular to the sensor loop plane.

Since we are interested in detecting the extremely small rotation, the method of reducing the so-called optical constant S_0 should be used to obtain a high sensitivity of the system. A theoretical investigation of the system sensitivity in the quantum noise limitation (Ostrzyżek 1989) shows that maximum sensitivity requires maximization of such parameters as: radius R of the loop, optical power P of the used source, length L of the used fibre: it also depends on wavelength λ and total losses of optical path σ. It should be noticed that the sensor loop length has the main influence on sensitivity. However, because with a growing fibre length and a decreasing radius of a loop, the losses increase too, the optimum length is evaluated at about 12–15 km for standard single-mode optical fibre at 1,285 nm as shown in Fig. 23.2 (Jaroszewicz and Wiszniowski 2008).

For the above reason optimized for high sensitivity AFORS-2 uses the 0.63 m diameter sensor loop made from special composite material with permalloy particles for shielding from magnetic field. A long length (15056 m, $\alpha = 0.450$ dB/km) of SMF-28 fibre has been winded in double-quadrupole mode (Dai et al. 2002) with 0.2 mm Teflon insulation between each fibre layers is for the thermal stabilization of the sensor's work. The system optimization made for AFORS-2 (among others, SLD with the high optical power $P = 20.8$ mW) allows for theoretical sensitivity equal to $2.47 \cdot 10^{-9}$ rad/s/Hz$^{1/2}$, for measured total optical loss $\sigma = 14.47$ dB.

Since the Sagnac phase-shift obtained directly in the interferometric system contains an unseparated noise component, the special signal processing is used. We have used the system based on the synchronous detection unit (Krajewski 2005). For AFORS-2 we used a new prototype of Autonomous Signal Processing Unit (ASPU), according to the scheme shown in the lower part of Fig. 23.1. The ASPU enables a detection of a rotation rate Ω from proper selection and processing the first ($A_{1\omega}$) and the second ($A_{2\omega}$) amplitude of the harmonic output signal, on basis of the following relation (Jaroszewicz et al. 2011):

$$\Omega = S_o \arctan\left[S_e \cdot \frac{A_{1\omega}(t)}{A_{2\omega}(t)}\right] \tag{23.2}$$

where S_e is electronic constants, related to parameters of used components, which are obtained during a sensor calibration based on the measurement of the Earth rotation for Warsaw, Poland i.e. $\Omega_E = 4.45 \cdot 10^{-5}$ rad/s (Jaroszewicz et al. 2011a; Krajewski et al. 2005). During the calibration the obtained constants for AFORS-2 were 0.059 s^{-1} and 0.013 for S_0 and S_e, respectively.

The measurement accuracy (Habel et al. 2009) for AFORS-2 checked in MUT laboratory located in Warsaw, Poland gives limited information about the system accuracy because of urban noises. Figure 23.3 summarizes these measurements. Because the ASPU allows for step changes of the detection frequency band in the range from 0.83 to 106.15 Hz, the obtained accuracy is at the level of $4.81 \cdot 10^{-9}$ to $6.11 \cdot 10^{-8}$ rad/s, respectively for lower and higher working frequency band. As one can see the obtained values are well correlated with Ω_{min} in quantum noise limitation. It should be noticed that the linear dependence of AFORS-2 sensitivity and the accuracy on the detected frequency band is the advantage of this system.

A new *FORS-Telemetric Server* with its main page shown in Fig. 23.4a (please use http://fors.m2s.pl with AFORSbook as login&password – for the open access to the system) is used for data storing and for monitoring the work of the AFORS-2. Because ASPU of the AFORS-2 contains GSM/GPS module and independent power supply for all electronic components of the system, hence it is as well as other AFORS systems, fully autonomous and mobile. In this moment, the 3 devices are managed via server: FORS-II, AFORS-1, AFORS-2 located in Ojców, Książ and Warsaw, respectively (Jaroszewicz et al. 2012) as it is shown in Fig. 23.4b.

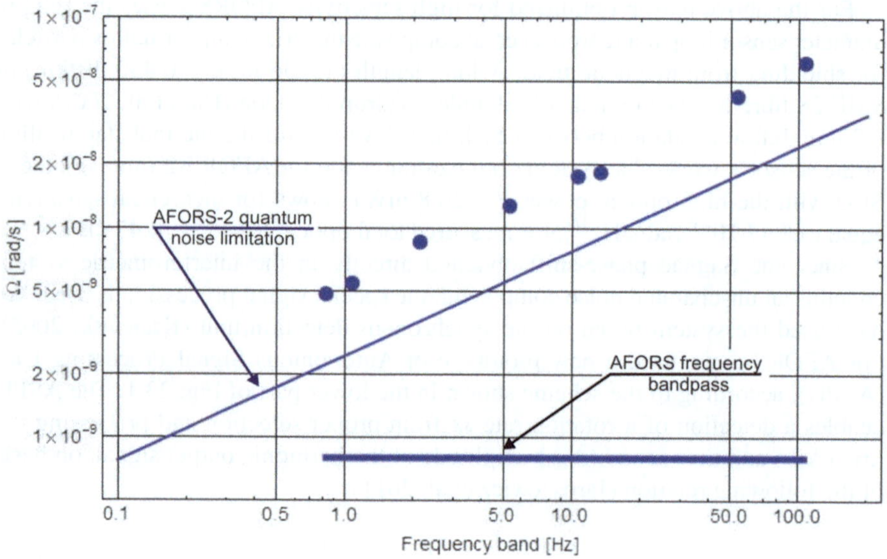

Fig. 23.3 The accuracy measured in Warsaw, Poland for the chosen detection bandpass for AFORS-2

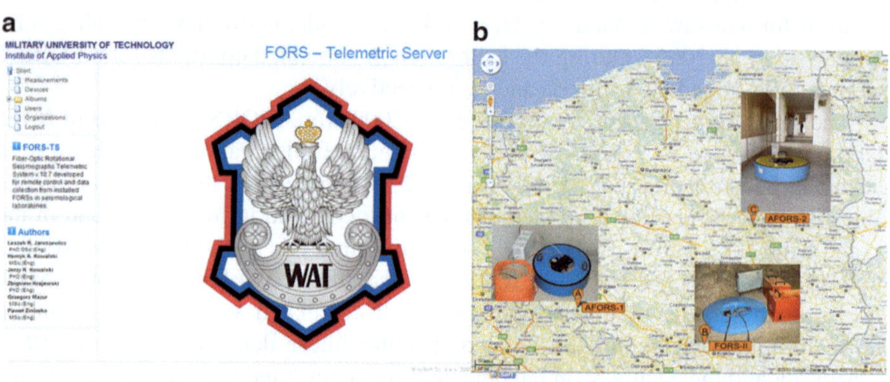

Fig. 23.4 The main page of FORS – Telemetric Server (**a**) and GOOGLE map with the devices localization (**b**)

23.3 Case Studies

The applied technology gives possibility for the remote (via Internet) controlling and changing of all electronic parameters of the ASPU for given AFORS, as presents the bookmark *Config* for the AFORS-2 in Fig. 23.5a. This remote control may comprise a software upgrade. Moreover, the bookmark *Data&Variables* (Fig. 23.5b) monitors

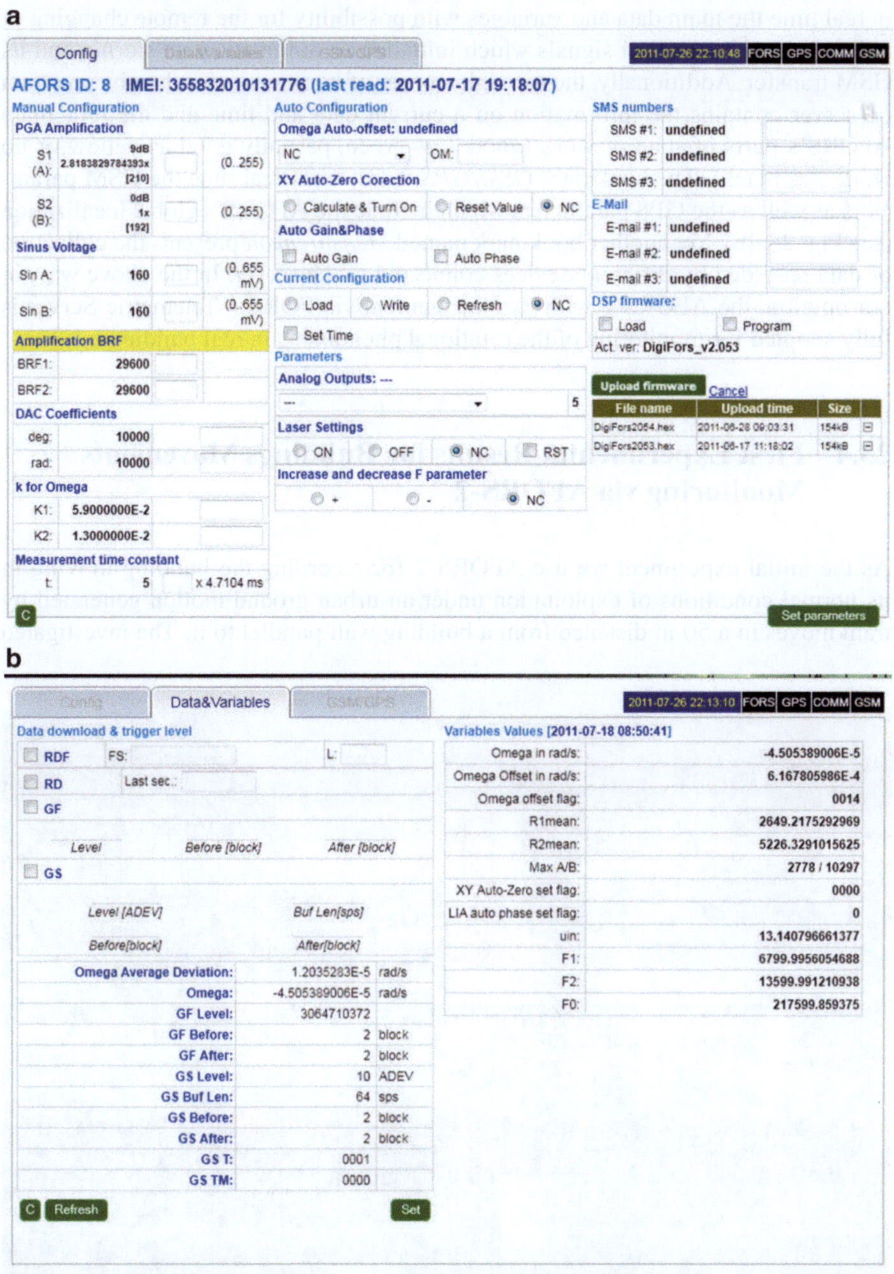

Fig. 23.5 The view of two main bookmarks: *Config* (**a**) and *Data&Variables* (**b**) for AFORS-2 at the FORS – Telemetric Server

in real time the main data and variables with possibility for the remote changing of a threshold – the level of signals which initialize an automatic data storing and its GSM transfer. Additionally, the top right corner of bookmarks for the given system on server contains the information on a current date and time and the four main AFORS's parts of state of work (good – as green, partially good as yellow or no work – as black). The bookmark *GSM/GPS* monitors in real time the GSM parameters as well as the GPS parameters which include the AFORS's global localization (see Fig. 23.4b). Yet another bookmark named *Measurement* presents the collection of data recorded by different devices connected to the server. In the above way, in our opinion, the AFORS-2 with its management via FORS – Telemetric Server is fully adopted for monitoring of the rotational phenomena in real buildings.

23.4 First Experimental Results for Buildings Movements Monitoring via AFORS-2

As the initial experiment we use AFORS-2 for recording the building moving in its normal conditions of exploitation under an urban ground motion generated by tram moves in a 50 m distance from a building wall parallel to it. The investigated

Fig. 23.6 The general view of the building used for monitoring of its rotation moving via AFORS-2. *Left picture* shows AFORS-2 mounted in 2nd and 1st building layers

building is a light construction (aluminium structure with sandwiches walls and ceilings) – see right picture in Fig. 23.6, where during the normal exploitation of a tram moves the building moves have been observed. The AFORS-2 has been installed, subsequently on the fifth, second and first floor in a hall, in the same vertical position (with accuracy of about 10 cm) as shown on the left pictures in Fig. 23.6. Since it is an old building with asbestos used as an inner wall isolation, now it is not utilize by the academy, so we expected that the recorder signals will be connected to an external perturbation, only. We expected the maximum signals for the lowest floor AFORS-2 installation, because light building steel frame construction should eliminate rotational vibration on the higher floors. For the last, fifth floor only perturbation signal has been recorded, which probably was connected to the extremely high temperature during this experiment (about 55–65 C). Figure 23.7 presents the building moves recorded on the second and first floors (difference about 3 m of height) for relatively the same ground motion generated by tram moves. Since these signals were recorded nearly at midnight on July 13th (AFORS-2 on the first floor) and July 14th (AFORS-2 on the second floor) i.e. in the night during summer holidays, the academy area was empty which had a direct influence on recorded signals and they were very clear.

As one can see in the above experiment the accuracy for AFORS-2 was $7.91 \cdot 10^{-6}$ and $3.15 \cdot 10^{-6}$ rad/s (see ADEV parameter in the left down corner of the two pictures in Fig. 23.7), for the chosen detection frequency band equal to 21.23 Hz. The amplitude of the detected rotation rate was about twice higher for the first floor, and was much higher than the system accuracy (more than ten orders).

Figure 23.8 presents a comparison of two floor extreme amplitudes of the normalized torsional rotations defined as torsional moves per meter with respect to the ground position. These data have been obtained for the correlated extreme vibrations presented in Fig. 23.7. As one can see, the rotational vibration muffling properties of the examined building are clearly observed.

The urban noise influence on the recorded signals can be observed on the data presented in Fig. 23.9, which have been obtained in the morning when the Academy opened for work. As one can see, the higher amplitude as well as frequency were observed in this time. Again, the much higher amplitude of torsional moves of the building is observed on the first floor of the building.

23.5 Conclusions and Final Remarks

According to the panel chaired by J. R. Evans during the First International Workshop on Rotational Seismology (Evans et al. 2007) with the summary recommendation reproduced by R. Cowsik (Cowsik et al. 2009), the engineering of a strong-motion seismology needs devices operating with a frequency range

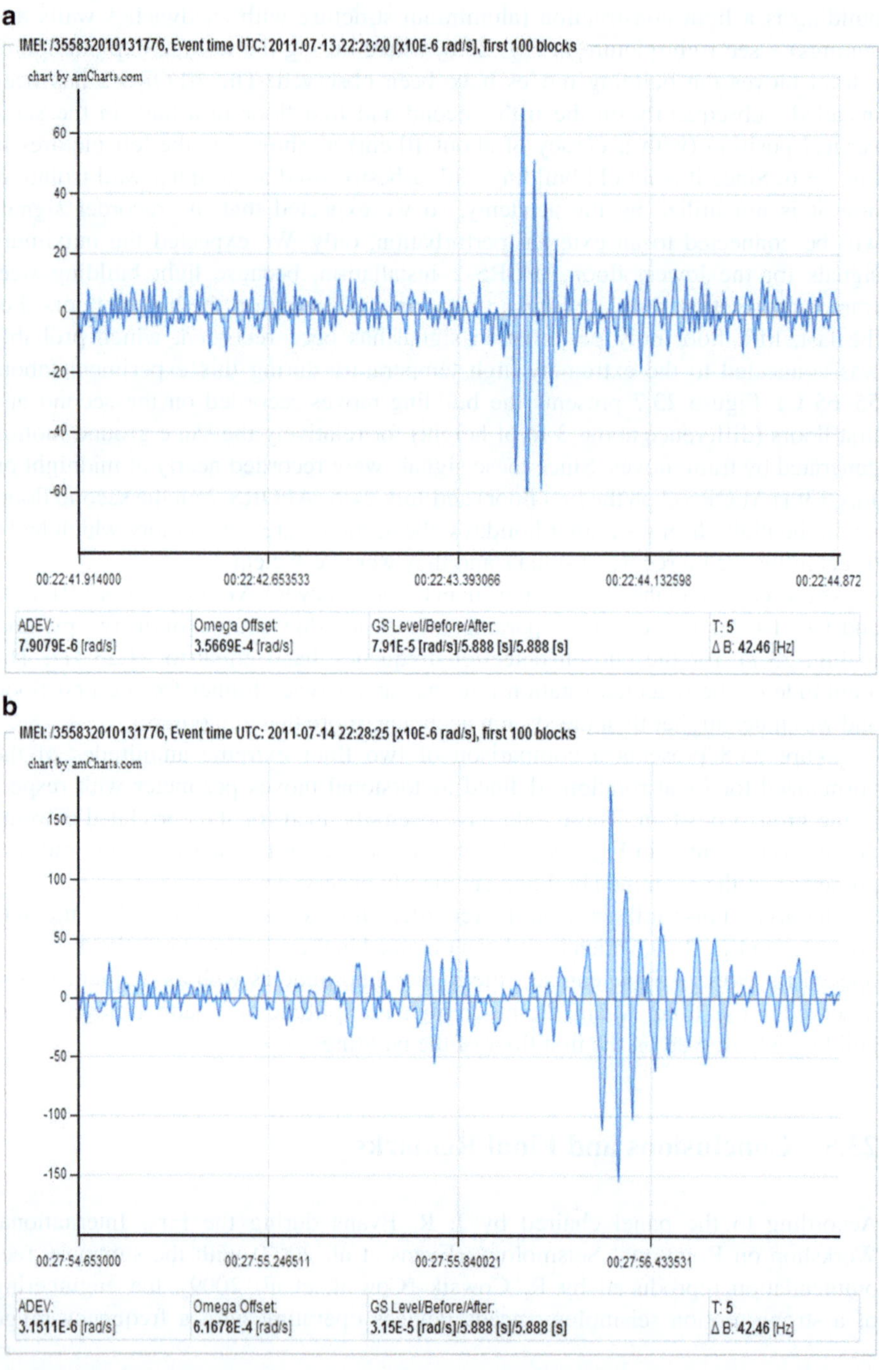

Fig. 23.7 The data recorded on second (*top*) and first (*down*) floor as response for ground moves after tram pass through street in distance about 50 m from and parallel to long building wall

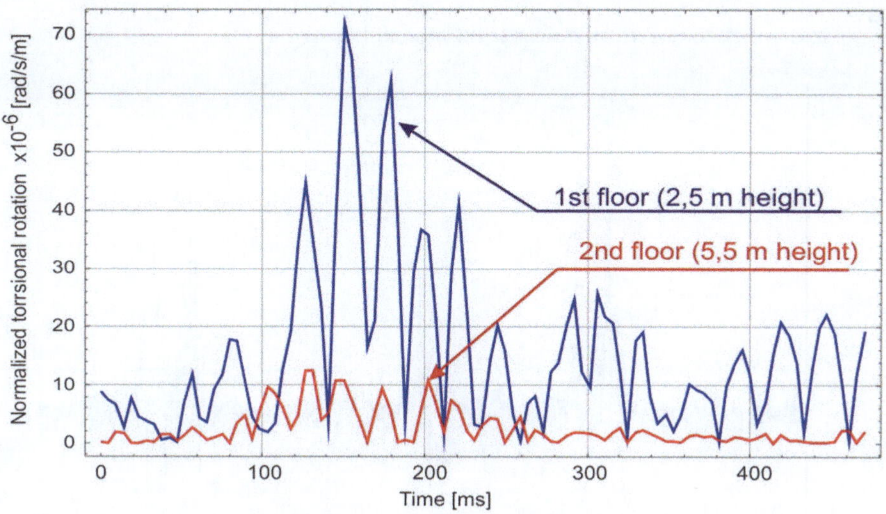

Fig. 23.8 The rotation vibration muffling properties of the examined building

0.05–100 Hz with resolution 10^{-6} to 10^{-1} rad/s/Hz$^{1/2}$. In the optical fibre technology the AFORS type device could be constructed with the total side equal to a flight hand luggage ($55 \times 40 \times 20$ cm size and up to 8 kg of weight). Such approach needs the sensor loop with diameter about 0.2 m as well as about 1000 m standard single mode fibre to achieve sensitivity in the level of $0.5 \cdot 10^{-7}$ rad/s/Hz$^{1/2}$. By using commercially available components, the final cost of such device may be below 5K Euro. We think that such system will be interesting mainly for its functionality with regard to the data management as well as for the remote control system.

The first results presented in this chapter show that the proposed approach can be useful for continuous monitoring of an engineering structure having some in-plane irregularity, for example multi-storey buildings with regard to the investigation of their torsional rotations as well as measuring interstory drifts. These measurements are made without any reference frame which is very important during earthquakes and may be made only by system based on the Sagnac effect. In comparison to the commercially available FOG instruments such as μFORS-1 (*Northrop-Gruman-LITEF GmbH*), the proposed system is designed for a direct measurement of a rotational rate, whereas any FOG measures change the angle which is written in their inner electronic system and difficult to direct changes. Additionally, our system prepared according to the AFORS technology has developed the software designed for the Internet system monitoring as well as the remote control which can manage a large number of such devices in a useful way for the operator.

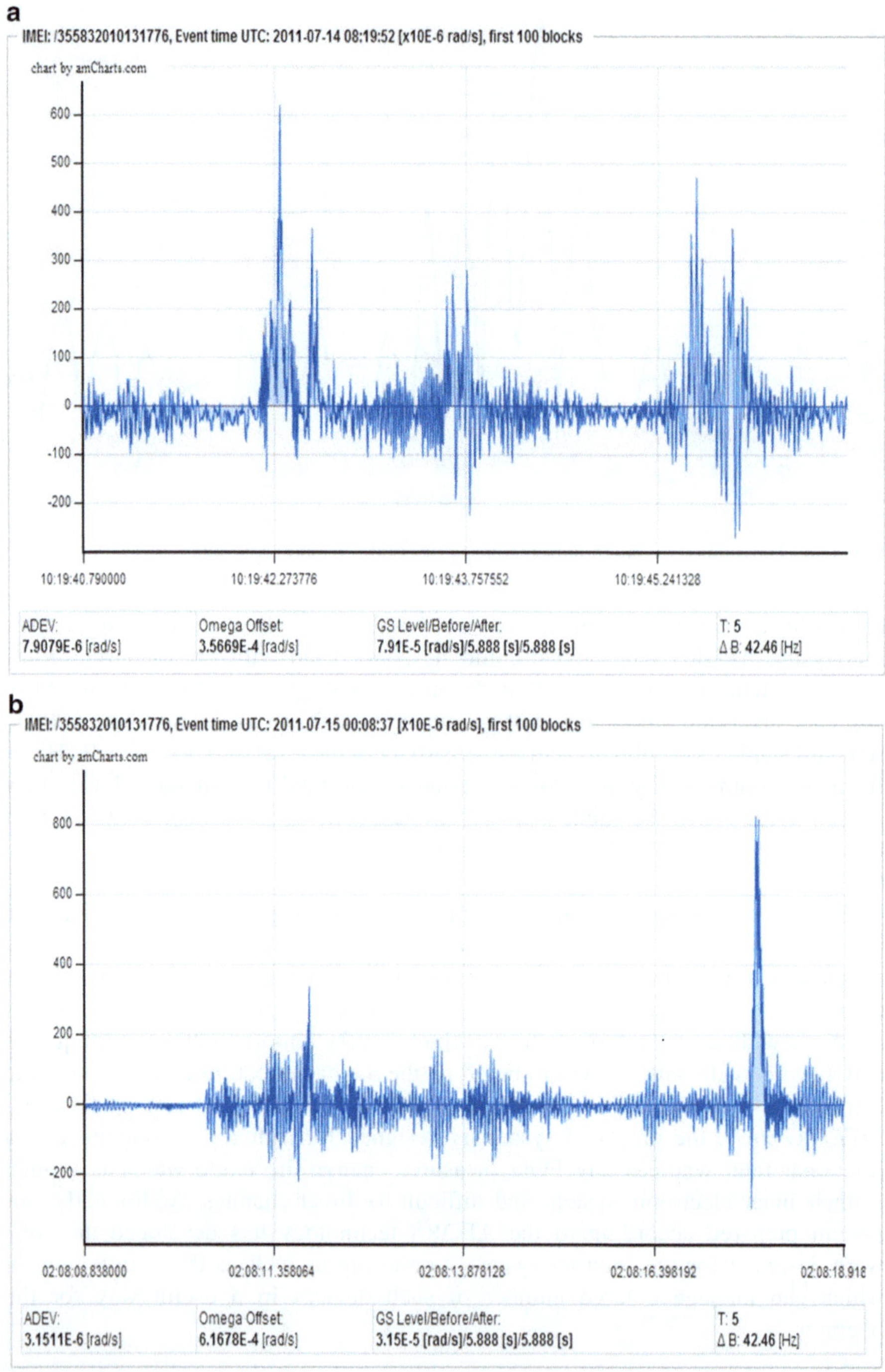

Fig. 23.9 The data recorded on the second (*top*) and the first (*down*) floor as a response for ground moves generated by morning intensity on the street in a distance about 50 m from and parallel to the long building wall

Acknowledgments This work was done in 2011 thanks to the financial support from the Polish Ministry of Science and Higher Education under Key Project POIG.01.03.01-14-016/08 "New photonic materials and their advanced application" and partially the Military University of Technology statutory activity No PBS-829.

References

Cowsik R, Madziwa-Nussinov T, Wagoner K, Wiens D, Wysession M (2009) Performance characteristics of a rotational seismometer for near-field and engineering applications. Bull Seismol Soc Am 99(2B):1181–1189

Dai X, Zhao X, Cai B, Yang G, Zhou K, Liu C (2002) Quantitative analysis of the Shupe reduction in a fiber-optic Sagnac interferometer. Opt Eng 41:1155–1156

Evans JR, Igel HI, Knopoff L, Lang TL, Trifunac MD (2007) Rotational seismology and engineering – online proceedings for the first international workshop. U.S. Geological Survey Open-File Report, 2007-1144, ver. 2.0, Appendix 2.5, pp 36–37

Franco-Anaya R, Carr AJ, Schreiber KU (2008) Qualification of fibre-optic gyroscope s for civil engineering applications. In: Proceedings of the New Zealand Society of Earthquake Engineering (NZSEE) conference, Available on CD-ROM, Wairakei, New Zealand

Habel WR, Baumann I, Berghmans F, Borzycki K, Chojetzki C, Haase K-H, Jaroszewicz LR, Kleckers T, Niklès M, Rothardt M, Schlüter V, Thévenaz L, Tur M, Wuilpart M (2009) Guidelines for the characterization and use of fibre optic sensors: basic definitions and a proposed standard for FBG-based strain sensors. Proc SPIE 7503:OF101–OF187

Jaroszewicz LR, Wiszniowski J (2008) Measurement of short-period weak rotation signals. In: Teisseyre R, Nagahama H, Majewski E (eds) Physics of asymmetric continuum: extreme and fracture processes. Spinger, Berlin, Chap. 2, pp 17–47

Jaroszewicz LR, Krajewski Z, Solarz L, Teisseyre R (2006) Application of the fibre-optic Sagnac interferometer in the investigation of seismic rotational waves. Meas Sci Technol 17(5): 1186–1193

Jaroszewicz LR, Krajewski Z, Kowalski H, Mazur G, Zinówko P, Kowalski JK (2011) AFORS autonomous fibre-optic rotational seismograph: design and application. Acta Geophys 59(3): 578–596

Jaroszewicz LR, Krajewski Z, Teisseyre KP (2012) Usefulness of AFORS – autonomous fibre-optic rotational seismograph for investigation of rotational phenomena. J Seismol 16(4): 573–586

Kao CG (1998) Design and shaking table tests for a four-storey miniature structure built with replaceable plastic hinges. Master's thesis, University of Canterbury

Krajewski Z (2005) Fiber optic Sagnac interferometer as device for rotational effect investigation connected with seismic events (in Polish). Doctoral thesis, Military University of Technology, Poland

Krajewski Z, Jaroszewicz LR, Solarz L (2005) Optimization of fiber-optic Sagnac interferometer for detection of rotational seismic events. Proc SPIE 5952:240–248

McGinnis (2004) Apparatus and method for detecting deflection of a tower. U.S. Patent application, 0107671 A1, 10 June 2004

Ostrzyżek A (1989) Accuracy analyze of the angle speed measurement in the fibre optic gyroscope (in Polish). Doctoral thesis, Military University of Technology, Poland

Post EJ (1967) Sagnac effect. Rev Mod Phys 39(2):475–493

Sagnac G (1913) L'ether lumineux demontre par l'effet du vent relatif d'ether dans un interferomfetre en rotation uniforme (in French). Comptes-rendus de l'Academie des Sciences 95:708–710

Schreiber KU, Velikoseltrsev A, Carr AJ, Franco-Anaya R (2009) The application fiber optic gyroscope for the measurement of rotations in structural engineering. Bull Seismol Soc Am 99(2B):1207–1214

Chapter 24
Seismic Monitoring of Linear and Rotational Oscillations of the Multistory Buildings in Moscow

Nataly Kapustian, Galina Antonovskaya, Vadim Agafonov, Kostantin Neumoin, and Maksim Safonov

Abstract The specially designed seismic monitoring system is deployed on high-rise buildings in Moscow. The system is capable of recording of the natural oscillations of the structures, with special attention to the angular motions, data processing to recover the natural motions from the micro seismic background noise and comparison of observed and theoretically predicted motion modes. Based on the comparison results the more accurate theoretical models and its software realizations for the building behavior can be developed. Both linear and rotational oscillations related to eigenmodes excited primarily by wind flow and oscillations of the building structure have been observed. The long-term permanent monitoring of the 44-store building in Moscow allowed to observe the creep of the concrete properties (the eigenmode frequencies changed by 20%) and seasonal variation (about 2% changes over a year).

N. Kapustian (✉)
Schmidt Institute of Physics of the Earth, Russian Academy of Science (RAS), IPE, 10 B. Gruzinskaya str., D-242, 123995 Moscow, Russia

Institute of Ecology Problems of the North UrB, Russian Academy of Science (RAS), 23 Naberezhnaya S. Dviny, 163000 Arkhangelsk, Russia
e-mail: nkapustian@gmail.com

G. Antonovskaya
Institute of Ecology Problems of the North UrB, Russian Academy of Science (RAS), 23 Naberezhnaya S. Dviny, 163000 Arkhangelsk, Russia
e-mail: galina.antonovskaya@gmail.com

V. Agafonov • K. Neumoin
Moscow Institute of Physics and Technology, State University, 9 Institutskii pereulok, 141700 Dolgoprudnuy, Moscow Region, Russia
e-mail: agvadim@yandex.ru

M. Safonov
LLC R-Sensors, 5A Zhukovskogo st, 141700 Dolgoprudnuy, Moscow Region, Russia
e-mail: r-sensors@mail.ru

O. Lavan and M. De Stefano (eds.), *Seismic Behaviour and Design of Irregular and Complex Civil Structures*, Geotechnical, Geological and Earthquake Engineering 24, DOI 10.1007/978-94-007-5377-8_24,

24.1 Introduction

The study of the eigenmodes of high-rise buildings has two main aspects: practical and scientific. The practical aspect consists of the possibility to estimate the condition of the building as a whole or to monitor the load-bearing constructions. The monitoring of the eigenmodes is technologically and cost-effective solution to control the changes of the principal load-bearing construction in real time. Indeed, the eigenmodes excited by wind or microseisms exist permanently, and using of high sensitive seismometers and modern digital data acquisition systems makes it possible to define the oscillations frequencies with practically any prescribed accuracy. On the other hand, the changes in load-bearing constructions properties or in "fixing" of the building on the ground manifest themselves in the changes of natural frequencies. Thus, we have a very sensitive instrument of monitoring. Moreover, only few observation points with the sensors installed are required, and in some cases it is possible to install the sensors on the ground in the vicinity of the building under study.

The scientific aspect consists of the fact that using the monitoring of the eigenmodes of a building it is possible to observe the processes in constructions in real scale – not on the models or construction fragments. As it's shown below, one succeeds in detecting the concrete creep effects, temperature changes, and to prove the correctness of theoretical conceptions on the constructions operation by means of comparison of the parameters obtained by calculating models and real measurements. Another important scientific aspect concerns the fact that the high-rise buildings are parts of artificial relief and are able to contribute some variations into the complex atmosphere-lithosphere interaction system. From some manifestations of this it is possible to obtain a practical result – using of the eigenmodes of high-rise buildings as instruments of the planet study; we'll illustrate that fact below using experimental examples.

24.2 The Instruments and Methods of Observations

In Russia the study of high-rise buildings oscillations induced by the effects of the wind pulsations began in 1950s by works of I.L. Korchinsky on several objects in Moscow (Korchinsky 1953), particularly on the main building of the Moscow State University (Fig. 24.1). The central 33-storey part of the building is detached from it's ells by movement joints and has steel load-bearing constructions, while the base of the building is the frozen ground. The feature of the building maintenance is that it has not been reconstructed for more than 50 years and the area in radius of 1 km has not been built on, that is the ground conditions have not being changed. From this point of view the building of the Moscow State University is unique and gives the possibility of the most pure observation of the effects of aging of the constructions. We reproduced the observations of I.L. Korchinsky 50 years later, exactly in the

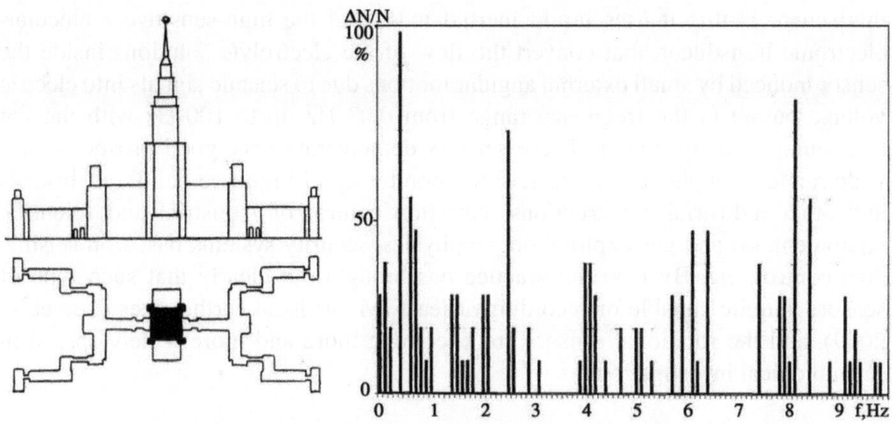

Fig. 24.1 Seismic diagnostic results for main building of Moscow State University: facade and plan (*to the left*) and peak reproducibility in spectra at observation points (*to the right*)

same points of the building; differences were only in the types of used seismometers and the way of data acquisition. The sensor created by Korchinsky did not allow to detect signals below 0.2 Hz, and the oscillographic detection technique defined the 0.05 Hz oscillation frequency measurement precision. Nowadays we have no such restrictions.

The registration of the building eigenmodes can be performed with both seismometers and deformometer, and seismometers are preferable at frequencies above 0.1 Hz. In seismometrical registration velocimeters or accelerometers are used, giving correspondingly velocity or acceleration of a point of the building. It is significant that the microseismic signals are being observed in a wide frequency range, and the following processing enhances the informative signals from this range. For that reason it is preferable to use velocimeters at lower storey of the high-rise building – here the long period motions is strong and observation technique is simpler.

Taking into account the fact that the eigenmodes of the high-rise buildings have distinct rotational component, it is rational to carry out the observations using both linear and rotational sensors. Surely, the rotational component declare itself in records of the linear sensors signals too, but to extract the rotational component in this case one should realize the special sensors disposal and its' exact synchronization that sometimes is not quite possible inside the building. In our experiments the 3-component (X, Y, Z) linear sensors were used: velocimeters models SM-3 (Russian standard sensor) and CMG-3TD, accelerometers model CMG-5T (Guralp ®Co.).

The direct registration of the rotational modes were carried out with the special sensors, installed side by side to the linear sensors and providing direct angular velocity data in XY, XZ and YZ planes. The rotational sensor model METR-03 produced by R-sensors, LLC® were used in the measurements. The features of this sensor is that it uses liquid inertial mass and the high-sensitive molecular-electronic transducer, that convert the flow of the electrolyte solutions inside the sensor induced by small external angular motions due to seismic signals into electric voltage output in the frequency range from 0.03 Hz up to 100 Hz with the flat to angular velocity output. These sensors demonstrate very good prospects in a wide range of applications like seismic monitoring of high-rise buildings, bridges and other industrial constructions, vibration control of industrial and scientific equipment, oil and gas exploration geophysics, security systems based on seismic area control, etc. By now the practice has brought out clearly that such type of sensors is quite capable of recording at least (M∼4) local earthquakes (Lee et al. 2009), and the rotational sensors are becoming more and more widely spread in seismological investigations.

24.3 Data Processing

Taking into account that the values of the eigenmode frequencies are unknown *a priori*, the oscillations in each point were recorded in the broad band – from DC up to 30 Hz. Than, the standard procedures of the spectrum analysis with the power spectral density calculation (including fitting of the analysis window, smoothing and averaging) and following selection of the stably occurring in signal narrow-band spectrum peaks were applied. These peaks were considered as possible manifestation of the eigenmodes. Than, the induced oscillations and interferences were thrown off taking into account the following criteria (Kapustyan and Rogozhin 2007). The eigenmodes must meet the stated below requirement:

- presence in practically all points of the building (except the standing-wave nodes),
- amplitude domination at horizontal components; in case of elongate in plan building the frequencies of oscillation differ along different axes of the plan,
- the amplitude of the horizontal component of the first mode grows on the height of the observation point,
- as a rule, the value of the eigenmode frequency is not a multiple of 50 Hz, otherwise that is the vibration of the electrical engine,
- at simultaneous observation in two points of a building, the oscillations are to be coherent,
- at simultaneous observation of seismic oscillations and atmosphere pressure pulsation, the eigenmode amplitude spikes and the wind pulsations are to be correlated.

In case the sensors of the observation system were deployed in a large number of points, it is possible to use the algorithm of extraction of standing waves which the eigenmodes consist of, that is based on the estimation of the coherentness of the signal records in the points (Emanov et al. 1998).

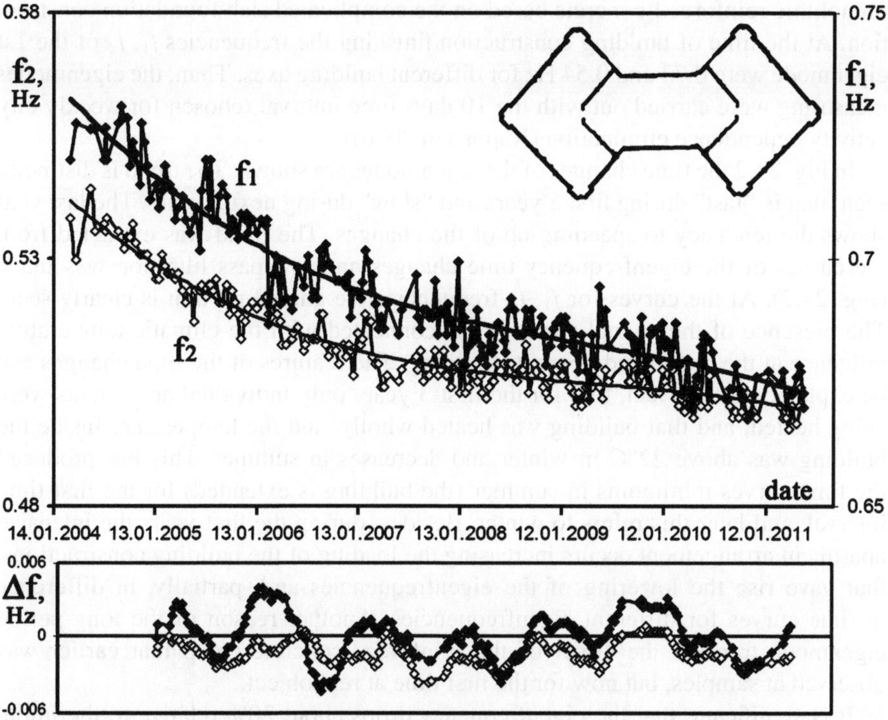

Fig. 24.2 The time changes of eigenmode frequencies of a high-rise building: floor plan and time evolution with trend are shown (*upper part*) and results of low-pass filtration (*lower part*)

24.3.1 The Main Types and Schemes of the Observations

24.3.1.1 The Space-Time Registration Systems. Monitoring of the Building Condition

Let's begin from the longest time period (50 years) for the Moscow State University main building (Fig. 24.1). The combination of the observation points at the building was passed one after another, and for each point the peaks of the spectrum were marked out. The diagram on the Fig. 24.1 shows the presence of the peaks (peaks repeatability) in the combination of the observation points for each frequency. From the diagram and presence of oscillations at certain points of the observation one can conclude that up to about 3 Hz peaks are characterizing the oscillations of the central part of the building cut by movement joints, range from 3 to 7 Hz corresponds to movement of the high-rise part of the building, and at upper frequencies – the oscillations of the spire and towers.

Let's take another example – monitoring of the 44-storey apartment building "Edelweiss" in Moscow that was being carried out from very finishing of the constriction for about 8 years. The building is compact-size in plan (Fig. 24.2), monolithic reinforced-concrete based on the complicated slab foundation construction. At the time of building construction finishing the frequencies f_1, f_2 of the 1st eigenmode were 0.73 and 0.54 Hz for different building axes. Than, the eigenmodes measuring were carried out with the 10 days time interval (chosen for weekly city activity dependence elimination (Kapustian 2000)).

In Fig. 24.2 the time changes of the eigenmodes are shown. The trend is distinctly seen, that is "fast" during first 3 years and "slow" during next 3 years. The last year shows the tendency to speeding-up of the changes. The trend was extracted from the curves of the eigenfrequency time changes and low-pass filtration was made (Fig. 24.2). At the curves for f_1, f_2 frequencies the annual rhythm is clearly seen. The presence of the annual variations is connected with the climatic-temperature influence at the reinforced-concrete building. The features of the time changes can be explained by the fact, that for the first 3 years only individual apartments were being heated, and that building was heated wholly, and the temperature inside the building was above 22°C in winter and decreases in summer. This has produced the time curves minimums in summer (the building is extended) for the first time interval, and later this refers to winter. Besides, during the first years the intensive apartment arrangement occurs increasing the loading of the building constructions, that gave rise the lowering of the eigenfrequencies and, partially, in difference in time curves for different eigenfrequencies. Another reason of the long-period eigenmode trend is the effect of the reinforced-concrete creep that earlier was observed at samples, but now for the first time at real object.

It is significant, that the eigenfrequency drops about 20% relative to the initial value (or, about 2% per year), and the annual variations are about 1%. Those are "natural" changes, and they are to be tacking into account when carrying out the monitoring of the building condition.

The interaction between different structures via ground is a well-known phenomenon for seismic interactions (Wong and Trifunac 1975). In a city environment the high-rise building are often situated at river banks and in neighborhood of bridges. Taking into account the close values of eigenfrequencies for such dissimilar constructions along with the simultaneous influence of the wind pulsations and big oscillations amplitudes of the bridges, we carried out some seismometrical observations on location.

The first example is the interaction of the 30-storey building "Tower-2000" and "Bagration" bridge in Moscow-City complex. The constructions are closely adjacent but constructively untied. Their eigenmodes are different, for the 1-st mode the most intensive oscillations are detected at 0.58 and 0.98 Hz for Tower and bridge, correspondingly. Nevertheless, the Tower has the rather strong oscillations at 0.98 Hz, and on the top floor the acceleration at this frequency exceeds the acceleration at eigenfrequency.

Another example refers to the construction interactions evaluation on the projecting stage of the high-rise building to be located in the vicinity of a bridge. The seismic observations made for the profile from the bridge to the site of the future building allowed to define signals going into the ground at eigenmode bridge oscillations and to select the most intensive modes and coincidence with the calcu-

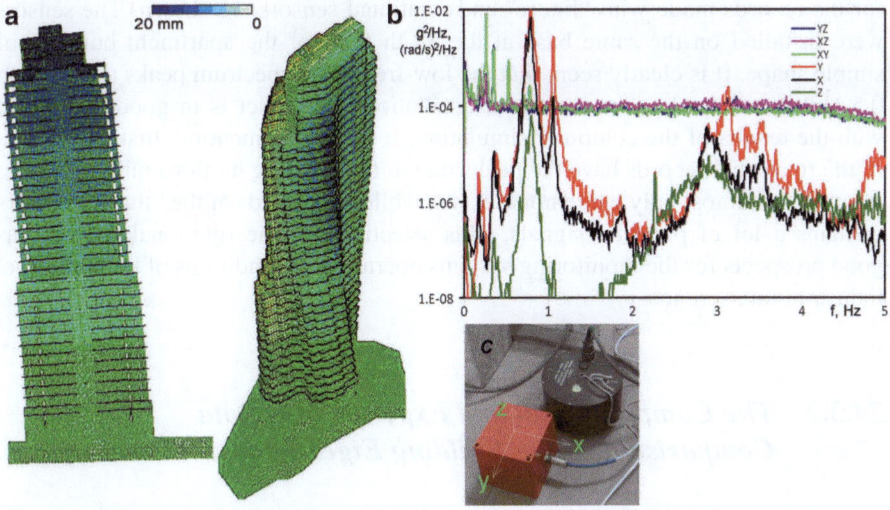

Fig. 24.3 The study of high-rise building oscillations: (**a**) calculations for linear (*left*) and rotational (*right*) vibrations, (**b**) power spectra for both types of sensors. Photo shows the accelerometer Guralp CMG-5T and the angular velocimeter METR-03 (*red*) installed at the 38-th floor of the building

lated eigenfrequencies. The real accelerogram obtained from the measurements was placed in the computational model, and the values of the displacementsat dynamic impacts were received. The calculations showed that for the building located 200 m from the bridge, the induced oscillations amplitudes (even in the case of resonance coincidence) are 1000 times less then induced by the wind influence. This example shows the possibility of the seismometry on a projecting stage.

24.3.1.2 Observations of the Rotational Oscillations of the High-Rise Buildings

The analysis of the dynamics of the high-rise buildings calculation models shows that the eigenmodes set besides the "linear" oscillations has quite intensive rotational components. Figure 24.3 shows the pictures of the rotational movements based on the calculation model for the high-rise apartment building; the maximum displacement amplitudes due to the rotational oscillations 20–100 times less than due to linear ones. Thus at seismic measurements on the buildings, part of the spectrum peaks potentially reflects the rotational motions.

For clarifying of the matter the series of the experiments with simultaneous registration of the seismic signals at the building using both linear and rotational sensors were performed. The "linear" oscillations were recorded using the accelerometer and velocimeters by Guralp®, and the rotational motions – by METR-03 sensors developed and produced by "R-sensors" LLC®. In Fig. 24.3 the power spectra for the records made with "linear" and rotational sensors are shown. The sensors were installed on the same base at the 38-th floor of the apartment building of simple shape. It is clearly seen, that the low-frequency spectrum peaks (lower than 0.5 Hz) correspond to the rotational oscillations. This fact is in good agreement with the results of the computer simulation. It should be mention, that the spectra of the rotational records have the peaks due to the building motions only – at other frequencies almost only white noise exists, while the records of the "linear" sensors contains a lot of parasitic signals. This selectivity of the rotational sensors has good prospects for the monitoring systems operating in conditions of the high level industrial interference.

24.3.2 The Computational and Experimental Data Comparison for the Building Eigenmodes

The calculations and measurements of the eigenmodes were carried during the construction when the height of the building was 8, 15, 22, 30, 37, 45 and 48 stories, and the obtained values of the eigenfrequencies were compared with each other. The following facts have been found:

- at relatively low number of stories (less 22) a lot of stably existing peaks is observed; the values of the observed characteristic frequencies coincide with the calculated frequencies rarely. This can be explained by the complicated spatial composition of the building, especially at the lower part of the building. Besides, the backfilling of a part of the basement construction was not made by that time, that gave the more complicated fixing of the building in the ground then in the computational model; this played a important role in formation of the eigenmodes,
- the higher the building is, the "simpler" integrally the building become in it's shape; the number of spectrum peaks decreases and it's values get to good agreement with the calculations. Nevertheless, there are some peak groups (for example, 0.6–0.7 Hz – the most intensive in the spectrum) that does not follow the calculation results (Fig. 24.4). This proves the need for the building model improvement.

24.3.3 The Computational and Experimental Data Comparison for the Building Eigenmodes

As it was shown above, at top stories of the buildings the eigenmodes amplitudes induced by wind pulsations are significantly greater than ones induced by microseisms via ground (including induction from the adjacent constructions). The

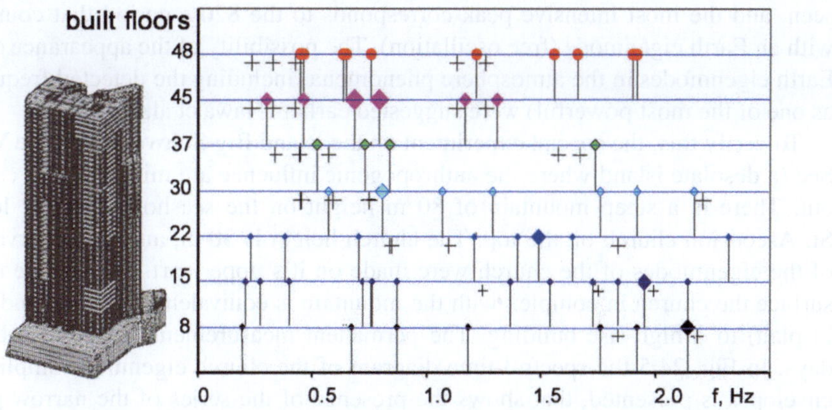

Fig. 24.4 Eigenmode frequency change during construction of "Continental" high-rise building: crosses – calculated values, other correspond to experimental detection of persistent peaks in a microseisms spectra of the building

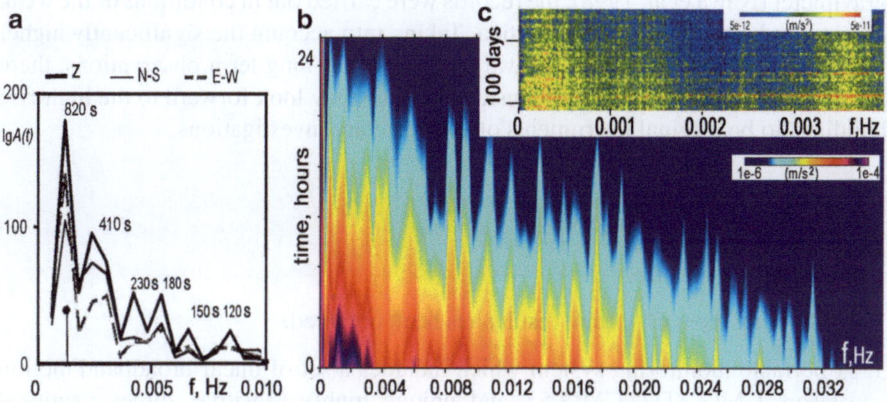

Fig. 24.5 Earth's characteristic oscillations observed as: a time variation of an envelope of high-rise buildings eigenmode oscillations (envelope spectra (**a**) and time-spectral diagrams (**b**)) and a direct measurement by supergravimeter (Nawa et al. 1998) (**c**)

simultaneous measurements of the pressure pulsations with microbarograph and mechanical oscillations clearly show the connection of these signals. The pressure variations are characterized by pulsations of different amplitude and time behavior. These signals modulate the amplitudes of the building eigenmodes. The long-term observations (for hours and days) of the eigenmodes amplitudes variations allow to reveal the modulating signal and to estimate it's properties.

The Fig. 24.5 contains the result of the many-hours recording of the high-rise building "Tower-2000" eigenmodes – the Fourier spectra of the envelopes for X-, Y- and Z-components of the seismic signal record. The series of peaks is clearly seen, and the most intensive peak corresponds to the 820 s period that coincides with an Earth eigenmode (free oscillation). The possibility of the appearance of the Earth eigenmodes in the atmosphere phenomena (including the detected frequency as one of the most powerful) were suggested earlier (Nawa et al. 1998).

To verify this, the special experiment on the island Big Solovetskiy in the White See (a desolate island where the anthropogenic influence are minimal) was carried out. There is a steep mountain of 80 m height on the seashore with the lonely St. Ascension church on the top. The church height is 30 m, and the observations of the eigenmodes of the church were made on it's upper part. So, relative to the surface the church in complex with the mountain is equivalent (in height and even in plan) to a high-rise building. The permanent measurements lasted for several days. In Fig. 24.5 the spectral-time diagram of the church eigenmode amplitudes envelopes is presented, that shows the presence of the series of the narrow peaks at the longperiod part of the spectrum, particularly the 820-s peak. Besides, in Fig. 24.5 for the comparison with the obtained results, the diagram demonstrates the presence of the free oscillations of the Earth during "quiet" (without powerful earthquakes) time intervals. The result is obtained using the unique superconducting gravimeter (Nawa et al. 1998); the records were carried out in conditions of the weak man-caused interference in Antarctica. Taking into account the significantly higher frequency resolution obtained (Nawa et al. 1998) at long-term observations, there is a satisfactory agreement of the results. So, one may look forward to the high-rise buildings to be original instruments of the planetary investigations.

24.4 Conclusions

The following most important results has been obtained.

1. A portable monitoring system which includes a set of linear broadband motion sensors CMG-3TD, CMG-5T and unique highly sensitive, direct rotational readout, molecular electronic sensors METR-03 has been designed and tested.
2. Both linear and rotational oscillations related to eigenmodes excited primarily by wind flow oscillations of the building structure have been observed.
3. Different allocation schemes of the sensors have been tested. It was found that most useful data can be obtained in case the sensors are located at some critical points which can be found a priory based on the numerical simulation of the structure response. The convenient and economical schemes for different types of buildings has been developed and tested.
4. The long-term (about 8 years) permanent monitoring during the course of construction of the 44-storey building in Moscow allowed to observe the long-term creep of the concrete properties (the natural frequencies were changed by 20% during the observation period) and seasonal variation of the elastic parameters of the building structure (about 2% changes in an year).

5. Based on the research results the following conclusions can be done:
6. The modeled and really observed parameters, especially for the lower eigenmodes, are very consistent for a simple or very high (>75 m) buildings.
7. The well pronounced rotational oscillations of the building structures can be detected using either differential technique and traditional highly sensitive 3-component seismic sensors or special direct rotational-readout seismic sensors.

References

Emanov AF, Seleznev VS, Kuzmenko AP et al (1998) Detailed engineering-seismological researches of buildings and structures. Methods of study structures and monitoring of the lithosphere. Novosibirsk, pp 142–153

Kapustian NK (2000) The price of progress. Sci Russ (Vol. 2):15–23

Kapustyan NK, Rogozhin EA (2007) Soil properties and building dynamic studies by microseismic methods. In: The international conference on modern trends in structural engineering and seismic design, University Center of Samaria, Ariel, 8–11 Oct 2007

Korchinsky IL (1953) High-rise building oscillations (in Russian). CNIISP 11:44, Moscow

Lee WHK, Igel H, Trifunac MD (2009) Recent advances in rotational seismology. Seismol Res Lett 80(3):479–490

Nawa K, Suda N, FukaoSato YT, Aoyama Y, Shibuya K (1998) Incessant excitation of the earth's free oscillations. Earth Planets Space 50:3–8

Wong HL, Trifunac MD (1975) Two-dimensional, antiplane, building-soil-building interaction for two or more buildings and for incident SH waves. Bull Seismol Soc Am 65(6):1863–1885

Index

A
AFORS. *See* Autonomous fibre-optic rotational seismograph (AFORS)
Ancient boulder structures, external impacts
eigenmode oscillations, 13
long-term impact, 12–13
methodology and dams description, 4–5
non-catastrophic earthquakes, 4
pulses
dam oscillation damping, 8
high-frequency microseisms, 6–8
lake dam experiment, 7–9
micro cracks, 7
nonlinear impact transformation, 9
seismic radiation anisotropy, 8
shearing deformation, 7
shot impact experiment, 6–7
smooth loads
external load processing, 12
Guttenberg-Richter law, 10
microseismic field, 11
seismic noise, 10
vibrations, 11
Solovetsky monastery, 4
strained rotations mechanism, 14
Angular motions, 356
"As-designed" buildings
ductility, 258
time history analyses, 258
torsionally flexible building, 260–262
torsionally stiff building, 259–260
translational period, 259
Asymmetric buildings
biaxial excitation, 190
displacement modification, 195
dynamic characteristics, 196–197
earthquake ground motions, 195
energy-based design method, 306
equivalent single-degree-of-freedom (E-SDOF), 190
horizontal seismic components, 193
inelastic deformations, 196
inelastic seismic analysis, 191
linear elastic response, 192
modal superposition, 198
multidirectional seismic effects, 194, 200
nonlinear response history analysis (NL-RHA), 198
plastic hinges, 196
Rayleigh damping, 193
roof displacement, 194
seismic components, 197
seismic response, 199
spatial structures, 201
static pushover analysis (SPA), 189
structural dynamics, 192
superposition principle, 191
uniaxial excitation, 199
vibration mode, 193
ATC40, 220
Autonomous fibre-optic rotational seismograph (AFORS)
automatic data storing, 346
autonomous signal processing unit (ASPU), 340
bookmarks, 344–346
continuous monitoring, 349
electronic parameters, 344
floor extreme amplitudes, 347, 349
in-plane irregularity, 349

O. Lavan and M. De Stefano (eds.), *Seismic Behaviour and Design of Irregular and Complex Civil Structures*, Geotechnical, Geological and Earthquake Engineering 24, DOI 10.1007/978-94-007-5377-8,

Autonomous fibre-optic rotational seismograph (AFORS) (*cont.*)
interstory drifts, 349
rotational components, 341
rotation rate amplitude, 347–348
specifications
FORS-Telemetric Server, 343
optical constant, 342
optical gyro configuration, 341
permalloy particles, 343
resolution *vs.* total optical length, 342
Sagnac phase-shift, 343
strong-motion seismology, 347
torsional effects, 340
urban ground motion, 346
urban noise influence, 347, 350

B
Bidirectional seismic effects, 200
Blocked media response, external impacts
long-term impact, 12–13
pulses, 6–9
smooth loads, 10–12
Body waves, 25
Bridge regularity
AASHTO LRFD code, 105
EC8 code, 105
elastic regularity index, 105–106
inelastic regularity index
elastic range, 106
lateral deformed configurations, 108
nonlinear static method, 107
normalized lateral displacement, 108
pushover analysis, 107

C
Capacity spectrum method (CSM), 163, 220, 222, 227, 229–231
Code provisions
concrete mechanical properties
coefficient of variation, 61
confidence factor, 61
design values, 62
intrinsic variability, 62
irregular industrial steel buildings, 73–79, 84
Concrete mechanical properties
analytical model, 64
code provisions
coefficient of variation, 61
confidence factor, 61
design values, 62
intrinsic variability, 62
concrete strength distribution, 63–64
elastic spectrum, 71
geometrical and mechanical characteristics, 62–63
inter-story drift, 71
quality standards, 72
reinforced concrete buildings, 60
seismic analysis
EC8 provisions *vs.* response domain, 67–69
FEMA provisions *vs.* response domain, 68–71
response domain, 65–67
spectral analysis, 64
seismic code, 60
seismic input, 64–65
seismic response, 60
Continuous monitoring, 349
Continuous reinforced concrete (RC) bridges. *See* Reinforced concrete (RC) bridges
Corrective eccentricities
bidirectional ground motions, 174
deck displacements, 177
elastic parameters, 183
elastic response
rigidity eccentricity, 177–178
torsionally balanced system, 178–179
inelastic response
strength eccentricity, 179–180
torsionally balanced system, 180
L-shaped building, 181
maximum dynamic displacements, 174
plan layout and member cross sections, 181–182
plastic base shear forces, 184
plastic range, 183
pushover and nonlinear dynamic analyses, 181
reinforced concrete frames, 176
seismic elastic and inelastic response, 175
seismic forces, 183–184
seismic response
force-displacement curve, 186
height-wise distribution, 185
in-plan distribution, 185
nonlinear dynamic analysis, 184
story drifts, 186
structural parameters, 175
CSM-FEMA440, 220, 222, 227, 229–231

D
Damper optimal placement, 304

Direct displacement-based design method (DDBD)
absolute and normalised frame top displacement, 279–280
beam reinforcement, 276–277
column reinforcement, 276–277
demand/capacity ratio, 279, 281
ductility, 269–270
earthquakes, 281
elastic analysis, 274
elastic stiffness, 270
force-based design method (FBD), 269
in-plan 3D multistory structure, 275–276
inter-story drifts, 279–280
nonlinear analyses
CANNY99, 277
cracking rotation, 278
elastic response spectrum, 279
hysteresis model, 278
mass displacements, 279
stress-strain diagram, 277
torsionally flexible system, 280
torsional response
asymmetric wall building, 271, 273
asymmetry in-plan structures, 270
code drift, 273
drift-controlled displacement, 273
plan eccentricities, 271–272
shear strength, 271
strength eccentricity, 271
torsional twist, 281

E
Earthquake excitation
base story columns, 245
deformed shape, 246–247
displacement response, 248
ground motion, 246
weakened columns, 247
Earthquake-resistant design
"as-designed" buildings
ductility, 258
time history analyses, 258
torsionally flexible building, 260–262
torsionally stiff building, 259–260
translational period, 259
ductility demand distribution, 267
earthquake-induced torsion, 254
Eurocode EC8, 266
Eurocode EC3, 256
inelastic range, 253
L-shaped torsionally flexible steel building, 254–255
L-shaped torsionally stiff steel building, 254–255
modification procedure
structural members, 261
torsionally flexible building, 263–267
torsionally stiff building, 262–265
modulus of elasticity, 257
natural eccentricity, 257
nonlinear dynamic analysis, 257–258
physical eccentricity, 266
semi-artificial motions, 256
three-story eccentric building, 254
torsional motions, 265
Energy-based design method
asymmetric building, 306
damper design, 301–302
earthquake, 297
energy-dissipation devices, 297
floor masses, 302–303
ground displacements and velocities, 300
hysteresis loops, 305–306
independent plane frames, 298
inter-story stiffness and resisting shear force, 303–304
linearizing coefficients, 300
mean dissipated hysteretic energy, 304
nonlinear differential equation, 299
probabilistic seismic response, 306
roof level, 305
shear forces, 299
spectral densities, 303
stiffness and energy dissipation capacity, 302
stochastic equivalent linearization method, 298
6-story steel building, 302
supplemental damping, 298
uncontrolled and controlled structures, 304–305
unit-impulse response functions, 301
Equivalent single-degree-of-freedom (E-SDOF), 190

F
FBD. *See* Force-based design method (FBD)
FEMA440, 220, 222, 227, 229–231
Force-based design method (FBD), 269, 270, 274, 276, 277, 279–281
Friction dampers
analysis/redesign algorithm, 289–290
brace stiffness, 292
damage indices (DIs), 286
design methodology, 290

Friction dampers (*cont.*)
design variables, 291, 293
earthquake excitation, 294
eight-story setback structure, 291
equations of motion, 287–288
formal optimization problem, 288–289
fully stressed design (FSD) characteristics, 289
ground motions, 294
inter-story drifts, 292
inter-story performance measure, 294
life safety, 285
performance measures, 288
peripheral envelope drifts, 293
plane symmetric frame, 287
seismic response, 286
structural engineering, 285
Fully stressed design (FSD)
bare structure frequency, 329
conjecture illustration, 328–329
floors, 328
friction dampers, 289
spectral densities, 328
trusses, 327

G
Geometric nonlinearity, 236, 238, 240, 248
Ground motions. *See* Seismic ground rotations
Ground rotations. *See* Seismic ground rotations
Guttenberg-Richter law, 10

H
High-rise buildings. *See* Multi-story buildings

I
Incremental dynamic analyses (IDA), 51, 113
Infilled and pilotis RC structures
beams and columns, 92
brittle failure, 89
collision shear walls, 88
ductility capacity medium (DCM), 93
flexural response, 100
infill panels, 90–92
inter-story structural pounding idealization, 89–90
non-linear dynamic analyses, 89
response of column
capacity demand, 94
flexural and shear capacity, 94
masonry infills, 96
seismic analysis, 96
shear demands, 95
shear forces, 96–97
shear strength, 95
time history ductility requirements, 94–95
response of infills and drifts
deformations, 96, 98
infilled frames, 96
interaction effect, 100
maximum inter-story drifts, 98–99
seismic performance, 98
special purpose joint element-model, 89
8-story frame structure, 92
structural damage, 88
structural system, 93
Infill panels
capacity curves and seismic response
bare frame, 126–127
inter-story drift, 128–129
inverse-triangular horizontal pattern, 126–127
mass-proportional horizontal pattern, 126
shear demands, constant horizontal pattern, 129
shear demands, inverse-triangular horizontal pattern, 129–130
shear limit state, 130
top displacement plan distribution, 127–128
eccentricity in structural system, 125–126
FeB38k steel, 121
geometrical and mechanical properties, 121–122
hospital complex, 120
limit states
inter-story drifts, 123–124
shear capacity, 124–125
mass and stiffness centers, 121
RC-framed buildings, 120
retrofit strategy, 131
rigidity, 132
seismic analysis, 122
seismic excitation, 122–123
spectral capacity *vs.* limit state, 131–132
torsional effects, 132
In-plane eccentric system, 138–139
In-plane irregularity, 349
Inter-story pounding. *See* Infilled and pilotis RC structures
Irregular industrial steel buildings
axial forces, braced frames, 80–81
column and brace forces, 82–83
3D dynamic time-history analysis, 79–80

elastic base shear, 81
elastic three-dimensional time history analyses, 74
equivalent static force method and response spectrum method
base shear force, 79
damping, 79
deformations, 77
dynamic analysis, 77
modal properties, 78
Rayleigh method, 78
equivalent static load method, 82
in-plane torsional effects, 82
NRCC/NBCC 2005, 73
Rayleigh method modal analysis, 84
seismic action
ductility system, 75
horizontal inertia forces, 76
large-capacity silos, 75
penthouse, 74–75
rigid diaphragm properties, 76
three-dimensional view and layout of columns, 74–75
seismic base shear, 80–81
seismic design, 73
seismic force-resisting system, 74
seismic response, 84
selection and scaling of ground motions, 76–77
story displacements, 80–81
Irregular tall structures' collapse analysis
deformations, 236
discrete Lagrangian formalism, 238–239
earthquake excitation
base story columns, 245
deformed shape, 246–247
displacement response, 248
ground motion, 246
weakened columns, 247
earthquake intensity, 249
geometric and material nonlinearity, 236
gravity loads, 239
inelastic deteriorations, 248
mixed Lagrangian formalism, 237–238
out-of-plane and torsional motion, 237
symmetric collapse mechanism
deformed shape of building, bird view, 241–242
deformed shape of building, plan view, 241, 243
displacement response, 241, 244
nonlinear dynamic analysis, 240
weakened base columns, 240–241
torsional collapse mechanism
base story weakened columns, 241, 244
column failure, 244
deformed shape, plan view, 242, 244–245
nonlinear static analysis, 241
torsional-lateral collapse, 249

J

Jin Mao Tower
rocking angles, 34–36
torsion damage, 36–37
translational displacements, 34–35

L

Linear dynamic response spectrum analysis
ATC40, 220
correction factor, 230
3D irregular structures, 219
dynamic properties, 226–227
eight-story building configuration, 222, 224
elasticity, 231
Eurocode 8, 230
extended CSM-FEMA440, 227
extended N2 method, 220
five-story building configuration, 222–223
ground motion intensity, 231
lateral transverse reinforcement, 224
normalised top displacements, 227
planar frames and bridges, 219
seismic assessment
seismic action, 225
structural analyses, 225–226
seismic intensity, 228
SeismoStruct, 223
three-story building configuration, 222–223
torsional correction factors
elastic dynamic analysis, 222
inelastic torsional response, 220
normalised roof displacement, 221
plastic deformations, 221
torsional effects, 228

M

Material nonlinearity, 236, 248
Mixed Lagrangian formulation (MLF). *See* Irregular tall structures' collapse analysis
Modal irregularity, 111
Modal pushover analysis (MPA)
asymmetric buildings (*see* Asymmetric buildings)

Modal pushover analysis (MPA) (*cont.*)
plan-asymmetric buildings (*see* Linear dynamic response spectrum analysis)
unsymmetric-plan buildings (*See* Unsymmetric-plan buildings)
Modal regularity, 108–109
Modal spectral analyses, 109
Multidirectional seismic effects, 194
Multiple tuned mass dampers (MTMDs)
allocation and sizing problem, 337
analysis/redesign algorithm, 329–330
equations of motion, 325–326
frame numbering, 334, 336
fully stressed design
bare structure frequency, 329
conjecture illustration, 328–329
floors, 328
spectral densities, 328
trusses, 327
ground motion, 334
multi-hazard mitigation, 324
natural frequencies, 324
performance-based design, 323
performance measures, 326
peripheral accelerations, 335
problem formulation, 326–327
proposed solution scheme
bare frame mass matrix, 332
Den-Hartog properties, 331
frequency-based spectrum, 330
modal damping, 332
power spectral density (PSD), 330
root mean square (RMS), 331
stiffness, 333
tuning frequencies, 333
retrofitted structure, 336–337
seismic excitations, 325
sequential search technique, 325
stiffness matrices, 335
8-story setback RC frame structure, 334
structural irregularities, 336
structural response, 323
Multi-story buildings
analyzed structure, 164–165
bidirectional loading, 170–171
continuous monitoring (*see* Autonomous fibre-optic rotational seismograph (AFORS))
corrective eccentricities (*see* Corrective eccentricities)
displacements, 169–170
earthquake-resistant design (*see* Earthquake-resistant design)
elastic hardening, 169
seismic input, 168–169
seismic monitoring (*see* Seismic monitoring, linear and rotational oscillations)
structural walls, 170
Takeda hysteretic behavior, 169
torsional sensitivity, 165–167
transformation to single-story structure, 166–168
unidirectional eccentricity, 171

N

National Building Code of Canada (NRCC/NBCC 2005), 73, 76
N2 method, 220, 222, 225, 227, 228, 230–231
Nonlinear analysis
"as-designed" buildings
ductility, 258
time history analyses, 258
torsionally flexible building, 260–262
torsionally stiff building, 259–260
translational period, 259
direct displacement-based design method (DDBD)
CANNY99, 277
cracking rotation, 278
elastic response spectrum, 279
hysteresis model, 278
mass displacements, 279
stress-strain diagram, 277
earthquake-resistant design, 257–258
Takeda hysteretic behavior, 169
Nonlinear response history analysis (NL-RHA), 198
Nonlinear static method, 107

O

One-story eccentric systems
Alpha method, 151
asymmetric systems, 137
base-isolated structures, 138
corner displacement amplifications, 138
eigenproblem, 141–142
equations of motion
differential equations, 140
lateral-torsional response, 139–140
natural frequency, 141
physical property, 141
stiffness, 140
free vibrations, 152
in-plane eccentric system, 138–139

lateral-torsional response, 137
maximum displacement and rotational response
damped seismic response, 151
undamped free vibration response, 147–150
maximum longitudinal centre mass displacement response
angular coefficient, 144
high-eccentric torsionally flexible systems, 145–146
low-eccentric torsionally flexible systems, 145
seismic excitation, 144
torsionally stiff systems, 144
maximum rotational response
damped seismic response, 146–147
undamped free vibration response, 145–146
modal contribution factors, 142–143
"period shifting" effect, 152
time-history analyses, 139
translational and torsional contributions, 139

P

Passive friction dampers. *See* Friction dampers
Period shifting effect, 152
Plan irregular structures
earthquake intensity, 172
multistory central core structure
analyzed structure, 164–165
bidirectional loading, 170–171
displacements, 169–170
elastic hardening, 169
seismic input, 168–169
structural walls, 170
Takeda hysteretic behavior, 169
torsional sensitivity, 165–167
transformation to single-story structure, 166–168
unidirectional eccentricity, 171
Romanian Code provisions, 172
simplified spectral analysis method (SESA), 156–157
single-story structure
analyzed structures, 157–158
capacity spectrum method, 163
code provisions, 164
ground motion, 162
horizontal elastic response spectrum, 161
parameters, 158–159
seismic design force, 160
seismic input, 158
seismic response, 159
serviceability limit state, 160
survivability limit state, 161–162
translational inelastic behavior, 163
ultimate limit state, 161
spectrum-compatible accelerograms, 156
ununiformed displacements, 156
Pushover analysis, 107
P wave reflection, 19–25

R

Reinforced concrete (RC) bridges
bridge regularity
AASHTO LRFD code, 105
EC8 code, 105
elastic regularity index, 105–106
inelastic regularity index, 106–108
damage characteristics, 116
elastic displacement, 104
evolutive modal spectral analyses, 109
geometric and mechanic characteristics, 112–113
hysteretic energy dissipation, 110
inelastic behaviour, 113
lateral deformed configuration, 117
modal irregularity, 111
modal regularity, 108–109
performance evaluation, 115–116
pier height relationship, 112
seismic demand, 113
seismic design codes, 104
seismic evaluation procedure, 105
seismic performance, 110–112, 117
stiffness degradation, 109
viaduct-type bridges, 114–115
Reinforced concrete (RC) framed buildings
beam and column dimensions and reinforcement, 49
C20/25 concrete and FeB38k steel, 49
collapse prevention, 54
concrete mechanical properties (*see* Concrete mechanical properties)
damage index, 56
direct displacement-based design method (*see* Direct displacement-based design method (DDBD))
ductility, 56
elastic response, 52
exceedance probability, 54
horizontal loads, 57
immediate occupancy limit state, 57

Reinforced concrete (RC) framed buildings (*cont.*)
infill panels (*see* Infill panels)
inter-story drift, 52–53
linearized fragility curve, 55
mechanical properties, 47
medium-low seismicity, 55
seismic performance level, 48
seismic response
assumed limit states, 51–52
incremental dynamic analysis, 51
seismic input, 51–52
statistical indexes, 54
structural irregularities, 49–50
top story displacement, 52–53
torsional effects, 48
Reinforced concrete frame-shear wall structure
rocking, 37–38
structure torsion, 37, 39
translational displacement, 37–39
Rotational ground motions. *See* Seismic ground rotations
Rotational motions, symmetric structures
Jin Mao Tower
rocking angles, 34–36
torsion damage, 36–37
translational displacements, 34–35
mass matrix and stiffness matrix, 33
30-m-height monument, 39–40
160-m reinforced concrete frame-shear wall structure
rocking, 37–38
structure torsion, 37, 39
translational displacement, 37–39
Rayleigh damping, 33
seismic motions, 34
seismic translational components, 32
Rotational oscillations, 359–360

S
Seismic analysis, 122
EC8 provisions *vs.* response domain, 67–69
FEMA provisions *vs.* response domain, 68–71
response domain
capacity curves, 65
ductility factor, 65
statistical parameters, 66
top story displacements, 65–66
spectral analysis, 64
Seismic control. *See* Multiple tuned mass dampers (MTMDs)
Seismic conversion efficiency, 9
Seismic demand, 209
Seismic design, 73
Seismic ground rotations
hypothetical rotational waves, 15
P wave reflection, 19–21
solid body mechanics, 17
spatial seismic effects, 16
spectral density derivation, P and SV waves joint reflections
accelerations, 20, 22
angles of incidence, 24–25
apparent velocities, 23
free surface reflection coefficient, 21
modulus of coefficients, 23–26
Poisson coefficient, 23
shear modulus, 23
Stieltjes-Fourier representation, 22
surface ground motion, 21
surface torsion, 16
SV wave reflection, 21
torsion, SH waves
coefficient of tortional component *vs.* angle of incidence, 19, 20
frequency domain, 19
principal coordinate system, 17
shear waves, 19
Stieltjes-Fourier integral, 18
wave propagation, 18
tortional and rocking acceleration, 25–26
unsymmetric-plan buildings, 208
Seismic monitoring, linear and rotational oscillations
building eigenmodes
eigenfrequencies, 360
frequency resolution, 362
microbarograph, 361
spectral-time diagram, 362
"Tower-2000" eigenmodes, 361
wind pulsations, 360
3-component seismic sensors, 363
eigenmode frequency, 356
high-rise buildings, 354
instruments and methods
angular motions, 356
high-sensitive molecular-electronic transducer, 356
linear sensors, 355
seismometers, 355
wind pulsations, 354
load-bearing constructions, 354
METR-03, 362
rotational oscillations, 359–360
space-time registration systems
accelerograms, 359

construction interactions evaluation, 358
low-pass filtration, 358
seismic diagnostic results, 355, 357
seismic observations, 358
44-story apartment building, 357–358
Seismic radiation, 1–14
Seismic response
concrete mechanical properties, 60
corrective eccentricities, multistory buildings
force-displacement curve, 186
height-wise distribution, 185
in-plan distribution, 185
nonlinear dynamic analysis, 184
story drifts, 186
damped systems, 146–147, 151
irregular industrial steel buildings, 84
single-story structure, 159
Shear requirements, 94–95
SH waves, 17–19
Single-story structure
analyzed structures, 157–158
capacity spectrum method, 163
code provisions, 164
ground motion, 162
horizontal elastic response spectrum, 161
parameters, 158–159
seismic design force, 160
seismic input, 158
seismic response, 159
serviceability limit state, 160
survivability limit state, 161–162
translational inelastic behavior, 163
ultimate limit state, 161
Static pushover analysis (SPA), 189
Steel buildings. *See* Irregular industrial steel buildings
Stochastic equivalent linearization method, 298
Structural control. *See* Friction dampers
Structural irregularities, 49–50
Supplemental damping
column sway mechanism, 310
damper design, 315
drift amplification factor, 312
energy-based design method (*see* Energy-based design method)
energy dissipation properties, 315
frame structures, 309
hysteretic dampers, 313–314
idealized shear structure, 311
irregularity limits
global displacement response, 318
ground motions, 316
regular base frames, 318
shear structures, 315–316
story strength capacity, 316
type L irregularities, 316–317
linear viscous damper, 322
modal participation factor, 311
performance spectra-based design method, 313
shear building model, 312
stiffness distribution, 314
9-story moment frame design
elastic frame, 319–320
ground motions, 318
inter-story drifts, 319, 321
nonlinear time-history analysis, 321
retrofit designs, 318
shear strength, 319
strength irregularities, 310
time-history analysis, 322
vertical irregularities, 310
viscous/viscoelastic dampers, 313–314
weak-beam-strong-column structural model, 312
Surface rocking. *See* P wave reflection; SV wave reflection
Surface torsion
definition, 16–17
SH waves
coefficient of tortional component *vs.* angle of incidence, 19, 20
frequency domain, 19
principal coordinate system, 17
shear waves, 19
Stieltjes-Fourier integral, 18
wave propagation, 18
SV wave reflection, 21–25
Symmetric structures
ANSYS software, 41–43
estimation method for rotational components
elastic wave method, 30
frequency dispersion effect, 31
plastic deformation, 30–31
translational components, 32
Eurocode8 method, 40–41
rotational motions
Jin Mao Tower, 34–37
mass matrix and stiffness matrix, 33
30-m-height monument, 39–40
Rayleigh damping, 33
reinforced concrete frame-shear wall structure, 37–39
seismic motions, 34
seismic translational components, 32

Symmetric structures (*cont.*)
seismic rotational loads, 43
seismic torsional damage, 30
structural response, 44
synthesized rotational components, 32–33

T
Torsional demands estimation. *See* Linear dynamic response spectrum analysis
Torsional effects, 48
direct displacement-based design method (DDBD)
asymmetric wall building, 271, 273
asymmetry in-plan structures, 270
code drift, 273
drift-controlled displacement, 273
plan eccentricities, 271–272
shear strength, 271
strength eccentricity, 271
infill panels, 132
Torsionally flexible steel building, 138, 141–143, 145, 150–152, 254–255, 260–267
Torsionally stiff steel building, 138, 141, 143, 144, 150–152, 254–255, 259–260, 262–265
Torsional sensitivity, 165–167

U
Unsymmetric-plan buildings
ASCE/SEI 41-06 nonlinear static procedure, 214
bending moments, 213
building evaluation standard, 205
elastic response spectrum analysis, 204
floor displacements and story drifts
IBC06 buildings, 209, 211
inelastic deformation, 209
linearly elastic systems, 212
seismic demand, 209
UBC85 buildings, 209–210
ground motions, 208, 214
internal forces, 213
mass-eccentric structural systems, 204
median story drifts, 214–215
natural vibration periods and modes
lateral displacements, 206
stiffness, 208
torsional rotations, 206
UBC85 buildings, 206–207
non-linear static procedure, 216
plastic hinge rotations, 212
steel moment-resisting frame buildings, 216
structural modeling, 206
structural systems, 205–206